Groups, Algebras and Identities

CONTEMPORARY
MATHEMATICS

726

Israel Mathematical Conference Proceedings

Groups, Algebras and Identities

Research Workshop of the Israel Science Foundation
Groups, Algebras and Identities
in Honor of Boris Plotkin's 90th Birthday
March 20–24, 2016
Bar-Ilan University and The Hebrew University of Jerusalem, Israel

Eugene Plotkin
Editor

AMS AMERICAN MATHEMATICAL SOCIETY

Tradition of Excellence
Bar-Ilan University
אוניברסיטת בר-אילן

2010 *Mathematics Subject Classification.* Primary 03Gxx, 08Axx, 08Bxx, 20Axx, 20Fxx, 16Sxx, 17Axx, 18Fxx.

Library of Congress Cataloging-in-Publication Data

Names: Research Workshop of the Israel Science Foundation on Groups, Algebras and Identities (2016 : Jerusalem) | Plotkin, Eugene, 1955– editor. | Plotkin, B. I. (Boris Isaakovich), honoree.

Title: Groups, algebras, and identities : Research Workshop of the Israel Science Foundation on Groups, Algebras and Identities held in honor of Boris Plotkin's 90th birthday, March 20-24, 2016, Bar Ilan University and the Hebrew University of Jerusalem, Israel / Eugene Plotkin, editor.

Description: Providence, Rhode Island : American Mathematical Society, [2019] | Series: Contemporary mathematics ; volume 726 | Series: Israel Mathematical Conference proceedings | Includes bibliographical references.

Identifiers: LCCN 2018040270 | ISBN 9781470437138 (alk. paper)

Subjects: LCSH: Group algebras–Congresses. | Geometry, Algebraic–Congresses. | Algebra, Universal–Congresses.

Classification: LCC QA174 .R47 2019 | DDC 512/.2–dc23 LC record available at https://lccn.loc.gov/2018040270

Contemporary Mathematics ISSN: 0271-4132 (print); ISSN: 1098-3627 (online)

DOI: https://doi.org/10.1090/conm/726

∞ The paper used in this book is acid-free and falls within the guidelines established to ensure permanence and durability.

Visit the AMS home page at https://www.ams.org/

10 9 8 7 6 5 4 3 2 1 24 23 22 21 20 19

Contents

Preface

"Groups, Algebras and Identities", a Research Workshop of the Israel Science Foundation in honor of Boris Plotkin's 90th birthday, took place at Bar-Ilan University and the Hebrew University of Jerusalem during March 20–24, 2016. It hosted about 50 participants from 10 countries and was devoted to various problems of Universal Algebra and related areas. These topics reflect the vast spectrum of scientific interests of Boris Plotkin, who was one of the main speakers.

These proceedings include the papers of some of the speakers. In some cases, they have chosen to submit manuscripts which depart from the texts of their lectures. They represent different areas of universal algebra and, of course, are related to the scientific interests and career of B. Plotkin. His paper "Seven Lectures on Universal Algebraic Geometry", which is reprinted here, was previously available only via the internet.

Professor Plotkin's scientific career started in the early 1940s at the Ural University of Sverdlovsk, USSR. He returned to this city after the battles of World War II, where he was seriously wounded during the liberation of his hometown of Bobruisk, Belorussia. His scientific advisor was a prominent group-theorist, P.G. Kontorovich. Since the early 1950s, B. Plotkin concentrated his efforts on difficult problems of abstract group theory. His results include the construction of groups with a normalizer condition; the invention of the locally-nilpotent radical, which became known as the Hirsch-Plotkin radical; a comprehensive study of radical groups, and many others. Also, in the early 1950s, he posted the famous Engel (nil) problem. This extremely difficult Burnside-type problem is still open. Numerous group-theoretic results were summarized in his survey papers and in the monograph *Groups of Automorphisms of Algebraic Systems.*

In 1960, B. Plotkin moved to Riga, Latvia, where he established the Riga Algebraic Seminar and developed a new theory called Varieties of Representations of Algebraic Systems. The main results of this theory were collected in the book *Varieties of Representations of Groups* (joint with S. Vovsi). Later on, in the 1980s, his interest in applied problems of universal algebra was awakened, including problems of theory of automata and theory of databases and knowledge. This resulted in the writing of the books *Universal Algebra, Algebraic Logic and Databases* and *Algebraic Structures in Automata and Databases* (joint with L. Greenglaz and A. Gvaramija).

A new period in the flourishing mathematical activity of B. Plotkin started after his move to Israel in 1993. In the late 1990s, working at the Hebrew University, he focused his efforts on the theory now commonly known as Universal Algebraic Geometry. An extension of this theory is called Logical Geometry.

We wish Boris Plotkin good health, which will enable him to continue his active and fruitful research in mathematics, in particular, in algebra and its applications.

We thankfully acknowledge the support provided for the Workshop by the following institutions:

Israel Science Foundation; Einstein Institute of Mathematics (Hebrew University); Institute of Advanced Studies (Hebrew University); Department of Mathematics (Bar-Ilan University); Gelbart Research Institute (Bar-Ilan University); Israel Mathematical Union.

We have the pleasure of thanking Miriam Beller, who served as technical editor of the volume.

The Editor

Bibliography of Boris Plotkin

BOOKS

(1) B.I. Plotkin, *Groups of Automorphisms of Algebraic Systems*, Moscow, Nauka (1966), 603pp; English version: Wolters-Noordhoff Publishing, Groningen, The Netherlands (1972), 502pp.

(2) B.I. Plotkin, *Algebraic Structures I*, Textbook, Riga, RVKIKU (1973), 114pp. (in Russian).

(3) B.I. Plotkin, *Algebraic Structures II*, Textbook, Riga, RVKIKU (1974), 160pp. (in Russian).

(4) B.I. Plotkin, S.M. Vovsi, *Varieties of Representations of Groups*, Riga, Zinatne (1983), 338pp. (in Russian).

(5) B.I. Plotkin, *Universal Algebra, Algebraic Logic and Databases*, Moscow, Nauka (1991), 447pp. (in Russian).

(6) B. Plotkin, L. Greenglaz, A. Gvaramija, *Algebraic Structures in Automata and Databases*, World Scientific Publ., Singapore-New Jersey (1992), 280pp.

(7) B. Plotkin, L. Greenglaz, A. Gavaramiya, *Elements of Algebraic Theory of Automata*, Moscow, Vyshaja Shkola (1994), 190pp. (in Russian).

(8) B. Plotkin, *Universal Algebra, Algebraic Logic and Databases*, Kluwer Acad. Publ. (1994), 438pp.

SELECTED SURVEY PAPERS

(9) B. Plotkin, *Generalized solvable and generalized nilpotent groups*, Uspekhi Mat. Nauk **13** (1958), no. 4, 89–172; translation in Amer. Math. Soc. Transl. **17** (1961), no. 2, 29–115.

(10) B. Plotkin, *General group theory*, Itogi Nauki, Algebra, Topology, Geometry, 1970 Moscow, VINITY (1971), 5–73 (in Russian).

(11) B. Plotkin, *Varieties of group representations*, Uspekhi Mat. Nauk **32** (1977), no. 5, 3–68 (in Russian).

(12) B. Plotkin, *Varieties of representations of finite groups. Locally stable varieties*, Uspekhi Mat. Nauk **34** (1979), no. 4, 65–95 (in Russian).

(13) B. Plotkin, *Algebraic logic, varieties of algebras and algebraic varieties*, Algebraic Systems, Proc. of Int. Conf. St. Petersburg-95, St. Petersburg (1999), 189–271.

(14) B. Plotkin, *Algebra, categories and databases*, In: Handbook of Algebra, **2**, Elsevier Publ., North-Holland, Amsterdam (2000), 79–148.

(15) B. Plotkin, *Seven lectures on universal algebraic geometry*, Contemporary Math., in this Volume.

PAPERS

(16) B. Plotkin, *On matrix, permutable with given one*, Ural University Press **1** (1949), 10–17 (in Russian).

(17) B. Plotkin, *Structures with the basis of decomposition I*, Fiz.-Mat. Bull. of Ural University **1** (1950), 3–8 (in Russian).

(18) B. Plotkin, *Structures with the basis of decomposition II*, Fiz.-Mat. Bull. of Ural University **1** (1950), 9–13 (in Russian).

(19) B. Plotkin, *On the theory of noncommutative groups without torsion*, Dokl. Akad. Nauk USSR **73** (1950), no. 4, 655–657 (in Russian).

(20) B. Plotkin, *On the theory of locally nilpotent groups*, Dokl. Akad. Nauk USSR (N.S.) **76** (1951), no. 5, 639–641 (in Russian).

(21) B. Plotkin, *On the theory of noncommutative groups without torsion*, Mat. Sb. **30(72)** (1952), no. 1, 197–212 (in Russian).

(22) B. Plotkin, *On the theory of solvable groups without torsion*, Dokl. Acad. Nauk USSR **84** (1952), no. 4, 665–668 (in Russian).

(23) B. Plotkin, *Some criteria for locally nilpotent groups*, Uspekhi Mat. Nauk **9** (1954), no. 3, 181–188 (in Russian).

(24) B. Plotkin, *On nilgroups*, Dokl. Akad. Nauk USSR **94** (1954), 999–1001 (in Russian).

(25) B. Plotkin, *Structure isomorphisms of solvable R-groups*, Dokl. Akad. Nauk USSR **95** (1954), no. 6, 1141–1144 (in Russian).

(26) B. Plotkin *On the nilradical of groups*, Dokl. Akad. Nauk USSR **98** (1954), no. 3, 341–344 (in Russian).

(27) P. Kontorovich, B. Plotkin, *Structures with additive bases*, Mat. Sb. **35(77)** (1954), no. 1, 187–192 (in Russian).

(28) B. Plotkin, *On radical in groups*, Uspekhi Mat. Nauk **10** (1955), no. 3, 179–180 (in Russian).

(29) B. Plotkin, *On the theory of solvable groups with finiteness conditions*, Dokl. Akad. Nauk USSR **100** (1955), no. 3, 417–420 (in Russian).

(30) B. Plotkin, *On the theory of solvable groups without torsion*, Mat. Sb. **36(78)** (1955), no. 1, 31–38 (in Russian).

(31) B. Plotkin, *Radical groups*, Mat. Sb. **37** (1955), no. 3, 507–526; translation in Amer. Math. Soc. Transl. (2), **17** (1961), 9–28.

(32) B. Plotkin, *Some questions of the theory of groups without torsion*, Ukrain. Math. J. **8** (1955), no. 3, 325–329 (in Russian).

(33) B. Plotkin, *On groups with finiteness conditions for abelian subgroups*, Dokl. Akad. Nauk USSR **107** (1955), no. 5, 648–651 (in Russian).

(34) B. Plotkin, *On some classes of generalized solvable groups and Lie algebras*, Uspekhi Mat. Nauk **11** (1955), no. 1, 257 (in Russian).

(35) B. Plotkin, *On some classes of infinite groups*, Uspekhi Mat. Nauk **11** (1955), no. 1, 259 (in Russian).

(36) B. Plotkin, *Radical and semisimple groups and Lie algebras*, Proc. of the 3rd All-Union Math. Session, I, Moscow (1955), 33 (in Russian).

(37) B. Plotkin, *Radical and semisimple groups*, Proc. Moscow Math. Soc., Moscow **6** (1957), 299–337 (in Russian).

(38) B. Plotkin, *Radical and nil elements in groups*, Izv. Vysh. Uchebn. Zaved. Mat. **1** (1958), 130–135 (in Russian).

(39) B. Plotkin, *On some classes of infinite groups*, Uspekhi Mat. Nauk **13** (1958), no. 1, 189–192 (in Russian).

(40) B. Plotkin, *Algebraic elements and nil elements in groups*, Uspekhi Mat. Nauk **13** (1958), no. 1, 211–212 (in Russian).

(41) B. Plotkin, *On algebraic sets of elements in groups and Lie algebras*, Uspekhi Mat. Nauk **13** (1958), no. 6, 133–138; translation in Siberian Math. J. **9** (1968).

(42) B. Plotkin, *Some types of abstract theoretical group features*, Uspekhi Mat. Nauk **14** (1959), no. 5(89), 211–212 (in Russian).

(43) B. Plotkin, *Abstract Sylow properties*, Proc. of Ural Railroad Institute, 2, Sverdlovsk, 7–14 (in Russian).

(44) B. Plotkin, *Radical groups, whose radical has ascending central series*, Ural Univ. Press **23** (1960), no. 2:3, 40–43 (in Russian).

(45) B. Plotkin, *On algebraic elements in abstract groups*, Ural Univ. Press **23** (1960), no. 2:3, 44–48 (in Russian).

(46) B. Plotkin, *On a radical of an automorphism group of a group with maximal condition*, Dokl. Akad. Nauk USSR **130** (1960), 977–980; translated in Soviet Math. Dokl. **1** (1960), 117–121.

(47) B. Plotkin, V.G. Vilyatser, *On the theory of locally stable groups of automorphisms*, Dokl. Akad. Nauk USSR **134** (1960), no. 3, 529–532; translation in Soviet Math. Dokl. **1** (1960), 1108-1111.

(48) B. Plotkin, *Preface to the Russian edition of the book: M. Sudzuki, The structure of the group and the structure of the lattice of its subgroups*, Moscow, (1960), 5–7.

(49) B. Plotkin, *Normal subgroup bounding the group*, Mat. Sb. **55** (1961), 344-352 (in Russian).

(50) B. Plotkin, *Locally stable groups of automorphisms*, Sibirsk. Mat. Z. **2** (1961), no. 1, 100–114 (in Russian).

(51) B. Plotkin, *Some properties of automorphisms of nilpotent groups*, Dokl. Akad. Nauk USSR **137** (1961), 1303–1306 (in Russian).

(52) B. Plotkin, *Radicals in group pairs*, Dokl. Akad. Nauk USSR **140** (1961), no. 5, 1019–1022 (in Russian).

(53) B. Plotkin, *Some radicals of automorphism groups*, Uspekhi Mat. Nauk **17** (1962), no. 4, 165–171 (in Russian).

(54) B. Plotkin, *Aussere locale Eigenschaften von Automorphismengruppen*, Arch. Math. (Basel) **13** (1962), 401–407 (in German).

(55) B. Plotkin, *Radicals connected with representations of groups*, Dokl. Akad. Nauk USSR **144** (1962), no. 1, 52–55 (in Russian).

(56) B. Plotkin, *Automorphisms of groups and group pairs*, Uspekhi Mat. Nauk **17** (1962), no. 6, 222 (in Russian).

(57) B. Plotkin, *Some remarks on group pairs*, Math. Zap. Ural Univ. **3** (1963), 63–69 (in Russian).

(58) B. Plotkin, *Some problems of mathematical education in VTUZ*, Proc. PVKIKU, Riga **3** (1963), 20–25 (in Russian).

(59) B. Plotkin, *Some problems of the general theory of representations of groups*, Izv. Akad. Nauk USSR, Ser. Mat. **27** (1963), no. 4, 855–882 (in Russian).

(60) B. Plotkin, *Generalized stable and generalized nilpotent groups*, Sibirsk. Mat. Z. **4** (1963), no. 6, 1389–1404 (in Russian).

(61) B. Plotkin, *Ω-semigroups, Ω-rings, and general representations*, Dokl. Akad. Nauk USSR **149** (1963), no. 5, 1037–1040 (in Russian).
(62) B. Plotkin, *On infinite-dimensional linear groups*, Dokl. Akad. Nauk USSR **154** (1963), no. 1, 42–45 (in Russian).
(63) B. Plotkin, *On conditions of finiteness in the theory of representations*, Proc. 5-th All-Union Conf., Novosibirsk (1963), 43–44 (in Russian).
(64) B. Plotkin, *On a theorem of Zassenhaus*, Tributes to N.G. Chebotarev, Izdat. Kazan Univ. (1964), 67–74 (in Russian).
(65) B. Plotkin, *On a method of constructing radicals in groups*, Sibirsk. Mat. Z. **6** (1965), no. 5, 1197–1201 (in Russian).
(66) A.I. Maltcev, B. Plotkin, *Petr Grigorievich (on the 60–th anniversary)*, Uspekhi Mat. Nauk **20** (1965), no. 4, 209–211 (in Russian).
(67) L. Berezina, B. Plotkin, et al., *Works of Soviet Latvia mathematicians in the last seven years*, Izv. Akad. Nauk Latv. SSR, Riga (1965), no. 3, 3–26 (in Russian).
(68) B. Plotkin, *Some problems of linear group theory*, Abstracts of Int. Math. Congress, Section 2, Moscow (1966), 51 (in Russian).
(69) P.S. Alexandrow, B. Plotkin, L.A. Skornjakov, *Alexander Gennadjevich Kurosh (on the 60–th anniversary)*, Uspekhi Mat. Nauk **23** (1968), no. 2, 219–228 (in Russian).
(70) B. Plotkin, L.A. Simonjan, *VIII All–Union Algebraic Colloquia*, Uspekhi Mat. Nauk **23** (1968), no. 2, 234–236 (in Russian).
(71) A.G. Kurosh, B. Plotkin, N.F. Sesekin, L.N. Shevrin, *Petr Grigorievich Kontorovich*, Uspekhi Mat. Nauk **20** (1968), no. 4, 239–240 (in Russian).
(72) B. Plotkin, *Radicals in groups, radical classes and operators of radical closure*, Abstracts of LX All–Union Colloquia, Gomel (1968), 154 (in Russian).
(73) B. Plotkin, *On semisimple groups*, Sibirsk. Mat. Z. **9** (1968), no. 3, 623–631 (in Russian).
(74) B. Plotkin, *Free representations of groups*, Proc. Riga Algebr. Seminar, Riga Univ. Press (1969), 191–208 (in Russian).
(75) B. Plotkin, *Stability, nilpotency and radicals in groups of automorphisms of modules*, Proc. Riga Algebr. Seminar, Riga Univ. Press (1969), 191–208 (in Russian).
(76) B. Plotkin, *On the semigroup of radical classes of groups*, Sibirsk. Mat. Z. **10** (1969), no. 5, 1091–1108; translation in Siberian Math. J. **10** (1969).
(77) B. Plotkin, *Functorials, radicals and coradicals in groups*, Math. Zap. Ural Univ. **7** (1970), no. 3, 150–183 (in Russian).
(78) B. Plotkin, *Varieties and quasivarieties, connected with representations* of *groups*, Dokl. Akad. Nauk USSR **196** (1971), no. 3, 527–530; translation in Soviet Math. Dokl. **12** (1971), no. 1, 192–196.
(79) B. Plotkin, *Triangular products of pairs. Certain questions of group theory*, Proc. Riga Algebr. Seminar, Riga Univ. Press **151** (1971), 140–170 (in Russian).
(80) B. Plotkin, *Radicals and varieties in group representations*, Latv. Math. Annual **10** (1972), 75–132 (in Russian).
(81) B. Plotkin, *On semigroup of varieties, associated with representations of groups*, Dokl. Akad. Nauk USSR **203** (1972), no. 2, 293–296; translation in Soviet Math. Dokl. **13** (1972), no. 2, 392–396.

(82) A.S. Grinberg, B. Plotkin, *Semigroups of varieties associated with group representations*, Sibirsk. Mat. Z. **13** (1972), no. 4, 841–858; translation in Siberian Math. J. **13** (1972), 585–596.

(83) B. Plotkin, *Group varieties and varieties of pairs associated with group representations*, Sibirsk. Mat. Z. **13** (1972), no. 5, 1030–1053; translation in Siberian Math. J., **13**, (1972), 713–729.

(84) P.S. Alexandrow, M. Baranovich, O. Golovin, B. Plotkin, *Alexander Gennadjevich Kurosh*, Uspekhi Mat. Nauk **27** (1972), no. 1, 211–222 (in Russian).

(85) B. Plotkin, *Radicals in groups, operations on classes of groups, and radical classes*, Selected Questions of Algebra and Logic, Nauka, Novosibirsk (1973), 205–244; translation in Amer. Math. Soc. Transl. (2) **119** (1983), 89–118.

(86) B. Plotkin, *Remarks on stable representations of nilpotent groups*, Trudy Moscow Mat. Obsc. **29** (1973), 191–206; translation in Trans. Moscow Math. Soc. **29** (1973), 185–200.

(87) B. Plotkin, *Ten years of Riga algebraic seminar*, Latv. Math. Annual **13** (1973), 3–18 (in Russian).

(88) L.E. Krop, B. Plotkin, *Magnus varieties of group representations*, Mat. Sb. **95** (1974), no. 4, 499–524 (in Russian).

(89) L.E. Krop, B. Plotkin, L.A. Simonjan, *Some connections between group varieties and varieties of representations of groups*, Dokl. Akad. Nauk USSR **219** (1974), no. 5, 1072–1075; translation in Soviet Math. Dokl. **15** (1974), no. 6, 1747–1751.

(90) E.M. Kublanova, Ts. Dididze, B. Plotkin, *On varieties of automata*, Dokl. Akad. Nauk USSR **221** (1975), no. 6, 1284–1287; translation in Soviet Math. Dokl. **16** (1975), no. 2, 537–541.

(91) L.Y. Gringlaz, B. Plotkin, *A note on powers of varieties of pairs and faithful group representations*, Latv. Math. Annual **16** (1975), 23–32 (in Russian).

(92) B. Plotkin, *Small varieties of pairs – representations of groups*, Abstracts of All–Union Algebraic Conf., Gomel (1975), 153 (in Russian).

(93) B. Plotkin, *Multiplicative systems of varieties of pairs – group representations*, Latv. Math. Annual **18** (1976), 143–170 (in Russian).

(94) K. Agadzanjan, B. Plotkin, *Methodology problems in mathematic education*, Proc. Riga Milit. Akad. (1976), 75–90 (in Russian).

(95) B. Plotkin, *Algebraic structures III. Linear systems*, Manuscript (1977), 120pp. (in Russian).

(96) E.M. Kublanova, Ts. Dididze, B. Plotkin, *Varieties of automata*, Cybernetica **1** (1977), 47–54 (in Russian).

(97) E.M. Kublanova, Ts. Dididze, B. Plotkin, *Varieties of automata II*, Cybernetica **3** (1977), 16–24 (in Russian).

(98) B. Plotkin, *Varieties of representations of groups and varieties of associative algebras*, Collection of Papers in Algebra, Riga (1978), 172–187 (in Russian).

(99) B. Plotkin, *Locally finite and locally finite–dimensional varieties of pairs – group representations*, Collection of Papers in Algebra, Riga (1978), 188–245 (in Russian).

(100) B. Plotkin, *Notes on small varieties of representations of groups*, Collection of Papers in Algebra, Riga (1978), 246–257 (in Russian).

(101) B. Plotkin, *Algebra*, Latvian Encyclopedia, Riga (1979) (in Russian).

(102) B. Plotkin, *Algebraic numbers*, Latvian Encyclopedia, Riga (1979) (in Russian).
(103) B. Plotkin, *On varieties of representations of groups over rings*, Abstracts of All–Union Algebraic Conf., Krasnojarsk (1979), 83 (in Russian).
(104) B. Plotkin, *Verbal ideals in group algebras of finite groups*, Abstracts of All–Union Algebraic Conf., Krasnojarsk (1979), 116 (in Russian).
(105) B. Plotkin, *Algebra of automata – some problems*, Vestnic of Moscow Univ. (1980) (in Russian).
(106) E. Kublanova, B. Plotkin, *On the algebra of varieties of group representations over an arbitrary commutative ring*, Latv. Math. Annual **24** (1980), 183–196 (in Russian).
(107) B. Plotkin, *Verbal ideals in group algebras of finite groups*, Latv. Math. Annual **24** (1980), 229–246 (in Russian).
(108) I. Peranidze, B. Plotkin, V. Shteinbuk, *Automata, representations and semigroups*, Latv. Math. Annual **25** (1981), 222–236 (in Russian).
(109) B. Plotkin, V. Shteinbuk, *Semigroups and automata*, Abstracts of Int. Conf. on Semigroups, Szeged, Hungary, Janos Bolyai Math. Soc. (1981), 20.
(110) B. Plotkin, *Many–sorted algebras – applied aspect*, Abstracts of All–Union Algebraic Conf., Leningrad (1981), 127–128 (in Russian).
(111) B. Plotkin, *Pseudoidentities in rings*, Abstracts of All–Union Algebraic Conf., Batumi (1981), 121.
(112) A. Kushkuley, B. Plotkin, *Identities of the regular representations finite groups*, Preprint, Riga (1981), 50pp. (in Russian).
(113) B. Plotkin, *Models and databases*, Proc. Comp. Centre of Akad. Nauk USSR, Tbilisi **22** (1982), no. 2, 100–130 (in Russian).
(114) B. Plotkin, *Biautomata*, Tbilisi Univ. Press, Math., Tbilisi **12** (1982), 123–159 (in Russian).
(115) B. Plotkin, *Groups*, Latvian Encyclopedia, Riga (1982) (in Russian).
(116) B. Plotkin, *Homomorphisms of groups*, Latvian Encyclopedia, Riga (1982) (in Russian).
(117) B. Plotkin *Rings*, Latvian Encyclopedia, Riga (1982) (in Russian).
(118) B. Plotkin, *Textbook on applied algebra. vol. I*, Reference in New Books Abroad, ser. A **4** (1982), 39–40.
(119) B. Plotkin, *Algebraic model of database – automaton*, Latv. Math. Annual **27** (1983), 216–232 (in Russian).
(120) *Radicals and verbals*, Proc. Int. Conf. Radical Theory, Eger, (1982), Colloq. Math. Soc. Janos Bolyai Soc., North–Holland, Amsterdam **31** (1983), 379–403.
(121) B. Plotkin *Databases and data types*, Abstracts of All–Union Conf. on databases, Tashkent (1983), 79–80 (in Russian).
(122) B. Plotkin, *Universal algebra and databases*, Abstracts of All–Union Algebraic Conf., Minsk (1983), 93 (in Russian).
(123) E. Maftsir, B. Plotkin, *Galois theory of databases*, Abstracts of All–Union Algebraic Conf., Lvov (1987), 22 (in Russian).
(124) A. Gvaramia, B. Plotkin, *Fuzzy and stochastic automata*, Preprint, Suhumi (1987), 21pp. (in Russian).
(125) B. Plotkin, *Galois theory of databases*, Lecture Notes in Math., Springer, Algebra–some Current Trends, Proc. Int. Alg. Conf. Varna **1352** (1988), 147–161.

(126) E. Maftsir, B. Plotkin, *The automorphism group of databases*, Ukrainian Math. J. **40** (1988), no. 3, 335–345; translation in Ukrainian Math. J. 40 (1988), no. 3, 286-294.

(127) L. Greenglaz, M. Gobechia, B. Plotkin, *Identities and varieties of bi–automata*, Latv. Math. Annual **34** (1992) (in Russian).

(128) E. Maftsir, B. Plotkin, *Groups of automorphisms in database theory*, Latv. Math. Annual **34** (1992) (in Russian).

(129) B. Plotkin, *Some algebraic aspects of database theory*, Proc. International Conference on Algebra, Part 3 (Novosibirsk, 1989), 219–239, Contemp. Math. **131**, Amer. Math. Soc., Providence, RI, 1992.

(130) A. Gvaramia, B. Plotkin, *The homotopies of quasigroups and universal algebras*, Universal Algebra and Quasigroup Theory, Heldermann, Berlin **19** (1992), 89–99.

(131) B. Plotkin, *Algebraic logic, algebra and databases*, Symbolic Logic-94, Abstracts, Clermont–Ferran (1994).

(132) L. Greenglas, A. Gvaramia, B. Plotkin, *Elements of algebraic theory of automata*, Moscow, Vyshaja Shkola (1994), 190pp.

(133) B. Plotkin, *Varieties of representations of groups, quasivarieties, and pseudovarieties of groups*, Proc. Int. Conf. Infinite Groups, Ravello-94, De Gruyter (1995), 215–228.

(134) B. Plotkin, *Model theoretical and deductive approach to databases from the point of view of algebraic logic and geometry*, Advances in Databases and Information Systems, Springer (1996), 161–170.

(135) B. Plotkin, T. Plotkin, S. Kraus, *Algebraic logic in the description of the state of databases*, Fundam. Prikl. Mat. **3** (1996), 875–910 (in Russian).

(136) B. Plotkin, *Algebraic geometry and universal algebra*, Int. Conf. on Universal Algebra, Abstracts, Szeged (1996), p. 43.

(137) B. Plotkin, *Varieties of algebras and algebraic varieties*, Israel J. Math. **96** (1996), no. 2, 511–522.

(138) B. Plotkin, *Varieties of algebras and algebraic varieties. Categories of algebraic varieties*, Siberian Advanced Mathematics, Allerton Press **7** (1997), no. 2, 64–97.

(139) B. Plotkin, *Some concepts of algebraic geometry in universal algebra*, Algebra i Analiz **9** (1997), no. 4, 224–248; translation in St. Petersburg Math. J. 9 (1998), no. 4, 859–879.

(140) B. Plotkin, *Logic and algebraic geometry in varieties of group representations*, Abstracts in CMS, Manitoba, Winnipeg (1997), 17–18.

(141) E. Koublanova, B. Plotkin, *Varieties and pseudovarieties of group representations*, Proc. Int. Conf. Groups-97 in Bath, England.

(142) B. Plotkin, *Algebraic geometry in universal algebra*, International Algebraic Conference St. Petersburg-97, Abstracts (1997), 99–100.

(143) B. Plotkin, T. Plotkin, S. Kraus, *Problems of equivalence, categoricity of axioms and states description in databases*, Studia Logica **61** (1998), 321–340.

(144) B. Plotkin, E. Plotkin, A. Tsurkov, *Geometrical equivalence of groups*, Comm. Algebra **27** (1999), no. 8, 4015–4025.

(145) B. Plotkin, *Algebraic logic, varieties of algebras and algebraic varieties*, International Algebraic Conference St. Peterburg -95, St. Petersburg (1999), 189–271.

(146) B. Plotkin, *Zero divisors in group-based algebras. Algebras without zero divisors*, Bul. Acad. Stiinte Repub. Mold. Mat. **2** (1999), 67–84.

(147) D. Nikolova, B. Plotkin, S. Kraus, *Some notes on universal algebraic geometry*, Algebra, Proc. International Conference on Algebra on the Occasion of the 90th Birthday of A. G. Kurosh, Walter De Gruyter Publ., Berlin (2000), 237–261.

(148) E. Koublanova, B. Plotkin, A. Yakovlev, *Pseudoidentities and pseudovarieties of representations of groups*, J. Group Theory **3** (2000), no. 2, 169–176.

(149) A. Berzinsh, B. Plotkin, E. Plotkin, *Algebraic geometry in varieties of algebras with the given algebra of constants*, Algebra 13, J. Math. Sci. (New York) **102** (2000), no. 3, 4039–4070.

(150) B. Plotkin, T. Plotkin, *Geometrical aspect of databases and knowledge bases*, Algebra Universalis **46** (2001), no. 1-2, 131-161.

(151) B. Plotkin, *Automorphisms and endomorphisms in universal algebra*, The Concise Handbook of Algebra, ed. G. Pilz (2002), 120–123.

(152) B. Plotkin, *Groups of automorphisms of algebraic systems*, The Concise Handbook of Algebra, ed. G.Pilz (2002), 434–437.

(153) G. Mashevitzky, E. Plotkin, B. Plotkin, *Automorphisms of categories of free algebras of varieties*, ERA of the American Math. Soc. **8** (2002), 1–10.

(154) B. Plotkin, *Some problems in nonclassical algebraic geometry*, Ukrainian Math. J. **54** (2002), no. 6, 841–846.

(155) B. Plotkin, *Algebras with the same (algebraic) geometry*, Tr. Mat. Inst. Steklova **242** (2003), Mat. Logika i Algebra, 176–207; translation in Proc. Steklov Inst. Math. 2003, no. 3(242), 165-196.

(156) B. Plotkin, *Infinitary quasi-identities and infinitary quasivarieties*, Proc. Latvian Akad. Sci., Section B **57** (2003), no. 3-4, 111–112.

(157) G. Mashevitzky, B. Plotkin, E. Plotkin, *Automorphisms of categories of free Lie algebras*, J. Algebra **282** (2004), no. 2, 490–512.

(158) B. Plotkin, *Problems in algebra inspired by universal algebraic geometry*, Fundam. Appl. Math. **10** (2004), no. 3, 181-197.

(159) B. Plotkin, *Algebraic geometry in First Order Logic*, Sovrem. Mat. Prilozh., no. 22, Algebra i Geometry (2004), 16–62; translation in J. Math. Sci. **137** (2006), no. 5, 5049–5097.

(160) B. Plotkin, A. Tsurkov, *Action type geometrical equivalence of representations of groups*, Algebra Discrete Math. **4** (2005), 48-79.

(161) B. Plotkin, *Notes on Engel groups and Engel elements in groups. Some generalizations*, Izv. Ural Univ. Ser. Mat. Mekh **36** (2005), no. 7, 153–166.

(162) B. Plotkin, T. Plotkin, *An algebraic approach to knowledge bases informational equivalence*, Acta Appl. Math. **89** (2005), 109–134.

(163) A. Gvaramia, B. Plotkin, *Action-type axiomatizable classes of group representations*, Fundam. Appl. Math. **11** (2005), no. 2, 73–85; translation in J. Math. Sci. (N.Y.) **142** (2007), no. 2, 1915-1922.

(164) B. Plotkin, G. Zhitomirski, *On automorphisms of categories of free algebras of some varieties*, J. Algebra **306** (2006), no. 2, 344–367.

(165) B. Plotkin, *Geometrical equivalence, geometrical similarity, and geometrical compatibility of algebras*, Zapiski Nauch. Sem. POMI, Vopr. Teor. Predst. Algebr i Grupp **330** (2006), no. 13, 201–222, 274; translation in J. Math. Sci. (N.Y.) **140** (2007), no. 5, 716-728.

(166) E. Katsov, R. Lipyanski, B. Plotkin, *Automorphisms of the categories of free modules, free semimodules and free Lie algebras*, Comm. Algebra **35** (2007), 931–952.

(167) B. Plotkin, *Logically separable algebras in varieties of algebras*, Bul. Acad. Stiinte Repub. Mold. Mat. (2007), no. 2, 33-42.

(168) B. Plotkin, G. Zhitomirski, *On automorphisms of categories of universal algebras*, Int. J. Algebra Comput. **17** (2007), no. 5-6, 1085-1106.

(169) G. Mashevitzky, B. Plotkin, *On automorphisms of the semigroup of endomorphisms of free universal algebras*, Internat. J. Algebra Comput. **17** (2007), no. 5–6, 1115-1132.

(170) B. Plotkin, *Some results and problems related to universal algebraic geometry*, Internat. J. Algebra Comput. **17** (2007), no. 5-6, 1133-1164.

(171) B. Plotkin, G. Zhitomirski, *Some logical invariants of algebras and logical relations between algebras*, Algebra i Analiz **19** (2007), no. 5, 214–245; translation in St. Petersburg Math. J. **19** (2008), no. 5, 829-852.

(172) B. Plotkin, T. Plotkin, *Categories of elementary sets over algebras and categories of elementary algebraic knowledge*, Pillars of Computer Science **4800**, Lecture Notes in Comp. Sci., Springer, Berlin (2008), 555-570.

(173) E. Aladova, B. Plotkin, *PI-groups and PI-representations of groups*, Fundam. Prikl. Mat **14** (2008), no. 7, 3–13; translation in J. Math. Sci. (N.Y.) **164** (2010), no. 2, 155-162.

(174) A. Gvaramia, B. Plotkin, *On linear representations of quasigroups*, Fundam. Prikl. Mat. **15** (2009), no. 3, 113–118; translation in J. Math. Sci. (N.Y.) **168** (2010), no. 1, 76-79.

(175) B E. Aladova, B. Plotkin, *Varieties of representations of groups and varieties of associative algebras*, Int. J. Algebra Comput **21** (2011), no. 7, 1149-1178.

(176) A. Gvaramia, B. Plotkin, *Logic in the representations of groups*, Algebra Logika **51** (2012), no. 1, 3–40, 150, 153; translation in Algebra Logic **51** (2012), no. 1, 1-27.

(177) B. Plotkin, *Sergei Nikolaevich Chernikov Memoirs*, Algebra Discrete Math **14** (2012), no. 1, C–H.

(178) E. Aladova, B. Plotkin, E. Plotkin, *Algebraic logic and logically-geometric types in varieties of algebras*, J. Algebra Appl. **12** (2013), no. 2, 125–146.

(179) B. Plotkin, *Algebraic logic and logical geometry. Two in one*, Vestnik St. Petersburg Univ. Math. **46** (2013), no. 1, 35-42.

(180) B. Plotkin, T. Plotkin, *Cascade connection and triangular product of linear automata*, Fundam. Prikl. Mat **17** (2011/12), no. 7, 175–186; translation in J. Math. Sci. (N.Y.) **197** (2014), no. 4, 565-572.

(181) B. Plotkin, *Algebraic logic and logical geometry in arbitrary varieties of algebras*, Group Theory, Combinatorics, and Computing, Contemp. Math. **611**, Amer. Math. Soc., Providence, RI (2014), 151-167.

(182) E. Plotkin, G. Zhitomirski, B. Plotkin, *Type of a point in universal geometry and in model theory*, Algebra Discrete Math. **19** (2015), no. 1, 87–100.

(183) E. Plotkin, B. Plotkin, *Multi-sorted logic and logical geometry: some problems*, Demonstr. Math. **48** (2015), no. 4, 578-619.

(184) E. Aladova, A. Gvaramia, B. Plotkin, E. Plotkin, T. Plotkin, *Logically-geometrical similarity for algebras and models with the same identities*, Tr. Inst. Mat. Belarus Akad. Nauk **23** (2015), no. 2, 112-122.

(185) E. Aladova, A. Gvaramia, B. Plotkin, T. Plotkin, *Multisorted logic, models and logical geometry*, Fundam. Prikl. Mat **19** (2014), no. 3, 5–22; translation in J. Math. Sci. (N.Y.) **214** (2016), no. 6, 742-754.

(186) G. Mashevitzky, B. Plotkin, E. Plotkin, *Action of endomorphism semigroups on definable sets*, Internat. J. Algebra Comput. **28** (2018), no. 8, 1585-1612.

Conference Program

Sunday, March 20, Hebrew University

10:00 - 10:50	Vladimir N. Remeslennikov *What is Universal Algebraic Geometry?*
11:20 - 12:10	Nikolay Romanovsky *Hilbert's Nullstellensatz in algebraic geometry over rigid solvable groups*
12:20 - 13:10	Zlil Sela *Word equations*
15:00 - 15:50	Avinoam Mann *Curiouser and Curioser, said Alice (partition formulae connected with abelian and non-abelian groups)*
16:10 - 16:40	Grigory Zhitomirsky *On logically-geometric types of algebras*
16:45 - 17:15	Arkady Tsurkov *Classification problem of quasi-varieties of the nilpotent class 2 torsion free groups is wild. How UAG works*
17:25 - 17:55	Boris Plotkin *How it was: glimpse of the past*

Monday, March 21, Hebrew University

10:00 - 10:50	Ilya Rips *Combinatorics of Engel relators and the geometry of the Cayley graph of the free Engel group*
11:20 - 12:10	Boris Plotkin *Algebra and logic in varieties of algebras*
12:20 - 13:10	Rostislav Grigorchuk *Groups, automata, and automatically generated sequences*
15:00 - 15:50	Boris Zilber *The geometrical semantics of algebraic quantum mechanics*
16:10 - 16:40	Vladislav Kharchenko *Quantum Lie theory*
16:45 - 17:15	Grigory Mashevitsky *The Galois correspondence between semigroups of transformations and logic formulas*

Tuesday, March 22, Bar-Ilan University

10:00 - 10:50	Efim Zelmanov *On groups with identities*
11:20 - 12:10	Ivan Shestakov *Simple finite dimensional noncommutative Jordan superalgebras*
12:20 - 13:10	Eli Aljadeff and Louis Rowen *Specht's conjecture*
14:40 - 15:10	Andrey Elishev *Independence of the B-KK isomorphism of infinite prime and Kontsevich conjecture*
15:15 - 15:40	Alexander Lashkhi *Modeling of ring geometry from von Neumann's point of view*
16:10 - 16:35	Alexander Pinus *Algebraic sets of universal algebras and algebraic closure operator*
16:40 - 16:55	Janis Cirulis *Models of a freeoid and algebraic logic*
17:10	Movie ["Eliyahu Rips: The Burning"] with commentary

Wednesday, March 23, Bar-Ilan University

10:00 - 10:50	Yuri Ershov *Continuous lattices and essentially complete topological spaces*
11:20 - 12:10	Mikhail Volkov *The finite basis problem for Kauffman monoids*
12:20 - 13:10	I. Ivanov-Pogodaev *Construction of infinite finitely presented nil-semigroup*
14:40 - 15:30	Stuart Margolis *CW-Decompositions, Leray numbers and the cohomology of left regular band algebras*
15:50 - 16:40	Leonid Bokut' *Composition-Diamond lemmas for associative (Lie) algebras and Lyndon-Shirshov (Lie) words*
16:45 - 17:10	Ruvim Lipyanski *Wild and superwild problems in algebra operator*
17:15 - 18:00	Discussions

Thursday, March 24, Hebrew University

10:00 - 10:50	Vahagn Mikaelian *On ascending series of nilpotent subvarieties in soluble varieties of groups*
11:20 - 12:10	Artem Shevljakov *Compactness conditions in universal algebraic geometry*
12:15 - 12:40	Elena Aladova *Syntax versus semantics in knowledge bases*
12:40 - 12:50	Evgeniia Mogilanskaia *About Tamura's problem*
15:50 - 16:40	Leonid Bokut' *Composition-Diamond lemmas for associative (Lie) algebras and Lyndon-Shirshov (Lie) words*
16:45 - 17:10	Ruvim Lipyanski *Wild and superwild problems in algebra operator*
17:15 - 18:00	Discussions

List of Participants

E. Aladova
Bar-Ilan University, Israel

Eli Aljadeff
Technion, Israel

T. Bauer
Bar-Ilan University, Israel

A. Belov-Kanel
Bar-Ilan University, Israel

L. Bokut'
Novosibirsk State University, Russia &
South China Normal University, China

S. Chobanyan
Georgian Technical University, Georgia

J. Cirulis
University of Latvia, Latvia

A.S. Dahari
Bar-Ilan University, Israel

A. Elishev
Sobolev Institute of Mathematics,,
Russia

Yu. Ershov
Moscow Institute of Physics and
Technology, Russia

R. Grigorchuk
Texas A & M University, U.S.A.

A. Gvaramia
Abkhazian State University, Abkhazia

I. Ivanov-Pogodaev
MFTI University, Russia

A. Juhasz
Technion, Israel

T. Kadeishvili
A.Razmadze Mathematical Institute,
Georgia

V. Kharchenko
Universidad Nacional Autónoma de
México, Mexico

E. Kleiman
Ashdod, Israel

E. Klimenko
Heinrich Heine University, Germany

E. Koublanova
Drexel University, U.S.A.

B. Kunyavskii
Bar-Ilan University, Israel

V. Kvaratskhelia
Institute of Computational
Mathematics, Georgia

A. Lashkhi
Georgian Technical University, Georgia

R. Lipyansky
Ben Gurion University, Israel

A. Lubotzky
Hebrew University, Israel

B. Mamuchishvili
A. Razmadze Mathematical Institute,
Georgia

A. Mann
Hebrew University, Israel

S. Margolis
Bar-Ilan University, Israel

G. Mashevitzky
Bar-Ilan University, Israel

E. Matzri
Bar-Ilan University, Israel

M. Menteshashvili
Georgian Technical University, Georgia

B. Mesablishvili
A.Razmadze Mathematical Institute, Georgia

V. Mikaelian
Yerevan State University, Armenia

E. Mogiljanskaja
Suai University, Russia

M. Muzychuk
Netanya College, Israel

A. Pinus
Novosibirsk State Technical University, Russia

B. Plotkin
Hebrew University, Israel

E. Plotkin
Bar-Ilan University, Israel

T. Plotkin
Bar-Ilan University, Israel

A. Ramishvili
Georgian Technical University, Georgia

V.N. Remeslennikov
Institute of Mathematics, Omsk, Russia

E. Rips
Hebrew University, Israel

N. Romanovsky
Institute of Mathematics, Novosibirsk, Russia

L. Rowen
Bar-Ilan University, Israel

Z. Sela
Hebrew University, Israel

I. Shestakov
University Sao Paulo, Brazil

A. Shevljakov
Institute of Mathematics, Omsk, Russia

G. Soifer
Bar-Ilan University, Israel

V. Tarieladze
Institute of Computational Mathematics, Georgia

A. Tsurkov
University Sao Paulo, Brazil

U. Vishne
Bar-Ilan University, Israel

M. Volkov
Ural Federal University, Russia

E. Zelmanov
University of California at San Diego, U.S.A.

G. Zhitomirskii
Bar-Ilan University, Israel

B. Zilber
University of Oxford, UK

Contemporary Mathematics
Volume **726**, 2019
https://doi.org/10.1090/conm/726/14605

Construction of a quotient ring of $\mathbb{Z}_2\mathcal{F}$ in which a binomial $1 + w$ is invertible using small cancellation methods

A. Atkarskaya, A. Kanel-Belov, E. Plotkin, and E. Rips

Abstract. We apply small cancellation methods originating from group theory to investigate the structure of a quotient ring $\mathbb{Z}_2\mathcal{F}/\mathcal{I}$ where $\mathbb{Z}_2\mathcal{F}$ is the group algebra of the free group $\mathcal{F}$ over the field $\mathbb{Z}_2$, and the ideal $\mathcal{I}$ is generated by a single trinomial $1 + v + vw$, where v is a complicated word depending on w. In $\mathbb{Z}_2\mathcal{F}/\mathcal{I}$ we have $(1 + w)^{-1} = v$, so $1 + w$ becomes invertible. We construct an explicit linear basis of $\mathbb{Z}_2\mathcal{F}/\mathcal{I}$ (thus showing that $\mathbb{Z}_2\mathcal{F}/\mathcal{I} \neq 0$). This is the first step in constructing rings with exotic properties.

1. Introduction

This paper describes the first step in a construction of a skew field with a finitely generated multiplicative group.

We will construct this skew field as a quotient ring of a group algebra of a finitely generated free group $\mathcal{F}$. The full construction of such a skew field would involve an iterative procedure, in which every step is similar to the one described in this paper, in a more complicated situation. The resulting injective limit will be either 0 or a skew field.

Our main objective is to show that the resulting skew field is non trivial, and, moreover, to obtain a skew field of infinite dimension over its center in which every non zero element is equal to a monomial, that is, an element of the free group $\mathcal{F}$. Hence, when $\mathcal{F}$ is finitely generated, the multiplicative group of this skew field is finitely generated. In order to show that the resulting skew field is non trivial, we need to develop its structure theory.

In this paper we consider the following problem. Let $\mathbb{Z}_2\mathcal{F}$ be the group algebra of the free group $\mathcal{F}$ over the field $\mathbb{Z}_2$, and let $1 + w$ be a binomial. We would like to find a quotient ring of $\mathbb{Z}_2\mathcal{F}$ in which the image of $1 + w$ is invertible, without losing control of the quotient ring relations.

Our way to deal with this problem is influenced by an analogue with small cancellation in groups. Namely, we equate $(1 + w)^{-1}$ to a complicated word

$$v = x^{\alpha}yx^{\alpha+1}y \cdots x^{\beta-1}y$$

with $|w| \ll \alpha \ll \beta$, where $|w|$ is a word length of w in $\mathcal{F}$. The monomial v exhibits small cancellation properties because every subword of v that contains at least two letters y appears in v only once ([**7**]).

2010 *Mathematics Subject Classification.* 20F67, 16S15.

In the case of groups we consider semicanonical monomials, namely, monomials that do not contain parts of the relations (or of their cyclic conjugates and inverses) that are too big. It turns out that semicanonical monomials representing the same element of the group are connected by a so-called one-layer diagram. The transition between semicanonical monomials representing the same element is done by substituting subwords of the relations by the inverses of their complements repeatedly (such transitions are called turns).

Here we imitate this process with inevitable complications. Since our relations are not binomial but polynomial, we substitute a subword by the sum of the rest of the monomials of the relation. We call such transitions multi-turns. The simple-minded picture goes like this

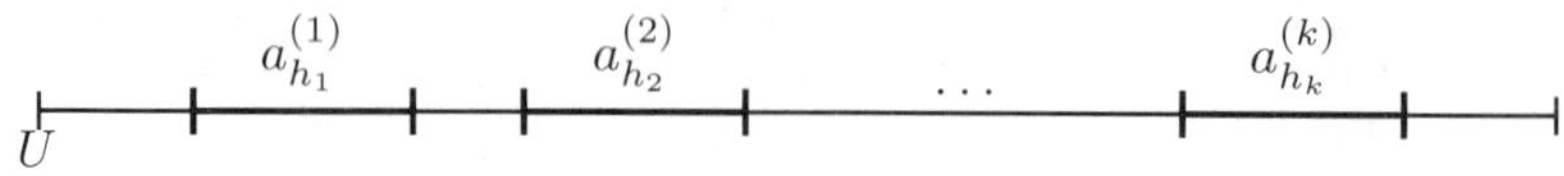

where U is a monomial, and we have polynomial relations

$$\sum_{j_i=1}^{l_i} a_{j_i}^{(i)} = 0,\ i = 1, \dots, k,\ h_i \in \{1, \dots, l_i\}.$$

Performing a number of multi-turns, we obtain sums of monomials of the form

We have $l_1 \cdot \ldots \cdot l_k$ monomials with linear dependencies induced by the multi-turns. In our (simple-minded) case we obtain a linear space of dimension $(l_1-1)\cdot \ldots \cdot (l_k-1)$.

The actual situation is more complicated, due to a number of factors. Let us list some of them:

- Some of the words $a_{j_i}^{(i)}$ might be short, causing certain degeneracies.
- There can be highly non-trivial interactions between the neighbour occurrences, so that the order of performing the multi-turns might matter.

This explains why pursuing this program turns out not entirely simple. We have to introduce a number of elaborate concepts to deal with the monomials and the subwords of the relations appearing in them (for this, see especially Section 3).

To control the degenerations, we introduce a (decreasing) filtration and study its properties (Section 4). Then we need to study the interaction between the linear dependencies and the filtration. Because of the properties like transversality and non-degeneracy, this interaction has nice properties (Section 4.2, Section 5). Our final result (Theorem 5.1) supplies an explicit linear basis for the quotient ring. In particular, the quotient ring does not collapse to 0.

Then we introduce a linear generating set of our quotient ring with particularly nice properties. For this generating set, the multiplication can be expressed as a sum of monomials forming thin triangles with the factors (Section 6, Theorem 6.1). Small cancellation groups are hyperbolic, so their multiplication is expressed by thin triangles. While we do not possess the concept of a hyperbolic ring (even for algebras over a field), the quotient ring we construct does display some features of what might be expected for a "hyperbolic ring".

2. Basic definitions

Consider the group ring $\mathbb{Z}_2\mathcal{F}$, where $\mathcal{F}$ is a free group with at least 4 free generators. We will call elements of $\mathcal{F}$ words or monomials. We deal only with reduced monomials unless the converse is explicitly stated. For our purposes we can use any field k, but we choose the field $\mathbb{Z}_2$ to simplify our calculations. Let us fix $w \in \mathcal{F}$, an arbitrary cyclically reduced primitive (not a proper power) word from $\mathcal{F}$. Fix positive integers α and β such that $|w| \ll \alpha \ll \beta$, where $|w|$ is a word length of w in $\mathcal{F}$. Our aim is inverting of the binomial $1+w$. Consider the word

$$v = x^{\alpha}yx^{\alpha+1}y\cdots x^{\beta-1}y \tag{1}$$

such that w does not start or end with the letters x, y, x^{-1}, y^{-1}. We are going to study a structure of the quotient ring $\mathbb{Z}_2\mathcal{F}/\mathcal{I}$, where

$$\mathcal{I} = \langle 1 + v + vw\rangle. \tag{2}$$

Clearly, in the quotient ring we have $v(1+w) = 1$. Multiplying it by v on the right side, we obtain $v(1+w)v = v$. Since v is a word, it is invertible in $\mathbb{Z}_2\mathcal{F}$, so, it is also invertible in $\mathbb{Z}_2\mathcal{F}/\mathcal{I}$ or the quotient ring is trivial. Multiplying the last equation by v^{-1}, we obtain $(1+w)v = 1$. Thus, $v^{-1} = 1+w$ in the ring $\mathbb{Z}_2\mathcal{F}/\mathcal{I}$.

DEFINITION 2.1. Suppose $M(v,w)$ is a non-commutative monomial over the words v, w. We call arbitrary subwords of monomials $M(v,w)$ *$(v,\ w)$-generalized fractional powers* (or simply *generalized fractional powers*).

Let v_1 be an arbitrary subword of v (or of v^{-1}). We define an additive (length) measure (aka Λ-measure) on subwords of v (or v^{-1}) in the following way

$$\Lambda(v_1) = \frac{m_y(v_1)}{\beta - \alpha}, \tag{3}$$

where $m_y(v_1)$ is the number of occurrences of the letter y (or of the letter y^{-1}) in v_1. Obviously, $\Lambda(v) = 1$ ($\Lambda(v^{-1}) = 1$). For any subword w_1 of w we put $\Lambda(w_1) = 0$. From now on, let us call Λ-measure of a generalized fractional power the sum of Λ-measure of its parts.

Let U be a word and U_1 be its subword. We call the triple that consists of U, U_1 and the position of U_1 in U *an occurrence of U_1 in U*.

An arbitrary word from $\mathcal{F}$ may contain occurrences of generalized fractional powers. While the notion of a maximal occurrence seems to be intuitively simple, it turns out that the notion we actually need has considerable subtleties.

So, we define a maximal occurrence in the following way. Let a be a generalized fractional power, that is, a is an occurrence in a monomial $M(v,w) = M_L a M_R$. Assume that $M_L = M_1(v,w)M_L'$, where $M_1(v,w)$ is a monomial in v, w, and M_L' has no initial subword equal to a monomial in v, w. In the same way, assume that $M_R = M_R' M_2(v,w)$, where $M_2(v,w)$ is a monomial in v, w, and M_R' has no final subword equal to a monomial in v, w. Then we consider a as an occurrence in the monomial

$$M_0(v,w) = M_1(v,w)^{-1}M(v,w)M_2^{-1}(v,w) = M_L' a M_R'.$$

Consider a set of monomials

$$\mathcal{A} = \{M(v,w) \mid M(v,w) = M_i(v,w)M_0(v,w)M_j(v,w), i,j \in \mathbb{N}\},$$

where $M_i(v, w)$ and $M_j(v, w)$ are arbitrary monomials in v, w. Then the occurrence of a in $M_0(v, w)$ can be considered as the occurrence of a in $M(v, w) \in \mathcal{A}$ with the same position in $M_0(v, w)$.

Suppose a is an occurrence in a word $U \in \mathcal{F}$, a letter a_1 prolongs a in U from the left side, a letter a_2 prolongs a in U from the right side. If a is an initial subword of U, a_1 is absent; if a is a final subword of U, a_2 is absent. If for any $M(v, w) \in \mathcal{A}$ the occurrence a can be prolonged neither by a_1 on the left nor by a_2 on the right in $M(v, w)$, then we call a *a maximal occurrence of a generalized fractional power in* U.

Let $U = U_L a U_R$. The above definition in fact means the following. The occurrence a is a maximal occurrence of a generalized fractional power in U if and only if it satisfies one of the conditions:

(L1) U_L is the empty word;
(L2) $U_L = U_L' a_1$, $M_L' = M_L'' a_1'$, $a_1 \neq a_1'$;
(L3) $U_L = U_L' a_1$, M_L' is the empty word, the last letter in each $M_i(v, w)$ is different from a_1;

and one of the conditions

(R1) U_R is the empty word;
(R2) $U_R = a_2 U_R'$, $M_R' = a_2' M_R''$, $a_2 \neq a_2'$;
(R3) $U_R = a_2 U_R'$, M_R' is the empty word, the first letter in each $M_j(v, w)$ is different from a_2.

DEFINITION 2.2. Let τ be some small value, U be an arbitrary word from $\mathcal{F}$. We call the set of maximal occurrences of generalized fractional powers in U of Λ-measure greater or equal to a given threshold τ *the chart of the word* U, and we call the corresponding occurrences of generalized fractional powers *members of the chart.*

Let us put $\varepsilon = \frac{1}{\beta - \alpha}$. Further we will assume that $\tau \geqslant 10\varepsilon$.

Any subword of v containing at least two letters y appears in a unique way in v. Moreover, a subword of v^l containing at least two letters y appears in v^l uniquely modulo the period v. Therefore, if the Λ-measure of some maximal occurrence of a generalized fractional power is greater or equal to 2ε, it can not be properly contained in another occurrence of a generalized fractional power (because otherwise it would not be maximal). In particular, given the chart of some word, one member of the chart can not be properly contained in another. However, members of the chart may have overlaps.

Consider an overlap of two members of the chart. If the overlap is a common part of two subwords of v^l, then according to the last remark, the Λ-measure of the overlap is not greater than ε. If the overlap contains a subword of w^k, it may have a more complicated form. Since $\alpha \gg |w|$, w does not contain subwords $y^{\delta_1} x^{\gamma} y^{\delta_2}$, where $|\gamma| \geqslant \alpha$. Hence, since the Λ-measure of subwords of w is equal to zero, the Λ-measure of the overlap still is not greater than ε for both members of the chart. In particular, since $\tau \geqslant 10\varepsilon$, we never have a situation when one member of the chart is fully covered by others.

DEFINITION 2.3. Consider oriented graphs with edges marked by generators of the group $\mathcal{F}$. Take such a graph of the form

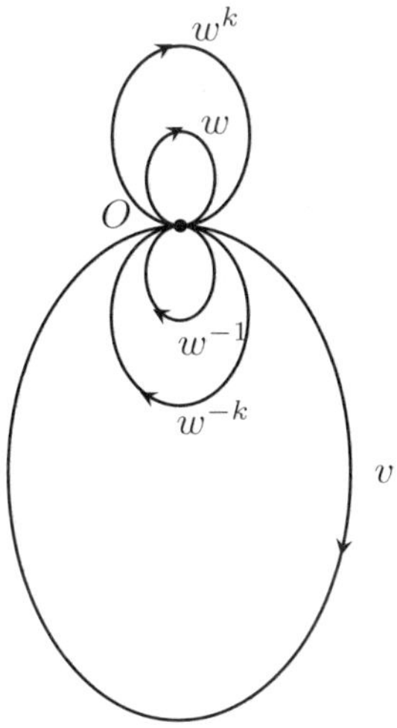

(4)

Here to each integer power of w corresponds a separate arc. That is, there are infinitely many arcs that correspond to different w^k. We call this graph a *v-diagram.*

Assume that we have an oriented path in the graph (4). When we go along this path, we can write down the mark of an edge if we pass the edge in the positive direction, and we can write down the inverse to the mark of an edge if we pass the edge in the negative direction. As a result, we obtain a generalized fractional power (possibly after cancellations if there are any).

It is easy to see that to each monomial over v, w there corresponds a path in the graph (4) with the initial and the final vertex O. By the definition, every generalized fractional power is a subword in a monomial over v, w. Hence, given a generalized fractional power M, one can specify two points I and F on the graph (4) such that M corresponds to the unique path starting at I and ending at F. For the sake of uniqueness, we assume that we always choose an arc with maximal absolute value of degree of w along the path. So, one can see that M corresponds to the path in the graph of one of the types (5)–(7).

Note that if a generalized fractional power is of Λ-measure strictly greater than ε, then the positions of the initial and the final points that belong to the v-arc are uniquely determined.

In what follows we use for v the notation $v = v_i v_m v_f$, where v_i is some initial part, v_m is some middle part and v_f is some final part of v (any part is allowed to be empty). Analogously to the notion of v-diagram, in this paper we represent all monomials (not only generalized fractional powers) as segments that consist of oriented edges marked by generators of the group $\mathcal{F}$; such segments we always read from left to right. So, for v we have

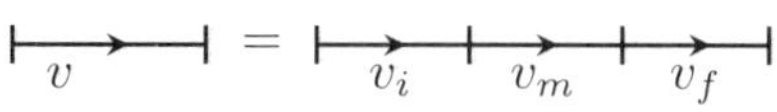

Similarly, for w we use the notation $w = w_i w_m w_f$, where w_i is some initial part, w_m is some middle part and w_f is some final part of w (any part is allowed to be empty).

Let us enumerate all possible positions of points I and F on the v-diagram. Thereby, we enumerate all possible types of generalized fractional powers. In every case, we explicitly write the corresponding forms of generalized fractional powers. As above, $M(v, w)$ is a monomial over v and w. First, both points I and F may lie on the v-arc. Then they divide the v-arc into three parts and, according to the notations introduced above, we denote them v_i, v_m and v_f and obtain the pictures

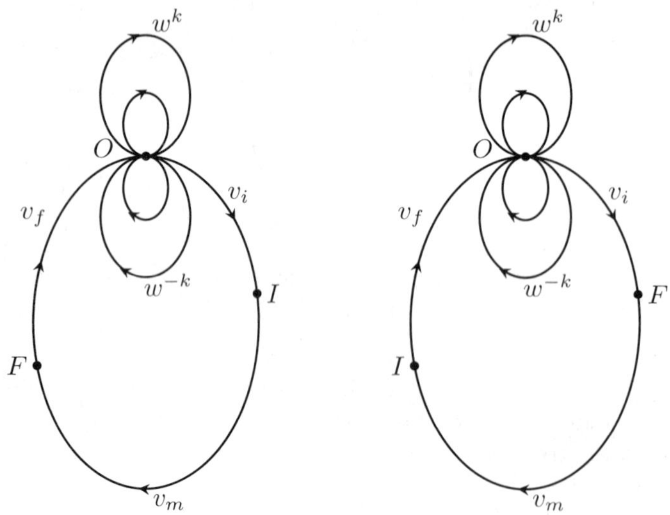

$$\begin{array}{ll} v_m v_f M(v,w) v_f^{-1}, & v_f M(v,w) v_i, \\ v_m v_f M(v,w) v_i v_m, & v_f M(v,w) v_f^{-1} v_m^{-1}, \\ v_i^{-1} M(v,w) v_i v_m, & v_m^{-1} v_i^{-1} M(v,w) v_i, \\ v_i^{-1} M(v,w) v_f^{-1}; & v_m^{-1} v_i^{-1} M(v,w) v_f^{-1} v_m^{-1}. \end{array} \tag{5}$$

The next configuration is when the point I lies on the v-arc and the point F lies on a w-arc or vice versa. Then, according to the above notations, we put $v = v_i v_f$, $w = w_i w_f$ and obtain the pictures

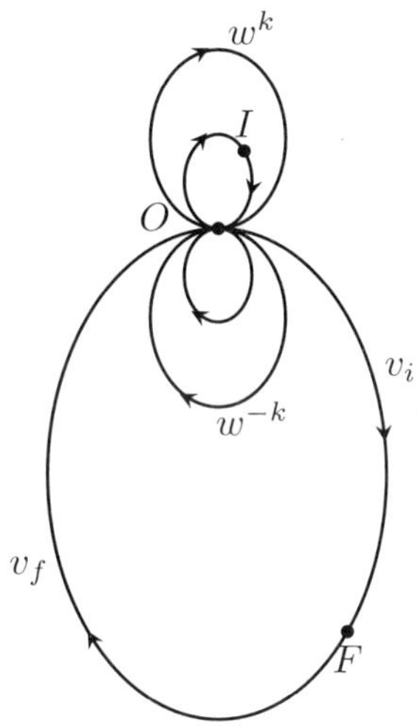

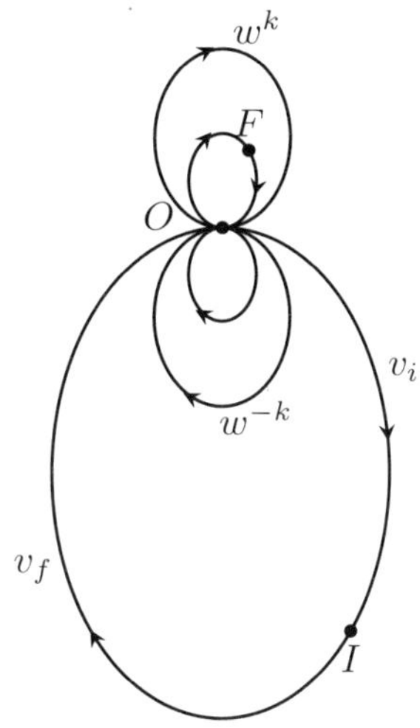

$w_f M(v,w) v_i,$	$v_f M(v,w) w_i,$
$w_f M(v,w) v_f^{-1},$	$v_i^{-1} M(v,w) w_i,$
$w_i^{-1} M(v,w) v_i,$	$v_f M(v,w) w_f^{-1},$
$w_i^{-1} M(v,w) v_f^{-1};$	$v_i^{-1} M(v,w) w_f^{-1}.$

(6)

In picture (6) points I and F lie on a positive w-arc, but they may lie on a negative w-arc within this type of paths as well.

The last configuration is when both points I and F lie on a w-arc. Then, according to the above notations, we put $w = w_i w_m w_f$ and obtain the pictures

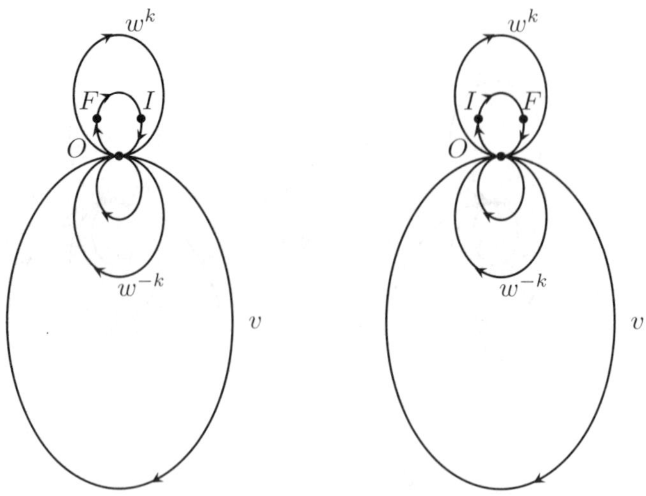

$$\begin{array}{ll} w_f M(v,w) w_i, & w_m w_f M(v,w) w_f^{-1}, \\ w_f M(v,w) w_f^{-1} w_m^{-1}, & w_m w_f M(v,w) w_i w_m, \\ w_m^{-1} w_i^{-1} M(v,w) w_i, & w_i^{-1} M(v,w) w_i w_m, \\ w_m^{-1} w_i^{-1} M(v,w) w_f^{-1} w_m^{-1}; & w_i^{-1} M(v,w) w_f^{-1}. \end{array} \tag{7}$$

In picture (7) points I and F lie on a positive w-arc, but each of them may lie on a negative w-arc within this type of paths as well. Also the point I and the point F may lie on different w-arcs.

Given a set of generalized fractional powers M_j, $j = 1, \ldots, k$, that correspond to the paths with the same initial and the same final point in the diagram of type (5)–(7), their sum $\sum_{j=1}^{k} M_j$ corresponds to the collection of these paths.

Let $\mathbb{Z}_2(w)$ be the field of rational functions in one variable w over $\mathbb{Z}_2$. Consider a non-commutative Laurent polynomial $P(x_1, x_2)$ over $\mathbb{Z}_2$ such that

$$P((1+w)^{-1}, w) = 0 \textit{ as an element of } \mathbb{Z}_2(w). \tag{8}$$

Notice that if $P(x_1, x_2)$ satisfies condition (8), then $P(v, w) \in \langle 1 + v + vw \rangle = \mathcal{I}$.

EXAMPLE 2.1. Let us give an example of such polynomials. The first example comes from the obvious equality in $\mathbb{Z}_2(w)$

$$w \cdot \frac{1}{1+w} = \frac{1}{1+w} \cdot w.$$

From this equality it follows that the polynomial

$$P(x_1, x_2) = x_1 x_2 + x_2 x_1$$

satisfies condition (8) and

$$P(v, w) = vw + wv \in \mathcal{I}.$$

Recalling the reduction of rational functions of the form $\frac{w^{\pm k}}{1+w}$ in $\mathbb{Z}_2(w)$ to elementary fractions, we obtain the following equalities in $\mathbb{Z}_2(w)$ for $k > 0$:

$$\begin{aligned}\frac{w^k}{1+w} &= \frac{w^k + w^{k-1} + w^{k-1} + \ldots + w + w + 1 + 1}{1+w} \\ &= w^{k-1} + w^{k-2} + \ldots + w + 1 + \frac{1}{1+w}\end{aligned}$$

and

$$\begin{aligned}\frac{w^{-k}}{1+w} &= \frac{w^{-k} + w^{-k+1} + w^{-k+1} + \ldots + w^{-1} + w^{-1} + 1 + 1}{1+w} \\ &= w^{-k} + w^{-k+1} + \ldots + w^{-1} + \frac{1}{1+w}.\end{aligned}$$

Hence, the polynomials

$$\begin{aligned}P_1(x_1, x_2) &= x_1 x_2^k + x_2^{k-1} + x_2^{k-2} + \ldots + x_2 + 1 + x_1, \\ P_2(x_1, x_2) &= x_1 x_2^{-k} + x_2^{-k} + x_2^{-k+1} + \ldots + x_2^{-1} + x_1\end{aligned}$$

satisfy condition (8) and

$$\begin{aligned}P_1(v, w) &= vw^k + w^{k-1} + w^{k-2} + \ldots + w + 1 + v \in \mathcal{I}, \\ P_2(v, w) &= vw^{-k} + w^{-k} + w^{-k+1} + \ldots + w^{-1} + v \in \mathcal{I}.\end{aligned}$$

According to types (5)–(7) of v-diagrams that correspond to possible forms of generalized fractional powers, we consider the list of expressions (9)–(14), where $P(x_1, x_2)$ is a non-commutative Laurent polynomial such that $P((1+w)^{-1}, w) = 0$ as an element of $\mathbb{Z}_2(w)$. We allow possibility of cancellations in monomials in (9)–(14). First, consider expressions

$$\begin{aligned}&v_f P(v, w) v_i, \\ &v_f P(v, w) v_f^{-1} v_m^{-1}, \\ &v_m^{-1} v_i^{-1} P(v, w) v_i, \\ &v_m^{-1} v_i^{-1} P(v, w) v_f^{-1} v_m^{-1}.\end{aligned} \tag{9}$$

To each expression of (9), there corresponds a collection of paths in the graph (5). Similarly, to each expression

$$\begin{aligned}&v_m v_f P(v, w) v_f^{-1}, \\ &v_m v_f P(v, w) v_i v_m, \\ &v_i^{-1} P(v, w) v_i v_m, \\ &v_i^{-1} P(v, w) v_f^{-1},\end{aligned} \tag{10}$$

there corresponds a collection of paths in the graph (5). To each expression

$$\begin{aligned}&w_f P(v, w) v_i, \\ &w_f P(v, w) v_f^{-1}, \\ &w_i^{-1} P(v, w) v_i, \\ &w_i^{-1} P(v, w) v_f^{-1},\end{aligned} \tag{11}$$

there corresponds a collection of paths in the graph (6). Similarly, to each expression

$$\begin{aligned} &v_f P(v,w) w_i, \\ &v_i^{-1} P(v,w) w_i, \\ &v_f P(v,w) w_f^{-1}, \\ &v_i^{-1} P(v,w) w_f^{-1}, \end{aligned} \tag{12}$$

there corresponds a collection of paths in the graph (6). Finally, consider the expressions

$$\begin{aligned} &w_f P(v,w) w_i, \\ &w_f P(v,w) w_f^{-1} w_m^{-1}, \\ &w_m^{-1} w_i^{-1} P(v,w) w_i, \\ &w_m^{-1} w_i^{-1} P(v,w) w_f^{-1} w_m^{-1}. \end{aligned} \tag{13}$$

To each expression of (13) there corresponds a collection of paths in the graph (7). Similarly, to each expression

$$\begin{aligned} &w_m w_f P(v,w) w_f^{-1}, \\ &w_m w_f P(v,w) w_i w_m, \\ &w_i^{-1} P(v,w) w_i w_m, \\ &w_i^{-1} P(v,w) w_f^{-1}, \end{aligned} \tag{14}$$

there corresponds a collection of paths in the graph (7). Since $P(v,w) \in \mathcal{I}$, the expressions (9)–(14) vanish in $\mathbb{Z}_2\mathcal{F}/\mathcal{I}$.

Every expression (9)–(14) is, in fact, a linear combination of generalized fractional powers. For any linear combination $\sum_{j=1}^{k} M_j$ of such type (where the cancellations in monomials are already performed) we call the monomials M_{j_1}, M_{j_2}, $j_1, j_2 \in \{1,\ldots,k\}$, *incident monomials.* Notice that the paths in the v-diagram corresponding to incident monomials always have the same initial points and the same final points. Clearly, one generalized fractional power has an infinite number of incident monomials because they may contain different powers of v and w.

Suppose we have a group G with a relator $R = M_1 M_2^{-1}$. Let $U = LM_1R$. Recall that the transition from $U = LM_1R$ to LM_2R representing the same element of G

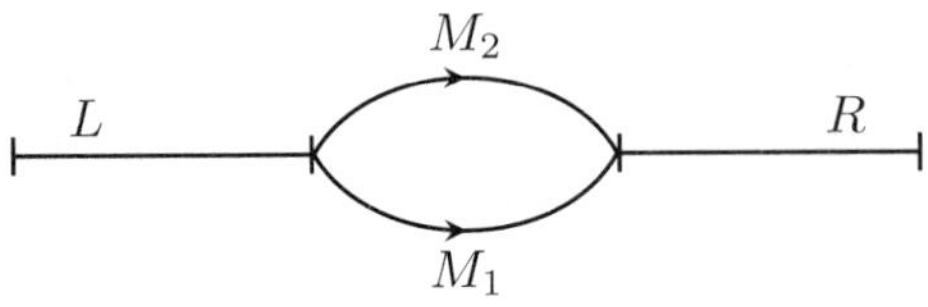

is called a *turn* of an occurrence of a subrelation M_1 (to its complement M_2). We generalize this notion to a multi-turn. Multi-turns will play a central role in our work.

Definition 2.4. Let $\sum_{j=1}^k M_j$ be one of the expressions (9)–(14) (where the cancellations in monomials are already performed). We call the transition

$$M_h \longmapsto \sum_{\substack{j=1\\ j\neq h}}^{k} M_j,$$

an *elementary multi-turn of* M_h.

Let $U_h = LM_hR$

$U_h =$ 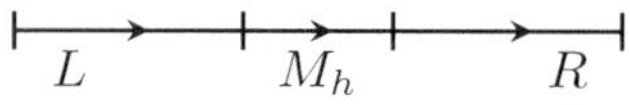

be a monomial in $\mathbb{Z}_2\mathcal{F}$, where M_h is a generalized fractional power, and $\sum_{j=1}^k M_j$ be one of the expressions (9)–(14), then the transition

$$U_h \longmapsto \sum_{\substack{j=1\\ j\neq h}}^{k} U_j,$$

where $U_j = LM_jR$,

$U_j =$ 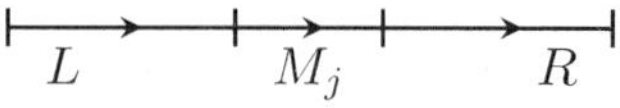

$j = 1, \ldots, k, j \neq h,$

is called a *multi-turn of the occurrence* M_h *in* U_h. Notice that since $\sum_{j=1}^k M_j \in \mathcal{I}$, $U_h = \sum_{\substack{j=1\\ j\neq h}}^{k} U_j$ in the ring $\mathbb{Z}_2\mathcal{F}/\mathcal{I}$.

We call the corresponding sum $\sum_{j=1}^k U_j$ *the support of the multi-turn.* Then the expressions (9)–(14) are all types of supports of elementary multi-turns.

3. Linear dependencies on $\mathbb{Z}_2\mathcal{F}$ induced by multi-turns. The description of the ideal $\mathcal{I}$ as a linear subspace of $\mathbb{Z}_2\mathcal{F}$

3.1. How multi-turns influence the chart. Let U be a word and a and b be members of its chart, i.e., maximal occurrences of generalized fractional powers. Since the monomial U is represented as a segment, it is natural to represent members of the chart of U as its subsegments. Let us describe possible configurations of the members a and b.

(1) We say that the members of the chart a and b are *separated* if there exists some non empty subword between them in the word U.

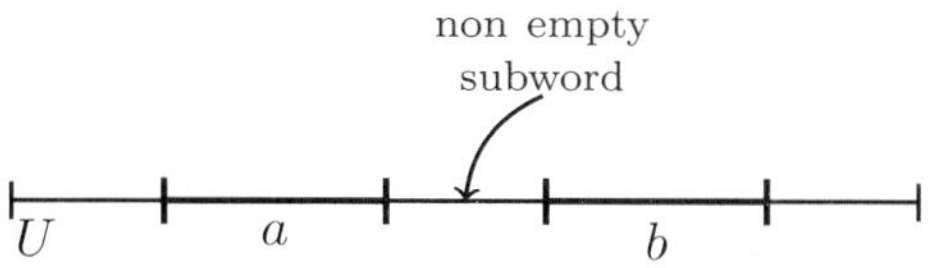

(2) We say that the members a and b *touch at a point* if a and b are adjacent and have no common non empty subword.

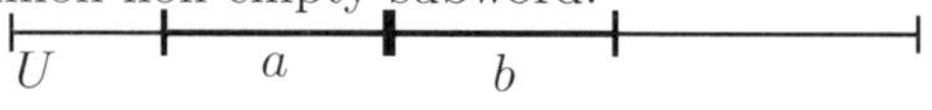

(3) We say that the members a and b *have an overlap* if the members have non empty common subword. Recall that the Λ-measure of the overlap is not greater than ε.

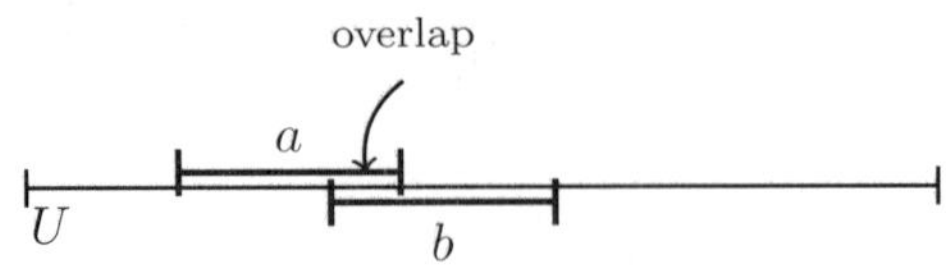

We will call the member of the chart of U closest to a from the left side *the left neighbour of a in the chart of U*. Similarly, we will call the member of the chart of U closest to a from the right side *the right neighbour of a in the chart of U*. Recall that maximal occurrences of generalized fractional powers are considered as members of the chart only when they have Λ-measure greater or equal to a given threshold τ and we assume that $\tau \geqslant 10\varepsilon$. The restriction on the Λ-measure allows us to avoid situations when one member of the chart is fully covered by other members. So, we avoid the situation when there exist two members that are not separated from a on one side. For example, we never obtain configurations like the following:

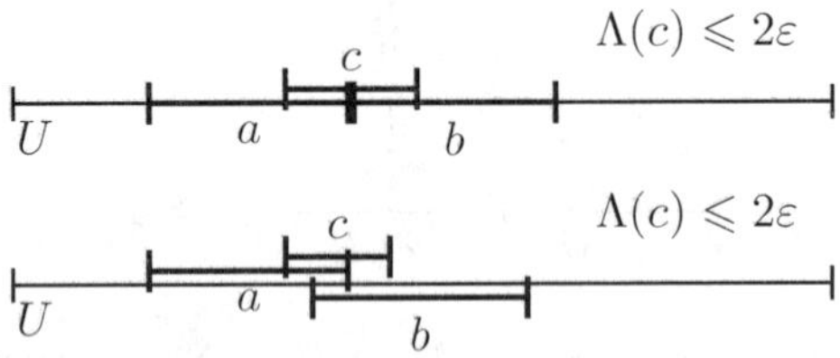

Suppose $U_h = La_hR$ is a word, a_h is a member of its chart, in particular $\Lambda(a_h) \geqslant \tau$. Suppose $U_h \mapsto \sum_{\substack{j=1\\ j\neq h}}^{k} U_j$ is a multi-turn of U_h that comes from an elementary multi-turn $a_h \mapsto \sum_{\substack{j=1\\ j\neq h}}^{k} a_j$. So, $U_h = \sum_{\substack{j=1\\ j\neq h}}^{k} U_j$ in the ring $\mathbb{Z}_2\mathcal{F}/\mathcal{I}$. Let us study how the chart of $U_j = La_jR$, $j = 1, \dots, k$, $j \neq h$, is related to the chart of $U_h = La_hR$. We distinguish three types of monomials in the sum $\sum_{\substack{j=1\\ j\neq h}}^{k} U_j$:

(1) La_jR, where $\Lambda(a_j) > \varepsilon$;
(2) La_jR, where $a_j = 1$.
(3) La_jR, where $\Lambda(a_j) \leqslant \varepsilon$ but $a_j \neq 1$;

Let $U_j = La_jR$ be a monomial of type 1, that is, $\Lambda(a_j) > \varepsilon$. Let us study the chart of La_jR. First of all notice that since a_h is a maximal occurrence of a generalized fractional power in La_hR, the monomial $U_j = La_jR$ has no cancellations and a_j is a maximal occurrence of a generalized fractional power in U_j. Clearly, all members of the chart of $U_h = La_hR$ that are separated from a_h remain unchanged in the chart of $U_j = La_jR$. Let b_h be the left neighbour of a_h in the chart of U_h. Assume b_h is not separated from a_h, and b'_h is its initial subword such that $L = L'b'_h$, $U_j = L'b'_h a_h R$. Then there is a corresponding occurrence of b'_h in U_j. Let b_j be the member of the chart of U_j that prolongs b'_h. Clearly, the member of the chart of U_h that prolongs b'_h is precisely b_h. The member b_j may differ from b_h when $j \neq h$. There are four possibilities:

1.1 The members b_h and a_h touch at a point in U_h. The members b_j and a_j have a non-empty overlap in U_j. Assume c is the overlap between b_j and a_j. In this case $b_j = b_hc$. For example, we may obtain this effect when a beginning of v_m is replaced by an end of v_i^{-1} or vice versa.

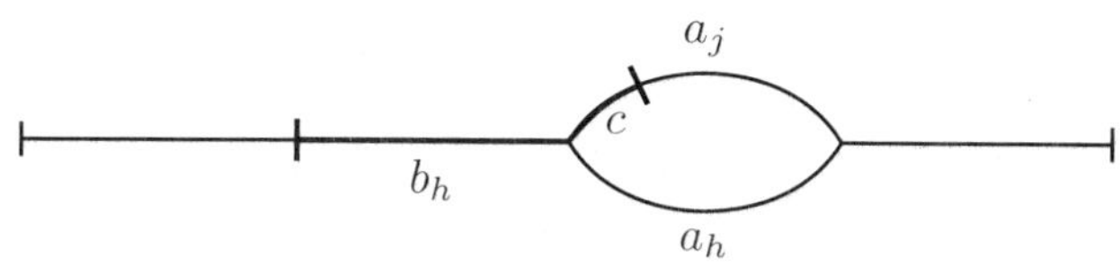

1.2 The members b_h and a_h have an overlap in U_h. The member b_h is enlarged in U_j with the use of a_j and the overlap between b_j and a_j increases in U_j. If the overlap increases by a piece c, then $b_j = b_h c$.

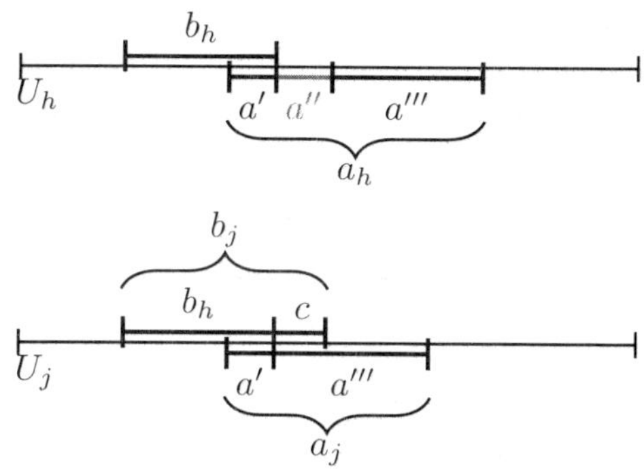

1.3 The members b_h and a_h have an overlap in U_h. The member b_h is shortened in U_j such that b_j and a_j touch at a point in U_j. For example, we obtain this effect when a beginning of v_m is replaced by an end of v_i^{-1} or vice versa. In this case $b_h = b_j c$.

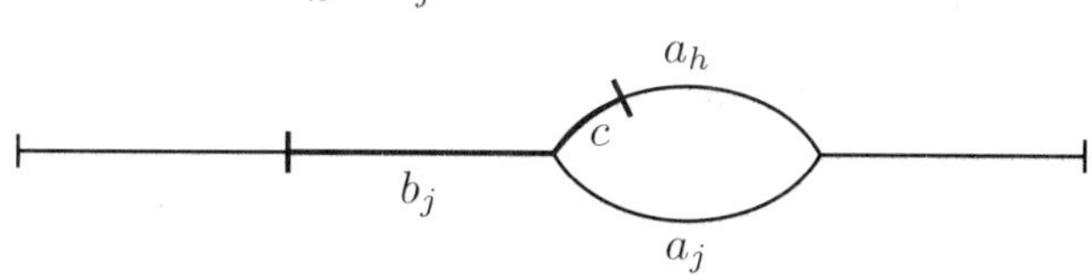

1.4 The members b_h and a_h have an overlap in U_h. The member b_h is shortened in U_j and the overlap between b_j and a_j becomes shorter in U_j. If the overlap decreases on a piece c, then $b_h = b_j c$.

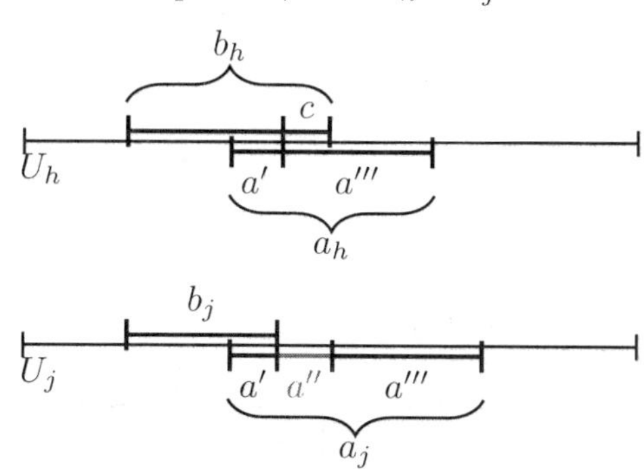

The essential observation is that either $b_h = b_j c$, or $b_j = b_h c$; in both cases the Λ-measure of c can not exceed ε. The possibilities for being changed for the right neighbour of a_h in the chart are the same.

The cases described in 1.1–1.4 may occur even if $\Lambda(b_h) < \tau$ or $\Lambda(b_j) < \tau$. Therefore, the chart of U_j also may change as a result of the following effect.

Assume $\tau - \varepsilon \leqslant \Lambda(b_h) < \tau$. Then we may obtain $\Lambda(b_j) \geqslant \tau$ in U_j, that is, b_j is counted as a member of the chart of U_j. So, a new member appears in the chart of U_j. It can also happen that $\Lambda(b_j) < \tau$, while $\Lambda(b_h) \geqslant \tau$.

Consider the chart of a monomial $U_j = LR$ of type 2. Cancellations between L and R may occur. Suppose $L = L'C$, $R = C^{-1}R'$ and $L'R'$ does not have further cancellations. Let P be the meeting point of L' and R'. Members of the chart of $L'R'$ that are separated from the point P remain unchanged. The possible configurations are as follows:

2.1 L' does not have a terminal subword equal to some generalized fractional power and R' does not have an initial subword equal to some generalized fractional power. Then the chart of $L'R'$ consists of members that lie left to P in L' and right to P in R'.

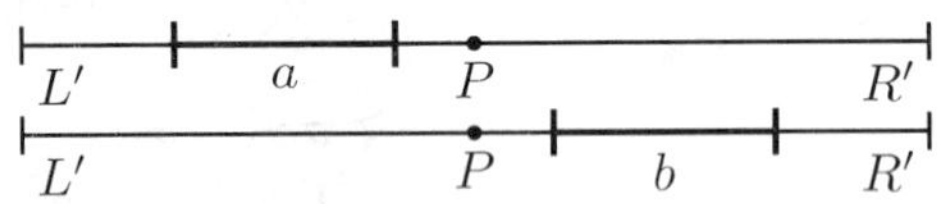

Suppose $L' = L'_1 a$, $R' = bR'_1$, a and b are maximal occurrences of generalized fractional powers in L' and in R', respectively, possibly with Λ-measure less than τ. We allow one of a or b to be equal to 1. The occurrence of a in $L'R'$ does not have to be maximal. Let us denote by a' the maximal occurrence of a generalized fractional power that prolongs a in $L'R'$. Similarly, let us denote by b' the maximal occurrence of a generalized fractional power that prolongs b in $L'R'$. Then interactions between a' and b' are as follows:

2.2 a' and b' touch at a point in $L'R'$. Then like in the case 2.1 the chart of $L'R'$ consists precisely of the members contained in L' and the members contained in R'.

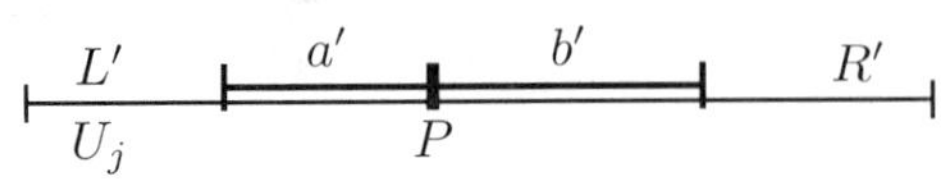

2.3 a' and b' in $L'R'$ have an overlap. Then a' is at most slightly enlarged with respect to a, and b' is at most slightly enlarged with respect to b. Since the Λ-measure of their overlap does not exceed ε, their total increase does not exceed ε.

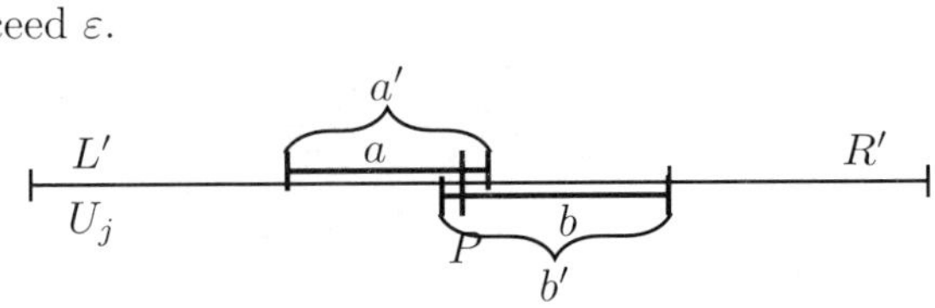

2.4 a' and b' merge to one generalized fractional power ab in $L'R'$.

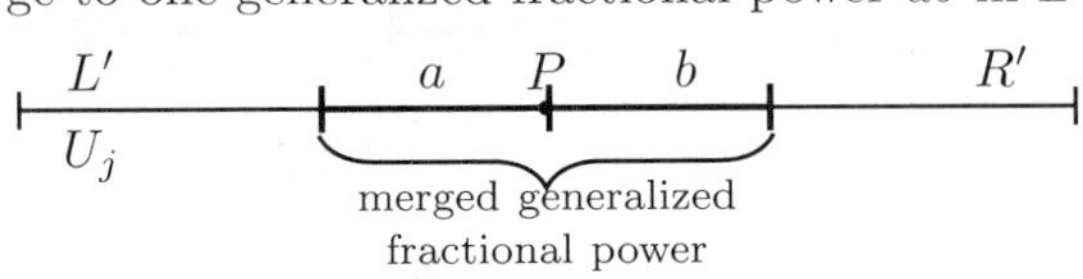

Consider the chart of monomials $U_j = La_jR$ of type 3. First of all notice that since a_h is a maximal occurrence of a generalized fractional power in U_0, the monomial $U_j = La_jR$ has no cancellations. The possible configurations are as follows:

3.1 L does not have a terminal subword equal to some generalized fractional power and R does not have an initial subword equal to some generalized fractional power. Then all members of the chart of La_jR are separated from a_j. In this case no new members are added; hence, the chart of U_j consists of occurrences that lie strictly left to a_j in L and strictly right to a_j in R.

Suppose $L = L_1a$, $R = bR_1$, a and b are maximal occurrences of generalized fractional powers in L and R, respectively, possibly with Λ-measure less than τ. We allow one of a or b to be equal to 1. As above, denote by a' the maximal occurrence of a generalized fractional power that prolongs a in La_jR, and denote by b' the maximal occurrence of a generalized fractional power that prolongs b in La_jR. We have the possibilities similar to the above 2.2–2.4.

3.2 a' remains equal to a and b' remains equal to b in U_j. Hence, a' and b' are separated in La_jR.

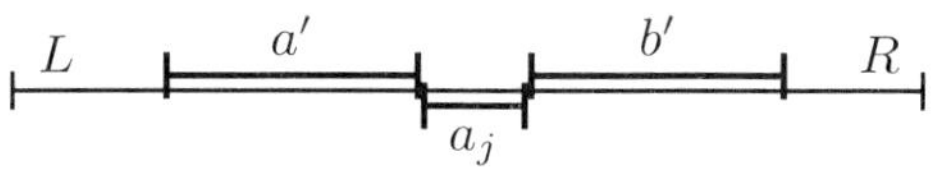

3.3 a' slightly enlarges from the right side with respect to a, and b' slightly enlarges from the left side with respect to b. Let us illustrate possible configurations of a' and b'.

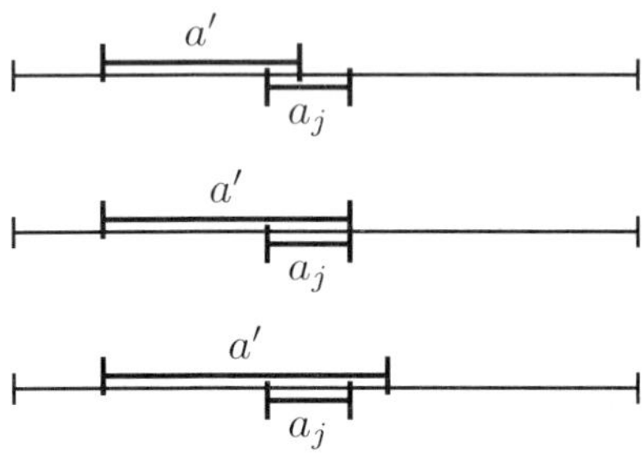

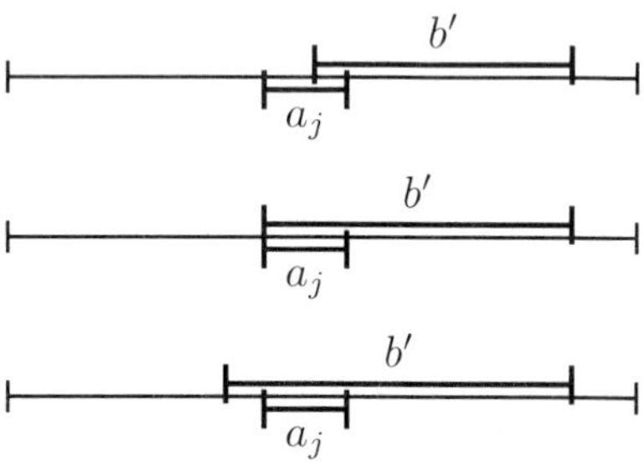

We may obtain a combination of any two configurations for a' and for b' in La_jR; hence, a' and b' may have any mutual position in La_jR (i.e., remain either separate, or touch at a point, or have an overlap). Let us illustrate one possibility as an example.

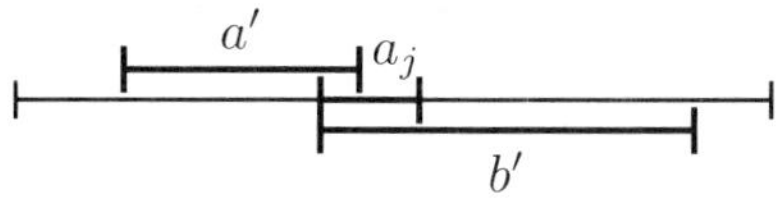

Since Λ-measure of a possible overlap of a' and b' does not exceed ε, the total increase of a' and b' with respect to a and b does not exceed 2ε. Hence, if Λ-measure of both a' and b' increases, Λ-measure of each of them increases at most by ε.

3.4 a' enlarges with respect to a, b' remains the same, or b' enlarges with respect to b, a' remains the same (the case when both of them enlarge is treated in 3.3). As above, a' and b' may have any mutual position in La_jR (i.e.,either remain separate, or touch at a point, or have an overlap). In this case Λ-measure of a' or b' increases at most by 2ε. Let us illustrate one important possibility, the case when Λ-measure of a' increases precisely by 2ε.

The most important case is the following:

3.5 a' and b' merge to one generalized fractional power aa_jb in U_j.

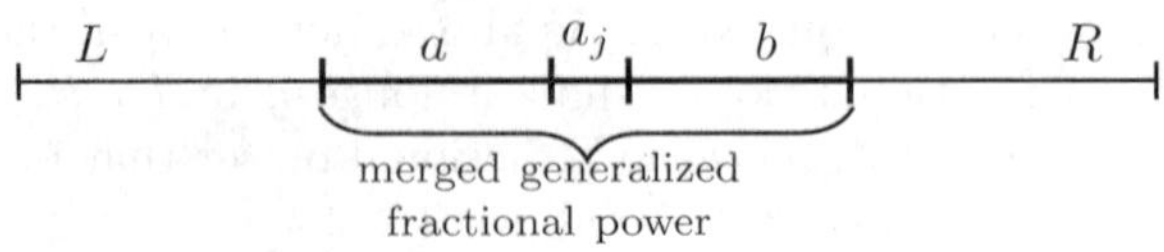

REMARK 3.1. Notice that, using the explicit description of generalized fractional powers (5)–(7) and the explicit description of elementary multi-turns (9)–(14), one can show that the case 3.5 may take place in at most one monomial of the corresponding sum $\sum_{\substack{j=1\\ j\neq h}}^{k} La_jR$.

3.2. Linear dependencies on $\mathbb{Z}_2\mathcal{F}$ induced by multi-turns. First we prove the important property of the supports of elementary multi-turns, that are the expressions (9)–(14). Then we will show that all the supports of multi-turns linearly generate the ideal $\mathcal{I}$. This fact plays the key role for the description of the linear structure of the quotient ring $\mathbb{Z}_2\mathcal{F}/\mathcal{I}$.

PROPOSITION 3.1 (Transversality). *Let $Q_1, \dots, Q_n \in \mathbb{Z}_2\mathcal{F}$ be linear combinations of generalized fractional powers of the form (9)–(14). Consider $\sum_{l=1}^{n} Q_l$. If in this sum, after additively cancelling out the identical monomials, we obtain nonzero, then at least one of the remaining monomials (generalized fractional powers) has Λ-measure $\geqslant \tau$.*

PROOF. Monomials of Q_l can be represented as paths in the corresponding v-diagram with the same initial and the same final point. If some initial or some final point lies on a w-arc, we fix it on the arc marked by the first power of w. Divide the sum $\sum_{l=1}^{n} Q_l$ into subsums corresponding to the same initial and final points (I, F) in the v-diagram

$$\sum_{l=1}^{n} Q_l = \sum_{(I,F)} \sum_{l=1}^{n(I,F)} Q_l^{(I,F)}.$$

Let a be a member of $\sum_{l=1}^{n} Q_l$ such that $\Lambda(a) \geqslant \tau$. Notice that since $\Lambda(a) \geqslant \tau > \varepsilon$, a has uniquely defined points I and F. So, a is a member of $\sum_{l=1}^{n(I,F)} Q_l^{(I,F)}$ and a is not a member of any $\sum_{l=1}^{n(I',F')} Q_l^{(I',F')}$ with $(I', F') \neq (I, F)$. Hence, if a is not additively cancelled out in the sum $\sum_{l=1}^{n(I,F)} Q_l^{(I,F)}$, then a is not additively cancelled out in the sum $\sum_{l=1}^{n} Q_l$. It follows that it is enough to prove Proposition 3.1 for a sum $\sum_{l=1}^{n(I,F)} Q_l^{(I,F)}$ with some fixed (I, F).

Assume the contrary. Suppose all members of $\sum_{l=1}^{n(I,F)} Q_l^{(I,F)}$ after additively cancelling out the identical monomials are of Λ-measure less than τ. Let us study possible forms of monomials in this sum. Suppose b is a monomial in this sum,

$\Lambda(b) < \tau$, I and F are the initial and the final points of the path corresponding to b. There are the following possible positions of I and F:

(1) The points I and F lie on the v-arc.
(2) The points I and F lie on a w-arc.
(3) The point I lies on the v-arc, the point F lies on a w-arc.
(4) The point I lies on a w-arc, the point F lies on the v-arc.

Since $\Lambda(b) < \tau$, the smallest path between I and F is necessarily of Λ-measure less than τ.

Consider case 1. Let us denote the smallest path from I to F by b', the smallest path from I to O by b_1, and the smallest path from O to F by b_2. Recall our notion that $v = v_i v_m v_f$. Then there are the following possibilities:

1.1 b' contains the point O;

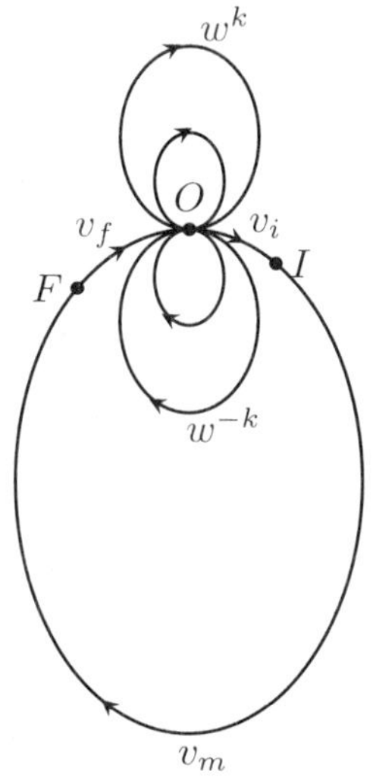

$b_1 = v_i^{-1}, b_2 = v_f^{-1}$,
$b' = v_i^{-1} v_f^{-1}$;

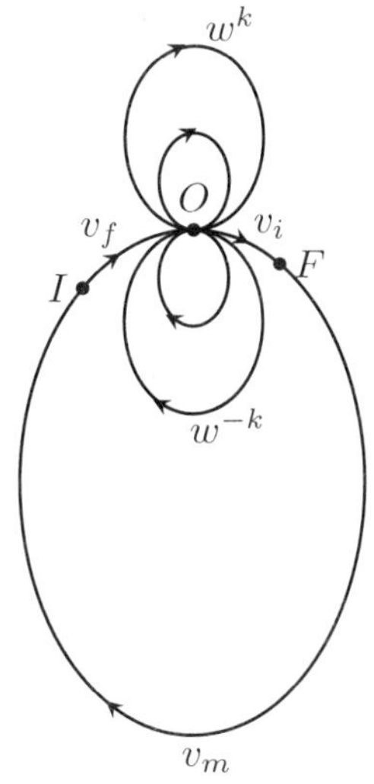

$b_1 = v_f, b_2 = v_i$,
$b' = v_f v_i$.

1.2 b' does not contain the point O and $\Lambda(b_1) + \Lambda(b_2) < \tau$

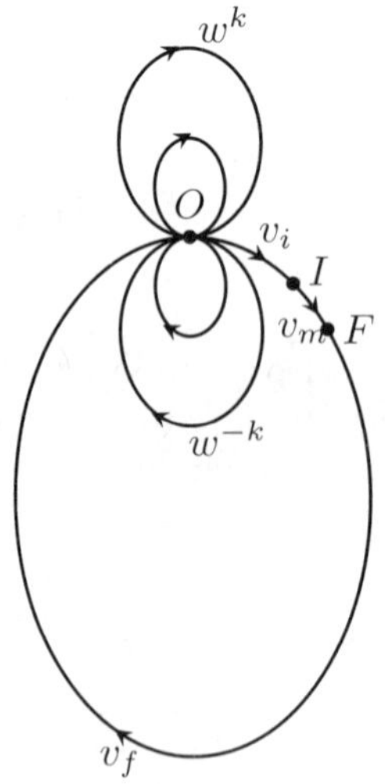

$b_1 = v_i^{-1}, b_2 = v_i v_m,$
$b' = v_m;$

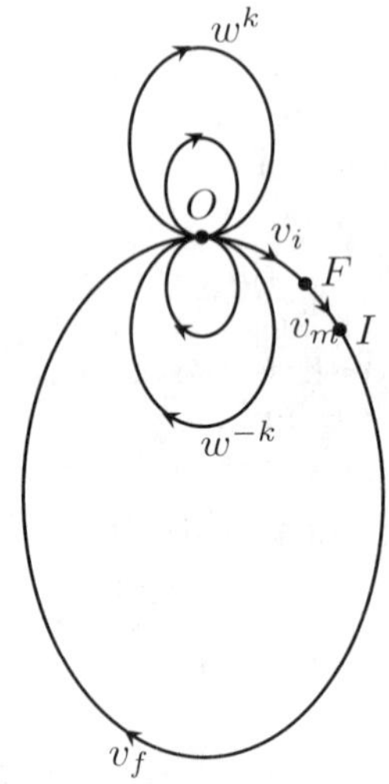

$b_1 = v_m^{-1} v_i^{-1}, b_2 = v_i,$
$b' = v_m^{-1};$

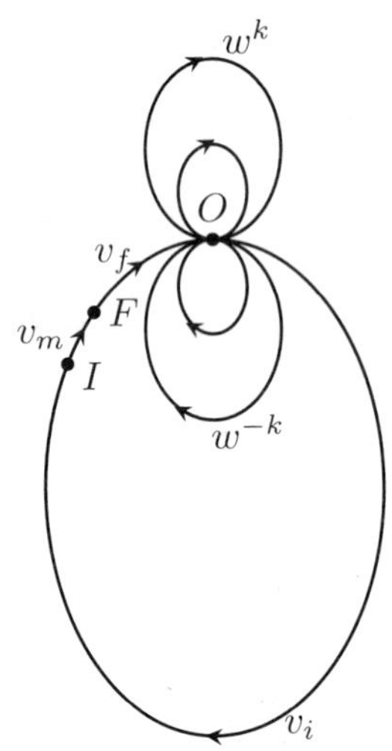

$b_1 = v_m v_f, b_2 = v_f^{-1},$
$b' = v_m;$

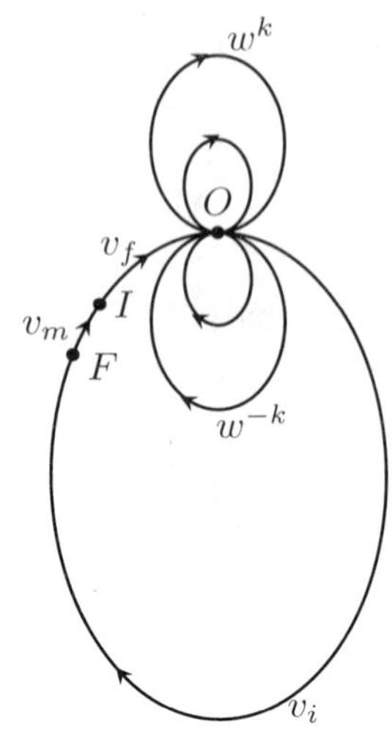

$b_1 = v_f, b_2 = v_f^{-1} v_m^{-1},$
$b' = v_m^{-1};$

1.3 b' does not contain the point O and $\Lambda(b_1) + \Lambda(b_2) \geqslant \tau$.

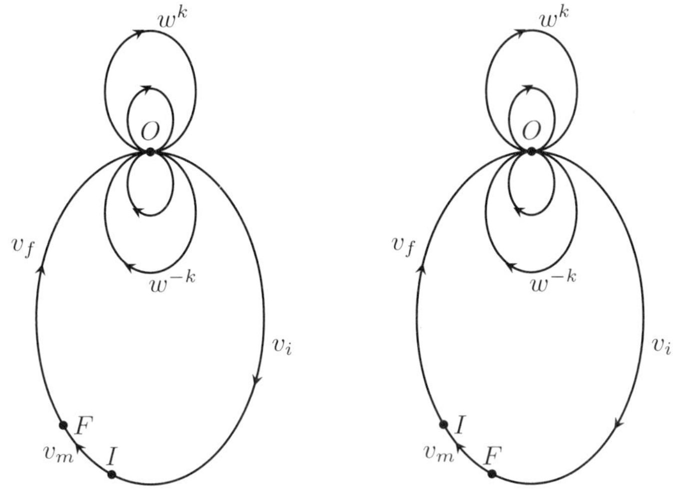

$b' = v_m$; $\qquad$ $b' = v_m^{-1}$;

In case 1.3 there is only one path of Λ-measure $< \tau$ ($b = b'$). Hence, if the sum $\sum_{l=1}^{n(I,F)} Q_l^{(I,F)}$ contains several different monomials of Λ-measure $< \tau$, then configuration 1.3 is not possible.

Consider configurations 1.1 and 1.2. Since all members of the sum $\sum_{l=1}^{n(I,F)} Q_l^{(I,F)}$ correspond to paths in the v-diagram with the initial point I and the final point F and of Λ-measure less than τ, the members are of the form

$$b_1 w^{k_j} b_2.$$

Notice that the monomial with $k_j = 0$ may have cancellations and equals to b' after cancellations.

Let us return to the consideration of case 1 in general. Recall that all $Q_l^{(I,F)}$ are of one of the types (9)–(14). So, if we multiply $\sum_{l=1}^{n(I,F)} Q_l^{(I,F)}$ by b_1^{-1} from the left side and by b_2^{-1} from the right side, we obtain the Laurent polynomial $P(v,w) = b_1^{-1} \sum_{j=1}^{n} Q_j b_2^{-1}$ such that $P(v,w)$ satisfies (8). But, on the other hand, in cases 1.1 and 1.2

$$P(v,w) = b_1^{-1} \sum_{l=1}^{n(I,F)} Q_l^{(I,F)} b_2^{-1} = \sum_{j=1}^{s} w^{k_l}.$$

In case 1.3, the polynomial consists of one monomial

$$P(v,w) = b_1^{-1} \sum_{l=1}^{n(I,F)} Q_l^{(I,F)} b_2^{-1} = b_1^{-1} b' b_2^{-1}.$$

Evidently, this polynomials does not satisfy (8). This contradiction completes the proof. The remaining cases are considered in the same way. □

Let us define a subspace of $\mathbb{Z}_2\mathcal{F}$ of linear dependencies induced by multi-turns. For every monomial from $\mathcal{F}$, we do all possible multi-turns of all members of the chart. As a result, we obtain a set of expressions

(15)

$$\mathcal{T} = \left\{ \sum_{j=1}^{k} U_j \mid U_h \in \mathcal{F}, U_h = La_hR, \textit{ where } a_h \textit{ is a member of the chart of } U_h, \right.$$

$$\left. U_h \mapsto \sum_{\substack{j=1\\ j\neq h}}^{k} U_j \textit{ is a multi-turn coming from an elementary multi-turn } a_h \mapsto \sum_{\substack{j=1\\ j\neq h}}^{k} a_j \right\}.$$

That is, $\mathcal{T}$ consists of the supports of all the multi-turns of members of the chart. We denote by $\langle \mathcal{T} \rangle$ the linear span of this set.

PROPOSITION 3.2. *The linear subspace $\langle \mathcal{T} \rangle \subseteq \mathbb{Z}_2\mathcal{F}$ is equal to the ideal $\mathcal{I} = \langle 1 + v + vw \rangle$.*

PROOF. First let us show that $\langle \mathcal{T} \rangle$ is an ideal in $\mathbb{Z}_2\mathcal{F}$. Assume that T is the support of a multi-turn of $M_h = La_hR$ that comes from an elementary multi-turn $a_h \mapsto \sum_{\substack{j=1\\ j\neq h}}^{k} a_j$, that is, $T = \sum_{j=1}^{k} U_j$, $U_j = La_jR$. We have to check that if Z is a monomial, then $ZT \in \langle \mathcal{T} \rangle$ and $TZ \in \langle \mathcal{T} \rangle$. Clearly, it is sufficient to check this property only for a monomial Z that consists of only one letter z. We will show an even stronger property, namely that $z(\sum_{j=1}^{k} U_j) \in \mathcal{T}$, $(\sum_{j=1}^{k} U_j)z \in \mathcal{T}$.

Consider $z(\sum_{j=1}^{k} U_j)$, $U_j = La_jR$. Suppose $\Lambda(a_h) < \tau$. From Proposition 3.1, it follows that in the sum $\sum_{j=1}^{k} a_j$ there exists a monomial $a_{h'}$ such that $\Lambda(a_{h'}) \geqslant \tau$. Then, by definition, $a_{h'}$ is a member of the chart of $La_{h'}R$. Hence, the sum $\sum_{j=1}^{k} U_j = \sum_{j=1}^{k} La_jR$ can be considered as the support of the multi-turn $M_{h'} = La_{h'}R \mapsto \sum_{\substack{j=1\\ j\neq h'}}^{k} La_jR$. So, in the sequel we can assume that $\Lambda(a_h) \geqslant \tau$.

First consider the case when L is not empty. Since $\Lambda(a_h) \geqslant \tau$, a_h is a member of the chart of $(zL)a_hR$ both when z does not cancel out with L, or when it does. Since $z(La_jR) = (zL)a_jR$, $\sum_{j=1}^{k} zU_j = \sum_{j=1}^{k} z(La_jR) = \sum_{j=1}^{k}(zL)a_jR$. Then, clearly, $\sum_{j=1}^{k} zU_j$ is the support of the multi-turn $(zL)a_hR \mapsto \sum_{\substack{j=1\\ j\neq h}}^{k}(zL)a_jR$. Hence, $\sum_{j=1}^{k} zU_j \in \mathcal{T}$.

Consider the case when L is empty, that is, $U_j = a_jR$. If z does not cancel with a_h and z does not prolong a_h to a generalized fractional power from the left, then a_h is a maximal occurrence of a generalized fractional power in $zU_h = za_hR$. Since $\Lambda(a_h) \geqslant \tau$, a_h is a member of the chart of za_hR. Hence, $za_hR \mapsto \sum_{\substack{j=1\\ j\neq h}}^{k} za_jR$ is a multi-turn of a member of the chart and its support $\sum_{j=1}^{k} za_jL = \sum_{j=1}^{k} zU_j$ belongs to $\mathcal{T}$.

Assume that z does not cancel with a_h and z prolongs a_h to a generalized fractional power from the left. Then for every $j \neq h$ either z does not cancel with a_j (and since $\Lambda(a_h) \geqslant \tau > \varepsilon$, z prolongs it to a generalized fractional power), or z cancels with a_j. Hence, $za_h \mapsto \sum_{\substack{j=1\\ j\neq h}}^{k} za_j$ is an elementary multi-turn (after the cancellations from the right hand side). Clearly, za_h is a maximal occurrence of a generalized fractional power in $zU_h = za_hR$. Since $\Lambda(za_h) \geqslant \Lambda(a_h) \geqslant \tau$, za_h is a member of the chart of za_hR. Therefore, $za_hR \mapsto \sum_{\substack{j\neq h\\ j=1}}^{k} za_jR$ is a multi-turn of a member of the chart and its support $\sum_{j=1}^{k} za_jR = \sum_{j=1}^{k} zU_j$ belongs to $\mathcal{T}$.

Assume that z cancels with a_h. Then for every $j \neq h$ either z does not cancel with a_j (and since $\Lambda(a_h) \geqslant \tau > \varepsilon$, z prolongs it to a generalized fractional power), or z cancels with a_j. Hence, $za_h \mapsto \sum_{\substack{j=1\\ j\neq h}}^{k} za_j$ is an elementary multi-turn (after the cancellations). Since $\Lambda(za_h) \leqslant \Lambda(a_h)$, we distinguish the following two possibilities:

(1) $\Lambda(za_h) \geqslant \tau$. Then za_h is a member of the chart of za_hR. Therefore, $za_hR \mapsto \sum_{\substack{j=1\\ j\neq h}}^{k} za_jR$ is a multi-turn of a virtual member of the chart and its support $\sum_{j=1}^{k} za_jR = \sum_{j=1}^{k} zU_j$ belongs to $\mathcal{T}$.

(2) $\Lambda(za_h) \geqslant \tau$. Then za_h is not a member of the chart of za_hR. Then we argue as at the beginning of the proof. From Proposition 3.1 it follows that in the sum $\sum_{j=1}^{k} za_j$ there exists a monomial $za_{h'}$ such that $\Lambda(za_{h'}) \geqslant \tau$. So, $za_{h'}$ is a member of the chart of $za_{h'}R$. Hence, the sum $\sum_{j=1}^{k} za_jR$ can be considered as the support of the multi-turn $za_{h'}R \mapsto \sum_{\substack{j\neq h'\\ j=1}}^{k} za_jR$.

Thus, the sum $\sum_{j=1}^{k} zU_j = \sum_{j=1}^{k} za_jR$ belongs to $\mathcal{T}$.

Summarising all of the above, we obtain $zT = z(\sum_{j=1}^{k}) \in \mathcal{T}$. Clearly, for the same reason we obtain $Tz = (\sum_{j=1}^{k})z \in \mathcal{T}$. Hence, $\langle\mathcal{T}\rangle$ is an ideal in $\mathbb{Z}_2\mathcal{F}$.

Evidently, every $\sum_{j=1}^{k} a_j$ of the form (9)–(14) belongs to $\mathcal{I}$. Hence, $\sum_{j=1}^{k} La_jR$ belongs to $\mathcal{I}$ for every $\sum_{j=1}^{k} a_j$ of the form (9)–(14). So, $\mathcal{T} \subseteq \mathcal{I}$ and $\langle\mathcal{T}\rangle \subseteq \mathcal{I}$. Since $vw + v + 1$ is the support of the multi-turn $vw \mapsto v + 1$ and $\Lambda(vw) = 1 \geqslant \tau$, $vw + v + 1 \in \mathcal{T}$. Therefore, since $\langle\mathcal{T}\rangle$ is an ideal in $\mathbb{Z}_2\mathcal{F}$, we obtain $\mathcal{I} \subseteq \langle\mathcal{T}\rangle$. Thus, $\mathcal{I} = \langle\mathcal{T}\rangle$. □

3.3. Virtual members of the chart. Recall that the chart of a word consists of maximal occurrences of generalized fractional powers with Λ-measure $\geqslant \tau$ (τ is our threshold). When we perform multi-turns, the measure of the occurrences may increase on the right or decrease on the right by at most ε in the resulting monomials of type 1. Similarly, it may increase on the left by at most ε or decrease on the left by at most ε. It follows that occurrences with measure $\geqslant \tau$ (above the threshold) may turn into occurrences $< \tau$ (below the threshold) and vice versa. Therefore we need to modify our notion of a member of the chart to a notion with more stable properties with respect to multi-turns. We call such a notion *a virtual member of the chart*. In this section, we define this notion and study its properties.

Namely, we may have the following effect. As above, let U_h be a monomial, a_h be a member of its chart, and $U_h = La_hR$. Let $U_h \mapsto \sum_{\substack{j=1\\ j\neq h}}^{k} U_j$ be a multi-turn of a_h that comes from an elementary multi-turn $a_h \mapsto \sum_{\substack{j=1\\ j\neq h}}^{k} a_j$, $U_j = La_jR$. Assume b_h is a maximal occurrence of a generalized fractional power in U_h different from a_h, $\Lambda(b_h) < \tau$, that is, b_h is not counted as a member of the chart of U_h. According to the previous section, the element b_h may be prolonged in U_j and the Λ-measure of the corresponding prolonged element b_j in U_j may increase and become $\geqslant \tau$. So, b_j may become a member of the chart of U_j and the number of members of the chart of U_j may become greater than the number of members of the chart of U_h. We may obtain this effect for both neighbours of a_h simultaneously.

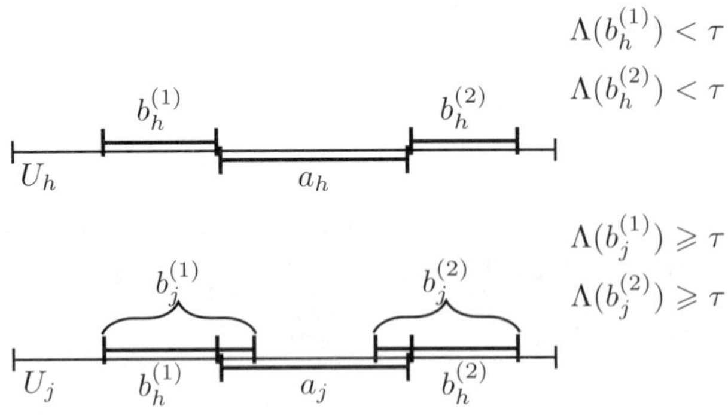

In this case, the number of members of the chart of U_j become greater than the number of members of the chart of U_h even if $\Lambda(a_j) < \tau$.

Assume that $U_j = La_jR$ is a resulting monomial of a multi-turn such that, roughly speaking, a_j is of small Λ-measure (for example $\Lambda(a_j) < \tau$, so a_j is not counted as a member of the chart). In Section 4, we will use an inductive argument for such resulting monomials of multi-turns. It looks natural to use induction by the number of members of the chart. But the example described above shows that the number of members of the chart may increase in U_j; hence, it is not an appropriate parameter for the induction. In order to prove that the induction is nevertheless finite, we refine a notion of a member of the chart in this section. After that we introduce a function that guarantees finiteness of the inductive process (see Corollary 3.6).

In this section, we consider the set $\mathcal{M}(U_h)$ of all maximal occurrences of generalized fractional powers in U_h, regardless of their Λ-measure, such that they are not properly contained in other occurrences of generalized fractional powers in U_h. There are two types of such occurrences: occurrences that are not fully covered by other occurrences from $\mathcal{M}(U_h)$ and occurrences that are fully covered by other occurrences from $\mathcal{M}(U_h)$. Denote the first set by $\mathcal{M}^{\mathbf{nfc}}(U_h)$ and the second set by $\mathcal{M}^{\mathbf{fc}}(U_h)$. Clearly, all maximal occurrences of a generalized fractional powers in U_h of Λ-measure greater than ε are contained in the set $\mathcal{M}(U_h)$. It is also clear that elements of $\mathcal{M}^{\mathbf{fc}}(U_h)$ are of Λ-measure not greater than 2ε.

For any maximal occurrence of a generalized fractional power in U_h we define *a set of its images* in a resulting monomial of a multi-turn.

DEFINITION 3.1. Let $U_h = La_hR$ be a monomial, and a_h be a maximal occurrence of a generalized fractional power such that $\Lambda(a_h) \geqslant \tau - 2\varepsilon$. Let a_j and a_h be incident monomials. Assume b_h is a maximal occurrence of a generalized fractional power in U_h that is different from a_h and is not properly contained in a_h. Since $\Lambda(a_h) \geqslant \tau - 2\varepsilon > \varepsilon$, a_h can not be properly contained in any other occurrence, particularly in b_h. To be precise, assume that the end of b_h lies strictly left to the end of a_h. We give the definition in the cases $a_j \neq 1$ and $a_j = 1$ separately.

(1) Assume $a_j \neq 1$, $U_j = La_jR$. Then we call an element of $\mathcal{M}(U_j)$ that contains a_j we call *an image of* a_h *in* U_j. We call the set of all these elements *the set of images of* a_h *in* U_j.

Denote by b'_h a subword of b_h that is an intersection of b_h and L. We have $L = L'_1 b'_h L'_2$, where L'_2 is possibly equal to 1. Hence,

$$U_h = La_hR = L'_1b'_hL'_2a_hR,$$
$$U_j = La_jR = L'_1b'_hL'_2a_jR.$$

We call an element of $\mathcal{M}(U_j)$ that contains b'_h *an image of* b_h *in* U_j. We call the set of all these elements *the set of images of* b_h *in* U_j.

(2) Assume $a_j = 1$. Then we say that *the set of images of* a_h *in* U_j *is empty.*

Let $L = L'C$, $R = C^{-1}R'$ and $L'R'$ has no further cancellations. If there is no non-empty subword of b_h that is contained in L', we say that *the set of images of* b_h *in* U_j *is empty.* Otherwise, let b'_h be a subword of b_h that is an intersection of b_h and L'. We have $L' = L'_1b'_hL'_2$, where L'_2 is possibly equal to 1. Hence,

$$U_h = La_hR = L'Ca_hC^{-1}R' = L'_1b'_hL'_2Ca_hC^{-1}R',$$
$$U_j = L'R' = L'_1b'_hL'_2R'.$$

We call an element of $\mathcal{M}(U_j)$ that contains b'_h *an image of* b_h *in* U_j. We call the set of all these elements *the set of images of* b_h *in* U_j.

Clearly, we have a similar correspondence for b_h when its beginning lies strictly right from the beginning of a_h.

If $\Lambda(b'_h) \leqslant \varepsilon$, it can be contained in several elements of $\mathcal{M}(U_j)$ and the set of images of b_h may contain several elements. If $\Lambda(b'_h) > \varepsilon$, then its prolongation in U_j is uniquely determined and the set of images of b_h consists of one element. In fact, in the previous section we described in detail all possible images of maximal occurrences of generalized fractional powers.

EXAMPLE 3.1. We observe the following effect for elements of $\mathcal{M}^{\mathbf{fc}}(U_j)$. Let a maximal occurrence b_h touch a_h at a point from the left side. Suppose $b_h = b_h^{(1)}c'_h$, where c'_h is a maximal occurrence of a generalized fractional power, $\Lambda(c'_h) \leqslant \varepsilon$. Suppose $a_j = c'_j a_j^{(1)}$, where c'_j is a maximal occurrence of a generalized fractional power, $\Lambda(c'_j) \leqslant \varepsilon$. Assume $c_j = c'_hc'_j$ is a maximal occurrence of a generalized fractional power in U_j, then we obtain a new element in $\mathcal{M}^{\mathbf{fc}}(U_j)$, which grows from two maximal occurrences $c'_h \notin \mathcal{M}(U_h)$ and $c'_j \notin \mathcal{M}(U_j)$.

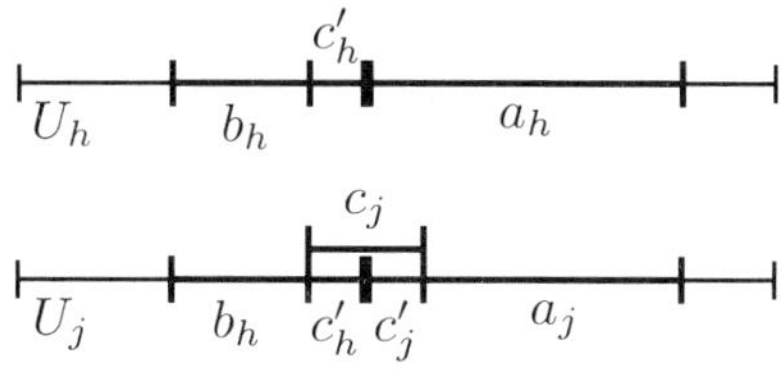

This example shows that we can not use just the size of $\mathcal{M}(U_j)$ as an inductive parameter. We will introduce special notions in order to control the behaviour of elements both from $\mathcal{M}^{\mathbf{nfc}}(U_j)$ and from $\mathcal{M}^{\mathbf{fc}}(U_j)$.

All elements of $\mathcal{M}(U_h)$ can be divided into sets such that every set covers a part of U_h and different parts are separated from each other. We denote them by $\mathcal{M}_1(U_h), \dots, \mathcal{M}_{k(U_h)}(U_h)$. We consider subcoverings of all parts, that are subsets of $\mathcal{M}_1(U_h), \dots, \mathcal{M}_{k(U_h)}(U_h)$, and call the union of this subcoverings *a covering of*

U_h. There are finitely many different coverings of U_h, we denote them by $\{\mathcal{C}_i(U_h) \mid i = 1, \dots, n(U_h)\}$. If Z is a subword of U_h and $\mathcal{C}_i(U_h)$ is a covering of U_h, we consider elements of $\mathcal{C}_i(U_h)$ that have non empty intersection with Z. We call the set of this elements *a covering of a subword* Z and denote it by $\mathcal{C}_i(Z, U_h)$.

Let us consider a subcovering of every set $\mathcal{M}_k(U_h)$ that consists of the smallest number of elements and denote this number by $N_k(U_h)$. Let us call the union of this subcoverings *a minimal covering of* U_h. Clearly, a minimal covering of U_h may not be uniquely defined. We define

$$N(U) = \sum_{k=1}^{k(U_h)} N_k(U_h),$$

that is, the number of elements in a minimal covering of U_h. Notice that all elements of $\mathcal{M}^{\mathbf{nfc}}(U_h)$ are contained in any covering of U_h. If some element of $\mathcal{M}^{\mathbf{fc}}(U_h)$ is fully covered by elements of $\mathcal{M}^{\mathbf{nfc}}(U_h)$, it is never contained in a minimal covering of U_h.

Let us study a general structure of positions of elements of $\mathcal{M}^{\mathbf{fc}}(U_h)$. Consider $c \in \mathcal{M}^{\mathbf{fc}}(U_h)$, that is, c is fully covered by other elements of $\mathcal{M}(U_h)$. Consider the set of all maximal occurrences $\{c_k\}$ from $\mathcal{M}(U_h)$ that have non empty intersection with c. Then for every c_k either its beginning point or its end point belong to c, since c is not properly contained in another occurrence. Consider c_{k_1} with the leftmost beginning with respect to the beginning of c and c_{k_2} with the rightmost beginning with respect to the beginning of c.

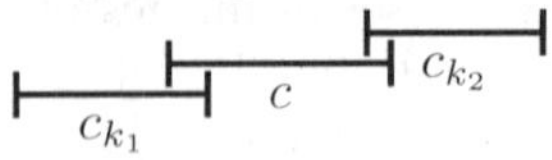

By the definition of $\mathcal{M}(U_h)$ neither of c_k is contained in another one. Hence, if $k \neq k_1$, then c_k begins after c_{k_1}, if $k \neq k_2$, then c_k ends before c_{k_2}. Therefore, if $k \neq k_1$, $k \neq k_2$, then c_k is fully covered by c, c_{k_1}, c_{k_2}, so, $c_k \in \mathcal{M}^{\mathbf{fc}}(U_h)$.

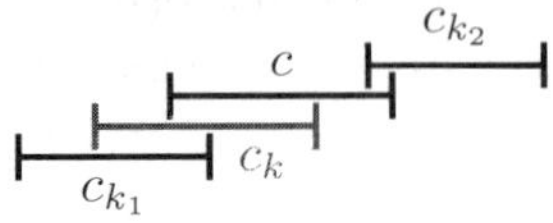

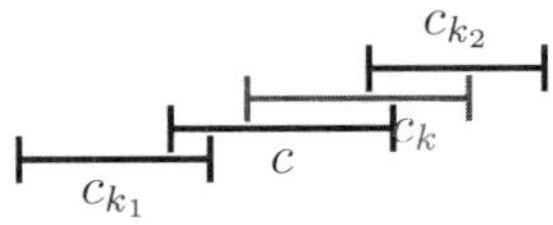

If c_{k_1} is also fully covered by other elements of $\mathcal{M}(U_h)$, we consider an element of $\mathcal{M}(U_h)$ that has the end at the leftmost side of c_{k_1} and repeat the argument. If c_{k_2} is fully covered by other elements of $\mathcal{M}(U_h)$, we consider an element of $\mathcal{M}(U_h)$ that has non-empty intersection with c_{k_2} with the rightmost beginning with respect to the beginning of c_{k_2} and repeat the argument. We continue the process until we find an element of $\mathcal{M}^{\mathbf{nfc}}(U_h)$ that has the beginning from the left of c and an element of $\mathcal{M}^{\mathbf{nfc}}(U_h)$ that has the end from the right of c. Denote the first element by d_1 and the second element by d_2. We see that all elements of $\mathcal{M}(U_h)$ that have the beginning after the beginning of d_1 and the end before the end of d_2 belong to $\mathcal{M}^{\mathbf{fc}}(U_h)$.

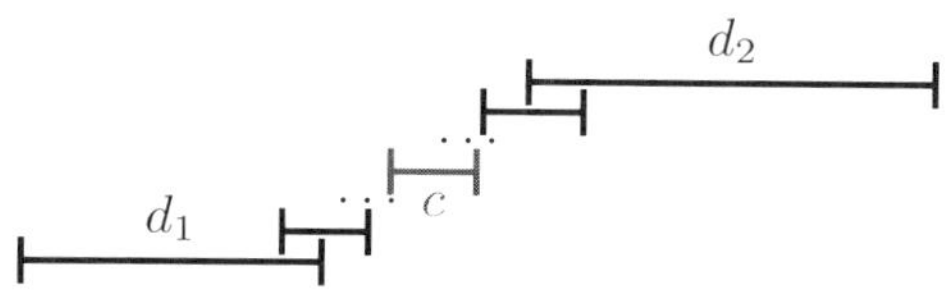

LEMMA 3.3. *Let U_h be a monomial, $a_h \in \mathcal{M}^{nfc}(U_h)$, $U_h = La_hR$. Assume a_h and a_j are incident monomials, $U_j = La_jR$. Then we have $N(U_j) \leqslant N(U_h)$. If, moreover, a_j is fully covered by images of elements of $\mathcal{M}^{nfc}(U_h) \setminus \{a_h\}$ (for instance, if $a_j = 1$), then $N(U_j) < N(U_h)$.*

PROOF. Let $\mathcal{C}_{i_1}(U_h)$ be a minimal covering of U_h. First consider the case $a_j \neq 1$. Since $a_h \in \mathcal{M}^{\mathbf{nfc}}(U_h)$, it necessarily belongs to $\mathcal{C}_{i_1}(U_h)$. Hence, the covering $\mathcal{C}_{i_1}(U_h)$ can be written as

$$\mathcal{C}_{i_1}(U_h) = \mathcal{C}_{i_1}(L, U_h) \sqcup \{a_h\} \sqcup \mathcal{C}_{i_1}(R, U_h),$$

where $\sqcup$ is a disjoint union. Denote by a'_j an image of a_h. Denote by $\mathcal{C}'$ the set of images of elements of $\mathcal{C}_{i_1}(L, U_h)$, by $\mathcal{C}''$ the set of images of elements of $\mathcal{C}_{i_1}(R, U_h)$. For elements that have more than one image, we take one arbitrary image. Every letter of L that is covered by $\mathcal{C}_{i_1}(U_h)$ is covered by $\mathcal{C}'$ in U_j. Every letter of R that is covered by $\mathcal{C}_{i_1}(U_h)$ is covered by $\mathcal{C}''$ in U_j. The occurrence of a_j in U_j is covered by a'_j. Therefore,

$$\mathcal{C}' \cup \{a'_j\} \cup \mathcal{C}'' \textit{ is a covering of } U_j.$$

Denote it by $\mathcal{C}_{i_2}(U_j)$.

Since we take one image for every element of $\mathcal{C}_{i_1}(U_h)$,

$$|\mathcal{C}'| = |\mathcal{C}_{i_1}(L, U_h)|, |\mathcal{C}''| = |\mathcal{C}_{i_1}(R, U_h)|,$$

where $|\cdot|$ is the number of elements in a set. Hence, $|\mathcal{C}_{i_2}(U_j)| \leqslant |\mathcal{C}_{i_1}(U_h)| = N(U_h)$, so we obtain $N(U_j) \leqslant N(U_h)$.

Let us show that $b_h \in \mathcal{M}^{\mathbf{nfc}}(U_h) \setminus \{a_h\}$ has one image in U_j. If b_h is separated from a_h, it is obvious. Suppose b_h is not separated from a_h from the left. Let b'_h be a subword of b_h that is an intersection of b_h and L. We have $L = L'_1 b'_h$. Since b_h is not fully covered by elements of $\mathcal{M}(U_h)$, b'_h is also not fully covered by elements of $\mathcal{M}(U_h)$. This means that b'_h can not be prolonged from the left neither in U_h nor in U_j. Hence, b'_h in U_j may be fully contained only in some element of $\mathcal{M}(U_j)$ that is contained in the word $b'_h a_j R$. Since b'_h is an initial subword of $b'_h a_j R$, there exists only one such element of $\mathcal{M}(U_j)$ that contains b'_h (possibly b'_h itself). Hence, b_h has one image in U_j. The case of b_h being not separated from a_h from the right is considered similarly.

Every element of $\mathcal{M}^{\mathbf{nfc}}(U_h)$ is contained in any covering of U_h. Therefore, every element of $\mathcal{M}^{\mathbf{nfc}}(U_h) \setminus \{a_h\}$ is contained either in $\mathcal{C}_{i_1}(L, U_h)$, or in $\mathcal{C}_{i_1}(R, U_h)$. Assume a_j is fully covered by images of elements of $\mathcal{M}^{\mathbf{nfc}}(U_h) \setminus \{a_h\}$. Since every element of $\mathcal{M}^{\mathbf{nfc}}(U_h) \setminus \{a_h\}$ has only one image, a_j is fully covered by $\mathcal{C}' \cup \mathcal{C}''$. Therefore, $\mathcal{C}' \cup \mathcal{C}''$ is a covering of U_j that has strictly less elements than $\mathcal{C}_{i_1}(U_h)$. Thus, we obtain $N(U_j) < N(U_h)$.

Now consider the case $a_j = 1$. Assume $U_j = LR = L'CC^{-1}R' = L'R'$, where $L'R'$ has no further cancellations. We have $U_h = L'Ca_hC^{-1}R'$, hence the covering $\mathcal{C}_{i_1}(U_h)$ can be written as

$$\begin{aligned}\mathcal{C}_{i_1}(U_h) = \mathcal{C}_{i_1}(L', U_h) \sqcup \{\mathcal{C}_{i_1}(L, U_h) \setminus \mathcal{C}_{i_1}(L', U_h)\} \sqcup \{a_h\} \\ \sqcup \{\mathcal{C}_{i_1}(R, U_h) \setminus \mathcal{C}_{i_1}(R', U_h)\} \sqcup \mathcal{C}_{i_1}(R', U_h).\end{aligned}$$

Denote by $\mathcal{C}'$ the set of images of elements of $\mathcal{C}_{i_1}(L', U_h)$, by $\mathcal{C}''$ the set of images of elements of $\mathcal{C}_{i_1}(R', U_h)$. Again, for elements that have more than one image, we take one arbitrary image. Every letter of L' that is covered by $\mathcal{C}_{i_1}(U_h)$ is covered by $\mathcal{C}'$ in U_j. Every letter of R' that is covered by $\mathcal{C}_{i_1}(U_h)$ is covered by $\mathcal{C}''$ in U_j. Hence,

$$\mathcal{C}' \cup \mathcal{C}'' \textit{ is a covering of } U_j.$$

Denote it by $\mathcal{C}_{i_2}(U_j)$.

Since we take one image for every element of $\mathcal{C}_{i_1}(U_h)$, we have

$$|\mathcal{C}'| = |\mathcal{C}_{i_1}(L', U_h)|, |\mathcal{C}''| = |\mathcal{C}_{i_1}(R', U_h)|.$$

Hence, $|\mathcal{C}_{i_2}(U_j)| < |\mathcal{C}_{i_1}(U_h)| = N(U_h)$, so we obtain $N(U_j) < N(U_h)$. □

Assume $a_h \in \mathcal{M}^{\mathbf{nfc}}(U_h)$. Consider the set of elements of $\mathcal{M}(U_h)$ that are not separated from a_h from the left side. Evidently, there exists at most one element of $\mathcal{M}^{\mathbf{nfc}}(U_h)$ in this set. Similarly, there exists at most one element of $\mathcal{M}^{\mathbf{nfc}}(U_h)$ that is not separated from a_h from the right side. Let us call them *the essential left neighbour of a_h and the essential right neighbour of a_h*, respectively.

We define a special sequence of transformations of monomials.

Definition 3.2. Let U be a word, $b \in \mathcal{M}^{\mathbf{nfc}}(U)$. We denote U by $U^{(1)}$ and b by $b^{(1)}$ and define a sequence of transformations inductively. Assume monomials $U^{(i)}$, maximal occurrence of a generalized fractional powers $b^{(i)}$ in $U^{(i)}$ and transformations $r_i : U^{(i-1)} \mapsto U^{(i)}$, $1 \leqslant i \leqslant k$, are already defined (we assume that the transformation r_1 is identical). Let $a_h^{(k)}$ be a maximal occurrence of a generalized fractional power in $U^{(k)} = L^{(k)} a_h^{(k)} R^{(k)}$ such that

(1) $a_h^{(k)}$ differs from $b^{(k)}$;
(2) $\Lambda(a_h^{(k)}) \geqslant \tau - 2\varepsilon$ (hence, $a_h^{(k)} \in \mathcal{M}^{\mathbf{nfc}}(U^{(k)})$);
(3) if the beginning of $b^{(k)}$ lies from the right of the beginning $a_h^{(k)}$, $a_h^{(k)}$ has the right essential neighbour; if the beginning $b^{(k)}$ lies from the left of the beginning $a_h^{(k)}$, $a_h^{(k)}$ has the left essential neighbour.

Suppose $a_h^{(k)}$ and $a_j^{(k)}$ are incident monomials. If $a_j^{(k)}$ in $L^{(k)} a_j^{(k)} R^{(k)}$ is not covered by images of $\mathcal{M}^{\mathbf{nfc}}(U^{(k)}) \setminus \{a_h^{(k)}\}$, then we say that

$$U^{(k+1)} = L^{(k)} a_j^{(k)} R^{(k)},$$
$$b^{(k+1)} \textit{ is an image of } b^{(k)} \textit{ in } U^{(k+1)},$$
$$r_{k+1} : U^{(k)} \mapsto U^{(k+1)} \textit{ is a replacement of } a_h^{(k)} \textit{ by } a_j^{(k)}.$$

Lemma 3.4. *Let U be a word, $b \in \mathcal{M}^{nfc}(U)$. Assume we have a sequence of transformations defined above that starts from U. If $U^{(K)}$ is a monomial in this sequence, then $b^{(K)}$ can increase or decrease at most by a piece of Λ-measure ε from the right side and at most by a piece of Λ-measure ε from the left side with respect to b.*

Proof. In this proof, we use notations from Definition 3.2. Let us prove the lemma by induction on K.

Assume $K = 2$, that is, the sequence of transformations has only one step r_1. If $a_h^{(1)}$ is separated from $b^{(1)}$, then the occurrence $b^{(2)}$ in $U^{(2)}$ is equal to $b^{(1)}$. Let $a_h^{(1)}$

be not separated from $b^{(1)}$. Since $b^{(1)} = b \in \mathcal{M}^{\mathbf{nfc}}(U)$, $b^{(1)}$ is an essential neighbour of $a_h^{(1)}$. According to the classification given in Section 3.1, $b^{(2)}$ can decrease at most by a piece of Λ-measure ε after the replacement $a_h^{(1)} \mapsto a_j^{(1)}$. Since $b^{(1)}$ is an essential neighbour of $a_h^{(1)}$, $a_j^{(1)}$ is not covered by the image of $b^{(1)}$, by Definition 3.2. Hence, according to the classification given in Section 3.1, $b^{(2)}$ can increase at most by a piece of Λ-measure ε after the replacement $a_h^{(1)} \mapsto a_j^{(1)}$.

Let us prove the step of the induction. Consider the replacement $r_2 : U^{(1)} \mapsto U^{(2)}$, $a_h^{(1)} \mapsto a_j^{(1)}$. Suppose $b^{(2)} \in \mathcal{M}^{\mathbf{fc}}(U)$. Then any element of $\mathcal{M}^{\mathbf{nfc}}(U)$ that is not separated from $b^{(2)}$ does not have the essential neighbour from the necessary side. Hence, there are no more possible replacements $U^{(i)} \mapsto U^{(i+1)}$, $i \geqslant 2$, of occurrences that are not separated from $b^{(i)}$. So, $b^{(i)}$ remains equal to $b^{(2)}$ after any further replacement.

Further we consider only the case $b^{(2)} \in \mathcal{M}^{\mathbf{nfc}}(U)$. If $a_h^{(1)}$ is separated from $b^{(1)}$, then the occurrence $b^{(2)}$ in $U^{(2)}$ is equal to $b^{(1)}$. The sequence of transformations starting from $U^{(2)}$ has a fewer number of steps. Hence, $b^{(K)}$ can increase or decrease at most by a piece of Λ-measure ε from the right and the left sides with respect to $b^{(2)}$ $(= b^{(1)})$ by the induction hypothesis.

Suppose $a_h^{(1)}$ is not separated from $b^{(1)}$. If $b^{(2)}$ remains equal to $b^{(1)}$, then as above $b^{(K)}$ can increase or decrease at most by a piece of Λ-measure ε from the right and the left sides with respect to $b^{(1)}$ by the induction hypothesis.

Assume $\Lambda(b^{(2)})$ changes with respect to $\Lambda(b^{(1)})$. Then there are the following cases:

(1) The beginning of $b^{(1)}$ lies from the left of the beginning of $a_h^{(1)}$, $\Lambda(b^{(2)}) < \Lambda(b^{(1)})$.
(2) The beginning of $b^{(1)}$ lies from the right of the beginning of $a_h^{(1)}$, $\Lambda(b^{(2)}) < \Lambda(b^{(1)})$.
(3) The beginning of $b^{(1)}$ lies from the left of the beginning of $a_h^{(1)}$, $\Lambda(b^{(2)}) > \Lambda(b^{(1)})$.
(4) The beginning of $b^{(1)}$ lies from the right of the beginning of $a_h^{(1)}$, $\Lambda(b^{(2)}) > \Lambda(b^{(1)})$.

Let us consider case 1. There are the following two configurations for $b^{(2)}$:

(α) $a_j^{(1)} \in \mathcal{M}^{\mathbf{nfc}}(U^{(2)})$

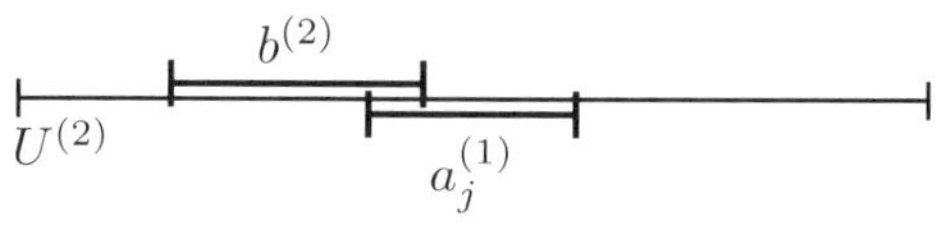

(β) $a_j^{(1)} \in \mathcal{M}^{\mathbf{fc}}(U^{(2)})$

$\Lambda(c) \leqslant 4\varepsilon$

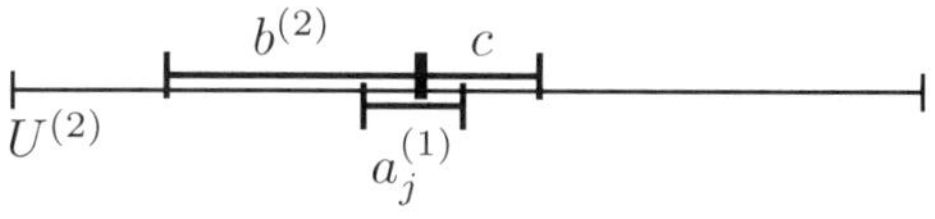

If $a_j^{(1)} \in \mathcal{M}^{\mathbf{fc}}(U^{(2)})$, then an element $c \in \mathcal{M}(U^{(2)})$ covers its terminal subword. Since $a_j^{(1)}$ is not covered by images of $\mathcal{M}^{\mathbf{nfc}}(U^{(1)}) \setminus \{a_h^{(1)}\}$, c is not an image of an element of $\mathcal{M}^{\mathbf{nfc}}(U^{(1)}) \setminus \{a_h^{(1)}\}$. Hence, c is an image of some generalized fractional power of Λ-measure not greater than 2ε. Then, by the classification from Section 3.1, $\Lambda(c) \leqslant 4\varepsilon$.

Let us study case (β) first. Since $\Lambda(c) \leqslant 4\varepsilon < \tau - 2\varepsilon$, c can not be used for the next replacement r_3. The sequence of replacements starting from $U^{(2)}$ has $K-1$ steps. Hence, if $c \in \mathcal{M}^{\mathbf{nfc}}(U^{(2)})$, then by the induction hypothesis the Λ-measure of an image of c can increase at most by 2ε with respect to $\Lambda(c)$ in every $U^{(i)}$, $3 \leqslant i \leqslant K$. So, an image of c is of less Λ-measure than $\tau - 2\varepsilon$ and can not be used for a replacement in any $U^{(i)}$, $3 \leqslant i \leqslant K$. Hence, all the next replaced occurrences are separated from $b^{(i)}$ from the right side. So, $b^{(i)}$ has no more changes from the right side with respect to $b^{(2)}$, $3 \leqslant i \leqslant K$.

Suppose $c \in \mathcal{M}^{\mathbf{fc}}(U^{(2)})$. Denote the element closest to c from the right such that it belongs to $\mathcal{M}^{\mathbf{nfc}}(U^{(2)})$ by d. Then d does not have the left essential neighbour. By the induction hypothesis, $b^{(i)}$ can decrease at most by a piece of Λ-measure ε from the left side. Hence, the image of d can not have the left essential neighbour in every $U^{(i)}$. Thus, d can not be used for a replacement in any $U^{(i)}$, $3 \leqslant i \leqslant K$. Therefore, all the next replaced occurrences are separated from the image of c from the right side. Hence, $b^{(i)}$ has no more changes from the right side with respect to $b^{(2)}$, $3 \leqslant i \leqslant K$.

Consider case (α). Consider the next replacement $r_3 : U^{(2)} \mapsto U^{(3)}$, $a_h^{(2)} \mapsto a_j^{(2)}$. With a slight abuse of language, we use the same indices in different incident monomials. By the induction hypothesis, $b^{(i)}$, $3 \leqslant i \leqslant K$, may increase or decrease from the left at most by a piece of Λ-measure ε with respect to $b^{(2)}$. This means that it remains to consider only replacements from the right of $b^{(2)}$. Therefore, without loss of generality, we can assume that the beginning $a_h^{(2)}$ lies from the right of the beginning of $b^{(2)}$. Then there are the following possibilities:

(1) $a_h^{(2)}$ is separated from $a_j^{(1)}$. Then we again obtain configuration (α) for $b^{(3)}$, that is, the image of $a_j^{(1)}$ belongs to $\mathcal{M}^{\mathbf{nfc}}(U^{(3)})$.

(2) $a_h^{(2)}$ coincides with $a_j^{(1)}$. In this case if $a_j^{(2)} \in \mathcal{M}^{\mathbf{nfc}}(U^{(3)})$, we obtain configuration (α) for $b^{(3)}$, possibly with a smaller overlap or without an overlap between $b^{(3)}$ and $a_j^{(2)}$. If $a_j^{(2)} \in \mathcal{M}^{\mathbf{fc}}(U^{(3)})$, we obtain configuration (β) for $b^{(3)}$.

(3) $a_h^{(2)}$ is not separated from $a_j^{(1)}$. If the image of $a_j^{(1)}$ belongs to $\mathcal{M}^{\mathbf{nfc}}(U^{(3)})$, we obtain configuration (α) for $b^{(3)}$. If the image of $a_j^{(1)}$ belongs to $\mathcal{M}^{\mathbf{fc}}(U^{(3)})$, we obtain configuration (β) for $b^{(3)}$.

So, we may obtain for $b^{(3)}$ either configuration (β) that was already considered, or again configuration (α). If we obtain configuration (α), $b^{(3)}$ increases or decreases at most by a piece of Λ-measure ε from the right side with respect to $b^{(1)}$.

Cases 2–4 are studied in the same way as case 1. So, if we continue the description for the next transformations r_i, we observe that possible changes of $b^{(i)}$ with respect to $b^{(1)}$ are restricted by a piece of Λ-measure not greater than ε from each side. $\square$

REMARK 3.2. In fact, in Lemma 3.4 we also proved the following. Assume that some maximal occurrence $a \in \mathcal{M}^{\mathbf{nfc}}(U)$ has an overlap of measure ε with its left essential neighbour. Then after a sequence of replacements from Definition 3.2, the Λ-measure of the image of a can not increase from the left with respect to the Λ-measure of a. Similarly, if a has an overlap of measure ε with its right essential neighbour, then after a sequence of replacements from Definition 3.2 the Λ-measure of the image of a can not increase from the right with respect to the Λ-measure of a.

DEFINITION 3.3. Let U be a word, $b \in \mathcal{M}^{\mathbf{nfc}}(U)$. If there exists a sequence of replacements constructed in Definition 3.2 such that $U^{(K)}$ is the last monomial in the sequence and $\Lambda(b^{(K)}) \geqslant \tau$, we call b *a virtual member of the chart of* U. We call the set of such maximal occurrences from $\mathcal{M}(U)$ *the virtual* τ*-chart of* U.

We denote the number of virtual members of the chart of U by $K_\tau(U)$.

From Lemma 3.4 it follows that if b is a virtual member of the chart of U, then $\Lambda(b) \geqslant \tau - 2\varepsilon$. Clearly, every member of the chart of U is also a virtual member of the chart of U.

From now on, when we speak about the chart of some word, we use only virtual members of the chart even if the qualification virtual is omitted.

Let us prove important properties of virtual members of the chart that we will use in the further argument.

LEMMA 3.5. *Let* U_h *be a monomial,* a_h *be a virtual member of its chart,* $U_h = La_hR$. *Assume* a_h *and* a_j *are incident monomials,* $a_j \neq 1$, $U_j = La_jR$. *If* a_j *is not fully covered by images of* $\mathcal{M}^{\mathbf{nfc}}(U_h) \setminus \{a_h\}$, *then* $K_\tau(U_j) \leqslant K_\tau(U_h)$. *If, moreover,* a_j *is not a virtual member of the chart of* U_j, *then* $K_\tau(U_j) < K_\tau(U_h)$.

PROOF. Let $U_h = La_hR$, $U_j = La_jR$. Suppose b is a virtual member of the chart of U_j different from a_j. Since a_j is not fully covered by images of $\mathcal{M}^{\mathbf{nfc}}(U_h) \setminus \{a_h\}$, a_j is not fully contained in b. To be precise, assume that the beginning of b lies from the left side of the beginning of a_j, that is, b is an occurrence in the subword La_j. Since b is a virtual member of the chart of U_j, there exists a sequence of transformations

$$U_j = U_j^{(1)} \mapsto \ldots \mapsto U_j^{(K)} \tag{16}$$

defined above such that the image of b in $U_j^{(K)}$ is of Λ-measure $\geqslant \tau$. Denote by b' the inverse image of b in U_h.

First assume that a_h in U_h has the left essential neighbour. Since a_j is not fully covered by images of $\mathcal{M}^{\mathbf{nfc}}(U_h) \setminus \{a_h\}$, the replacement $a_h \mapsto a_j$ can be used in the definition of a virtual member of the chart. Hence, we can add the transformation $U_h \mapsto U_j$ to the beginning of the sequence (16). Therefore, we obtain the sequence that starts from U_h and defines b' as a virtual member of the chart of U_h.

Now assume that a_h in U_h does not have the left essential neighbour. Then it is possible that there exist elements of $\mathcal{M}^{\mathbf{fc}}(U_h)$ that are not separated from a_h from the left side. The other possibility is that there are no elements of $\mathcal{M}(U_h)$ that are not separated from a_h from the left side. Let us study the first case. Since a_h in U_h does not have the left essential neighbour, b' is separated from a_h.

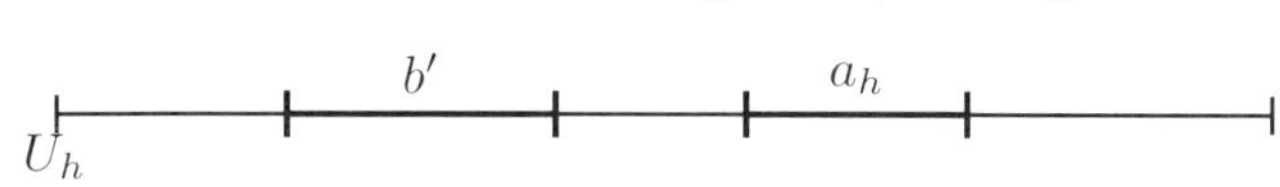

Consider the first possibility, namely, there exist elements of $\mathcal{M}^{\mathbf{fc}}(U_h)$ that are not separated from a_h from the left side. Let us show that there exists an element of $\mathcal{M}(U_j)$ of Λ-measure $\leqslant 5\varepsilon$ such that its beginning lies between the beginning of b and the beginning of a_j and it is not fully covered by its essential neighbours. Let $d' \in \mathcal{M}(U_h)$ be not separated from a_h from the left and has the leftmost beginning among such elements. Let d be an image of d' in U_j. Since a_h does not have the left essential neighbour, $d' \in \mathcal{M}^{\mathbf{fc}}(U_h)$. Hence, $\Lambda(d') \leqslant 2\varepsilon$, so, $\Lambda(d) \leqslant 5\varepsilon$. If d is not covered by its essential neighbours, then d is the necessary element.

Suppose d is covered by its essential neighbours. In particular, this means that d has both the left and the right essential neighbour. First assume that a_j is contained in d. Then a_j is covered by essential neighbours of d. Since a_j is not covered by images of $\mathcal{M}^{\mathbf{nfc}}(U_h) \setminus \{a_h\}$, at least one essential neighbour of d is not equal to an image of the element of $\mathcal{M}^{\mathbf{nfc}}(U_h) \setminus \{a_h\}$. Hence it is of Λ-measure $\leqslant 5\varepsilon$ and it satisfies necessary conditions.

Assume that a_j is not contained in d. Since d' has the rightmost beginning among elements that are not separated from a_h from the left, its image d in U_j has the same beginning position. Therefore, the left essential neighbour of d in U_j is the image of the left essential neighbour of d' in U_h. Hence, the left essential neighbour of d is separated from a_j, since a_h does not have the left essential neighbour. So, d can not be covered by its left essential neighbour and a_j. Then consider the right essential neighbour of d and denote it by c. Since d is not covered by its left essential neighbour and a_j, the occurrence c is not equal to the occurrence a_j. If some element of $\mathcal{M}(U_j)$ has the beginning from the right of the beginning of d and the end from the left of the end of a_j, it is covered by d and a_j. Hence, it does not belong to $\mathcal{M}^{\mathbf{nfc}}(U_j)$. Therefore, since $c \in \mathcal{M}^{\mathbf{nfc}}(U_j)$, a_j is contained in c. Since a_j is not covered by images of $\mathcal{M}^{\mathbf{nfc}}(U_h) \setminus \{a_h\}$, c is not equal to an image of an element of $\mathcal{M}^{\mathbf{nfc}}(U_h) \setminus \{a_h\}$. Since the occurrence c is not equal to the occurrence a_j, $\Lambda(c) \leqslant 5\varepsilon$ and it satisfies the necessary conditions.

We denote the obtained element by g. We put $g = g_l g_m g_r$, where g_l is an overlap with the left essential neighbour (if there is any) and g_r is an overlap with the right essential neighbour (if there is any). Since g is not covered by its left and right essential neighbours, g_m is not empty.

We put $U_j = U_j^{(1)} = L^{(1)} g_m R^{(1)}$. Then b is an occurrence in the subword $L^{(1)}$. From Lemma 3.4 it follows that the Λ-measure of an image of g in every $U_j^{(i)}$ is not greater than $7\varepsilon < \tau - 2\varepsilon$. Therefore, an image of g can not be used in any replacement $U_j^{(i)} \mapsto U_j^{(i+1)}$ in (16). Then every replacement in (16) is of the form

$$L^{(i)} g_m R^{(i)} \mapsto L^{(i+1)} g_m R^{(i+1)},$$
$$\textit{where either } L^{(i+1)} = L^{(i)}, \textit{ or } R^{(i+1)} = R^{(i)}.$$

The possibility of a transformation $L^{(i)} g_m R^{(i)} \mapsto L^{(i+1)} g_m R^{(i+1)}$, where $R^{(i+1)} = R^{(i)}$, does not depend on $R^{(i)}$. Since b is an occurrence in the subword $L^{(1)}$, it is sufficient to do only transformations $L^{(i)} g_m R^{(i)} \mapsto L^{(i+1)} g_m R^{(i+1)}$, where $R^{(i+1)} = R^{(i)}$, in a sequence that defines b. So, we obtain the sequence

$$U_j^{(1)} = L^{(1)} g_m R^{(1)} \mapsto L^{(i_1)} g_m R^{(1)} \ldots \mapsto L^{(i_s)} g_m R^{(1)}$$

that also defines b. We have $U_h = L^{(1)}R'$. Thus, the same replacements can be done starting with U_h, so, we obtain the sequence

$$U_h = L^{(1)}R' \mapsto L^{(i_1)}R' \ldots \mapsto L^{(i_s)}R' \tag{17}$$

such that the image of b' in $L^{(i_s)}R'$ is of Λ-measure $\geqslant \tau$. Therefore, b' is a virtual member of the chart of U_h.

The case when there are no elements of $\mathcal{M}(U_h)$ that are not separated from a_h from the left side is considered in a similar (but easier) way.

So, we have proved that every virtual member of the chart of U_j is the image of a virtual member of the chart of U_h. Hence, $K_\tau(U_j) \leqslant K_\tau(U_h)$. If a_j is not a virtual member of the chart of U_j, then the image of a_h is not a virtual member of the chart of U_j, because a_j is not fully covered by images of $\mathcal{M}^{\mathbf{nfc}}(U_h) \setminus \{a_h\}$. Thus, in this case $K_\tau(U_j) < K_\tau(U_h)$. □

COROLLARY 3.6. *Let U_h be a monomial, a_h be a virtual member of its chart, $U_h = La_hR$. Assume a_h and a_j are incident monomials, $U_j = La_jR$. If a_j is a virtual member of the chart of U_j, then $N(U_j) = N(U_h)$. If a_j is not a virtual member of the chart of U_j, then either $N(U_j) < N(U_h)$, or $N(U_h) = N(U_j)$ and $K_\tau(U_j) < K_\tau(U_h)$.*

PROOF. From Lemma 3.3 it follows that $N(U_j) \leqslant N(U_h)$. Let a_j be a virtual member of the chart of U_j, then $a_j \in \mathcal{M}^{\mathbf{nfc}}(U_j)$. Hence, we can apply Lemma 3.3 to the opposite replacement $a_j \mapsto a_h$ and obtain $N(U_h) \leqslant N(U_j)$. Thus, $N(U_j) = N(U_h)$.

Assume a_j is not a virtual member of the chart of U_j. From Lemma 3.3 it follows that if a_j is fully covered by images of $\mathcal{M}^{\mathbf{nfc}}(U_h) \setminus \{a_h\}$, then $N(U_j) < N(U_h)$. Therefore, if $N(U_h) = N(U_j)$, then a_j is not fully covered by images of $\mathcal{M}^{\mathbf{nfc}}(U_h) \setminus \{a_h\}$. Hence, from Lemma 3.5 it follows that $K_\tau(U_j) < K_\tau(U_h)$. □

COROLLARY 3.7. *Let U_h be a monomial, a_h be a virtual member of its chart, $U_h = La_hR$. Assume a_h and a_j are incident monomials, a_j is a virtual member of the chart of $U_j = La_jR$. Assume $b_h \in \mathcal{M}^{nfc}(U_h)$, b_h is different from a_h and b_j is the image of b_h in U_j. Then b_h is a virtual member of the chart of U_h if and only if b_j is a virtual member of the chart of U_j. In particular, in this case $K_\tau(U_h) = K_\tau(U_j)$.*

PROOF. Since a_j is a virtual member of the chart of U_j, $\Lambda(a_j) \geqslant \tau - 2\varepsilon$ and $a_j \in \mathcal{M}^{\mathbf{nfc}}(U_j)$. Hence, the conditions of Lemma 3.5 hold for the replacement $U_h \mapsto U_j$, $a_h \mapsto a_j$. In the proof of Lemma 3.5 we actually show that if b_j is a virtual member of the chart of U_j, then b_h is a virtual member of the chart of U_h.

Since a_j is a virtual member of the chart of U_j, we can consider the opposite transformation $U_j \mapsto U_h$, $a_j \mapsto a_h$. Then, obviously, the image of b_j in U_h is b_h. If we apply Lemma 3.5 in this case, we obtain that if b_h is a virtual member of the chart of U_h, then b_j is a virtual member of the chart of U_j. □

In the same way as (15) we define the set of all the supports of multi-turns of virtual members of the chart

$$\mathcal{T}' = \Bigg\{ \sum_{j=1}^{k} U_j \mid U_h \in \mathcal{F}, U_h = La_hR, \quad \textit{where } a_h \textit{ is a virtual member of the chart of } U_h, \quad U_h \mapsto \sum_{\substack{j=1\\ j\neq h}}^{k} U_j \textit{ is a multi-turn that comes from an elementary multi-turn } a_h \mapsto \sum_{\substack{j=1\\ j\neq h}}^{k} a_j \Bigg\}. \tag{18}$$

PROPOSITION 3.8. *The linear subspace $\langle \mathcal{T}' \rangle \subseteq \mathbb{Z}_2\mathcal{F}$ is equal to the ideal $\mathcal{I} = \langle 1 + v + vw \rangle$.*

PROOF. From Proposition 3.1 it easily follows that $\mathcal{T}' = \mathcal{T}$. Hence, by Proposition 3.2, $\langle \mathcal{T}' \rangle = \mathcal{I}$. □

3.4. Sequences of transformations of a given monomial. Let U be a monomial, a_h and b_h be virtual members of the chart of $U = L_a a_h R_a = L_b b_h R_b$. To be precise, assume that the beginning of a_h lies from the left of the beginning of b_h. Assume a_h and a_j, b_h and b_j are incident monomials and consider transformations $a_h \mapsto a_j$, $b_h \mapsto b_j$. If a_j or b_j is equal to 1 and the resulting monomial $L_a R_a$ or $L_b R_b$ have cancellations, we do not perform them right after the replacement.

Let $\widetilde{a}_h$ be the image of a_h in $L_b b_j R_b$. If $b_j = 1$ and $L_b R_b$ has cancellations, we say that $\widetilde{a}_h$ is the intersection of a_h and L_b (notice that this slightly differs from Definition 3.1). Then $\widetilde{a}_h = a_h d_1$, where d_1 is a generalized fractional power that cancels or prolongs a_h (d_1 may be empty). Let $\widetilde{b}_h$ be the image of b_h in $L_a a_j R_a$. If $a_j = 1$ and $L_a R_a$ has cancellations, we say that $\widetilde{b}_h$ is the intersection of b_h and R_a. Then $\widetilde{b}_h = d_2 b_h$, where d_2 is a generalized fractional power that cancels or prolongs b_h (d_2 may be empty).

Since the monomials a_h and a_j are incident, the monomials $a_h d_1 = \widetilde{a}_h$ and $a_j d_1$ are also incident. Hence, we consider the replacement $\widetilde{a}_h \mapsto \widetilde{a}_j$ in $L_b b_h R_b$, where $\widetilde{a}_j = a_j d_1$. Similarly, since the monomials b_h and b_j are incident, the monomials $d_2 b_h = \widetilde{b}_h$ and $d_2 b_j$ are also incident. So, we consider the replacement $\widetilde{b}_h \mapsto \widetilde{b}_j$ in $L_a a_h R_a$, where $\widetilde{b}_j = d_2 b_j$. The next lemma states an important property of these transformations.

LEMMA 3.9. *The result of the replacement $\widetilde{a}_h \mapsto \widetilde{a}_j$ in $L_b b_j R_b$ and the result of the replacement $\widetilde{b}_h \mapsto \widetilde{b}_j$ in $L_a a_j R_a$ are equal.*

PROOF. First assume that a_h and b_h are separated, $U = L_a a_h M b_h R_b$. Then we obtain two sequences of transformations

$$U = L_a a_h M b_h R_b \mapsto L_a a_j M b_h R_b \mapsto L_a a_j M b_j R_b$$

and

$$U = L_a a_h M b_h R_b \mapsto L_a a_h M b_j R_b \mapsto L_a a_j M b_j R_b.$$

Obviously, the results are equal.

Assume a_h and b_h touch at a point, $U = L_a a_h b_h R_b$. After the replacement $a_h \mapsto a_j$ we obtain the monomial $L_a a_j b_h R_b$. The image of b_h in $L_a a_j b_h R_b$ may have different forms depending on a_j (see classification in Section 3.1). But in any case the replacement $\widetilde{b}_h \mapsto \widetilde{b}_j$ in $L_a a_j b_h R_b$ can be represented as a replacement $b_h \mapsto b_j$ and the further cancellations (if there are any). So, we obtain

$$U = L_a a_h b_h R_b \mapsto L_a a_j b_h R_b \mapsto L_a a_j b_j R_b.$$

After the replacement $b_h \mapsto b_j$ we obtain the monomial $L_a a_h b_j R_b$. Similarly, the replacement $\widetilde{a}_h \mapsto \widetilde{a}_j$ in $L_a a_h b_j R_b$ can be represented as a replacement $a_h \mapsto a_j$ and the further cancellations (if there are any). So, we obtain

$$U = L_a a_h b_h R_b \mapsto L_a a_h b_j R_b \mapsto L_a a_j b_j R_b$$

and results of both sequences are equal.

Assume a_h and b_h have an overlap c. Denote by a'_h the intersection of a_h and L_b, by b'_h the intersection of b_h and R_a, then $U = L_a a'_h c b'_h R_b$. After the replacement $a_h \mapsto a_j$ we obtain the monomial $L_a a_j b'_h R_b$. As above, the replacement $\widetilde{b}_h \mapsto \widetilde{b}_j$ in $L_a a_j b'_h R_b$ can be represented as a replacement $b'_h \mapsto c^{-1} b_j$ and the further cancellations (if there are any). So, we obtain

$$U = L_a a'_h c b'_h R_b \mapsto L_a a_j b'_h R_b \mapsto L_a a_j c^{-1} b_j R_b.$$

After the replacement $b_h \mapsto b_j$ we obtain the monomial $L_a a'_h b_j R_b$. As above, the replacement $\widetilde{a}_h \mapsto \widetilde{a}_j$ in $L_a a'_h b_j R_b$ can be represented as a replacement $a'_h \mapsto a_j c^{-1}$ and the further cancellations (if there are any). So, we obtain

$$U = L_a a'_h c b'_h R_b \mapsto L_a a'_h b_j R_b \mapsto L_a a_j c^{-1} b_j R_b$$

and results of both sequences are equal. □

LEMMA 3.10. *Let U be a monomial, a_h and b_h be virtual members of the chart of U, $U = L_a a_h R_a = L_b b_h R_b$. Let a_h and a_j be incident monomials, b_h and b_j be incident monomials, and consider transformations $a_h \mapsto a_j$, $b_h \mapsto b_j$. Assume a_j is a virtual member of the chart of $L_a a_j R_a$, b_j is a virtual member of the chart of $L_b b_j R_b$. Denote by $\widetilde{b}_h$ the image of b_h in $L_a a_j R_a$ and consider the transformation $\widetilde{b}_h \mapsto \widetilde{b}_j$, obtained by multiplying of $b_h \mapsto b_j$ by the corresponding generalized fractional power. Let us apply this replacement to the monomial $L_a a_j R_a$ and denote the result by $U_{a,b}$. Then $\widetilde{b}_j$ is a virtual member of the chart of $U_{a,b}$.*

PROOF. To be precise, assume that the beginning of a_h lies from the left of the beginning of b_h. We will consider only the most interesting case when a_h and b_h have an overlap c. The cases when a_h and b_h are separate or touch at a point are considered similarly.

Denote by a'_h the intersection of a_h and L_b, that is, $a_h = a'_h c$. Denote by b'_h the intersection of b_h and R_a, that is, $b_h = c b'_h$.

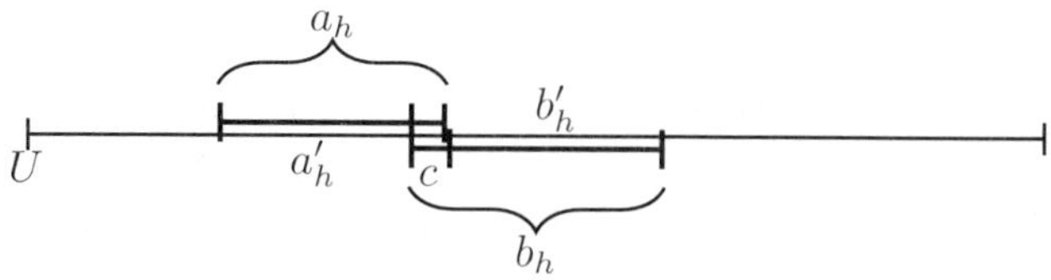

Let us perform the replacement $a_h \mapsto a_j$ in $U = L_a a_h R_a$. Then the image of b_h in $L_a a_j R_a$ is equal to $c_1 b'_h$, where c_1 is an overlap with a_j (possibly empty). We put $a_j = a'_j c_1$ and $\widetilde{b}_h = c_1 b'_h$.

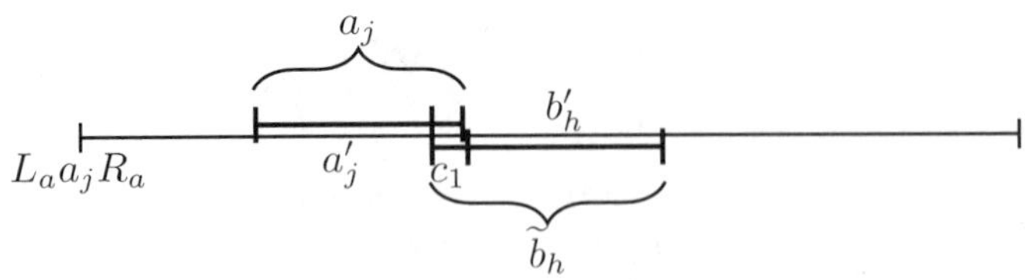

Since a_j is a virtual member of the chart of $L_a a_j R_a$, from Corollary 3.7 it follows that $\widetilde{b}_h$ is also a virtual member of the chart of $L_a a_j R_a$. The monomials b_h and b_j are incident. Hence, multiplying them by $c_1 c^{-1}$ from the left, we obtain incident monomials $c_1 b'_h$ and $c_1 c^{-1} b_j$. So, we have the replacement $c_1 b'_h = \widetilde{b}_h \mapsto \widetilde{b}_j$, where $\widetilde{b}_j = c_1 c^{-1} b_j$.

Let us perform the replacement $b_h \mapsto b_j$ in $U = L_b b_h R_b$. Then the image of a_h in $L_b b_j R_b$ is equal to $a'_h c_2$, where c_2 is an overlap with b_j (possibly empty). We put $b_j = c_2 b'_j$ and $\widetilde{a}_h = a'_h c_2$.

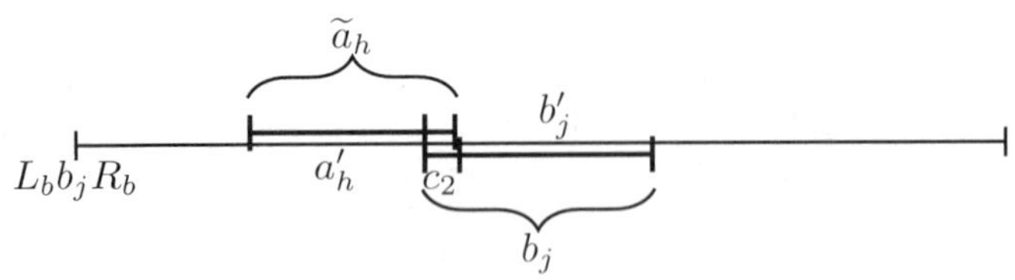

Since b_j is a virtual member of the chart of $L_b b_j R_b$, from Corollary 3.7 it follows that $\widetilde{a}_h$ is also a virtual member of the chart of $L_b b_j R_b$. The monomials a_h and a_j are incident. Hence, multiplying them by $c^{-1} c_2$ from the right, we obtain incident monomials $a'_h c_2$ and $a_j c^{-1} c_2$. So, we have the replacement $a'_h c_2 = \widetilde{a}_h \mapsto \widetilde{a}_j$, where $\widetilde{a}_j = a_j c^{-1} c_2$.

From Lemma 3.9 it follows that the sequences of replacements

$$U = L_a a_h R_a \mapsto L_a a_j R_a \mapsto L_a a'_j \widetilde{b}_j R_b$$

and

$$U = L_b b_h R_b \mapsto L_b b_j R_b \mapsto L_a \widetilde{a}_j b'_j R_b$$

give the same final result. So, we have

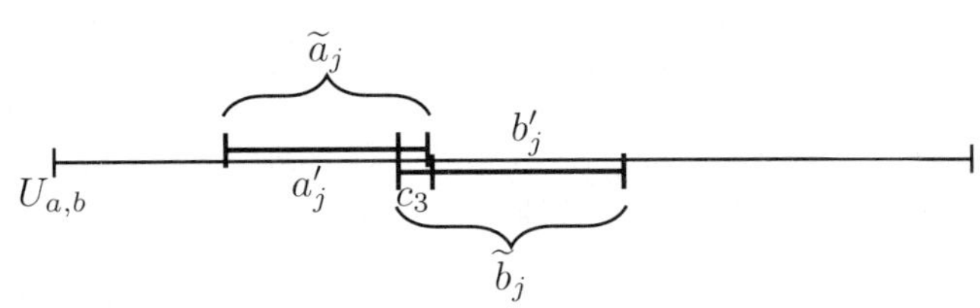

where the overlap c_3 is possibly empty. One can easily calculate that $c_3 = c_1 c^{-1} c_2$.

Recall that the Λ-measure of any overlap of two maximal occurrences of generalized fractional powers is equal either to ε, or to zero. Since $a_j = a'_j c_1$ is a virtual member of the chart of $L_a a_j R_a$, $\Lambda(a_j) \geqslant \tau - 2\varepsilon$. If $\Lambda(c_1) = \varepsilon$, then, according to

Remark 3.2, $\Lambda(a_j) \geqslant \tau - \varepsilon$. Hence, $\Lambda(a'_j) \geqslant \tau - 2\varepsilon$. If $\Lambda(c_1) = 0$, then, clearly, $\Lambda(a'_j) = \Lambda(a_j) \geqslant \tau - 2\varepsilon$.

We have $\Lambda(\widetilde{a}_j) \geqslant \Lambda(a'_j) \geqslant \tau - 2\varepsilon$. Since b_j is a virtual member of the chart of $L_b b_j R_b$, there exists a sequence of replacements from Definition 3.2 starting from $L_b b_j R_b$ such that the Λ-measure of the image of b_j in the last monomial of the sequence is not less than τ. Since $\Lambda(\widetilde{a}_j) \geqslant \tau - 2\varepsilon$, we can consider the replacement $\widetilde{a}_j \mapsto a'_h c_2 = \widetilde{a}_h$ in the monomial $U_{a,b}$. Then we obtain the transformation $U_{a,b} \mapsto L_b b_j R_b$. We add this transformation to the beginning of the sequence. Since the image of $\widetilde{b}_j$ in $L_b b_j R_b$ is equal to b_j, as a result we obtain the sequence of replacements starting from $U_{a,b}$ such that the Λ-measure of the image of $\widetilde{b}_j$ in the last monomial is not less than τ. Hence, $\widetilde{b}_j$ is a virtual member of the chart of $U_{a,b}$. □

Applying Lemma 3.9 and Lemma 3.10, we obtain the following statement, that we will use in the next section.

COROLLARY 3.11. *Let U be a monomial.*

(1) *Assume we have a sequence of replacements starting from U such that every replacement transforms a virtual member of the chart into a virtual member of the chart. Then any replacement can be moved to any position in the sequence and the final result remains the same. Moreover, after changing of the order of the replacements, every replacement in the obtained sequence still transforms a virtual member of the chart into a virtual member of the chart.*

(2) *Let $a_h^{(1)}, \dots, a_h^{(n)}$ be virtual members of the chart of U. Consider a number of replacements $a_h^{(i)} \mapsto a_j^{(i)}$ in U such that each $a_j^{(i)}$ is a virtual member of the chart of the resulting monomial. Suppose these transformation are applied consecutively. Then every transformation in the chain also transforms a virtual member of the chart into a virtual member of the chart. Moreover, we can apply the replacements in any order.*

4. The structure of certain subspaces of $\mathbb{Z}_2\mathcal{F}/\mathcal{I}$: filtration, grading and tensor products

First let us introduce the notion of derived monomials.

DEFINITION 4.1 (**derived monomials**). Let U be a monomial. Consider the following transformations of U:

(1) Replacements of a virtual member of the chart by an incident monomial non-equal to 1. Recall that in this case the result is always a reduced monomial.

(2) Replacements of a virtual member of the chart by incident monomial equal to 1 and further cancellations (in order to obtain the reduced monomial).

Starting with a certain monomial U we consecutively apply transformations (1), (2). All the monomials that we obtain after some sequence of transformations (1), (2) (including the monomial U itself) are called *derived monomials of U*.

DEFINITION 4.2. Let $\{U_i\}$ be either finite or countable set of monomials. By $\langle U_1, \dots, U_k, \dots\rangle_d$, we denote a subspace of $\mathbb{Z}_2\mathcal{F}$ generated by all the derived monomials of the monomials $\{U_i\}$.

REMARK 4.1. Assume U is a monomial and $U \in \langle U_1, \dots, U_k, \dots \rangle_d$, where U_i are monomials. Then $U = \sum_{j=1}^{l} \widetilde{U}_{i_j}$, where $\widetilde{U}_{i_j}$ is a derived monomial of U_{i_j}. Since monomials form a basis of $\mathbb{Z}_2\mathcal{F}$, we obtain $U = \widetilde{U}_{i_{j_0}}$. So, we obtain that $U \in \langle U_1, \dots, U_k, \dots \rangle_d$ if and only if U is a derived monomial of some U_i.

We have shown that $\mathcal{I} = \langle \mathcal{T}' \rangle$. Consider $\sum_{j=1}^{l} U_j = T \in \mathcal{T}'$, where $U_h \mapsto \sum_{\substack{j=1 \\ j \neq h}}^{k} U_j$ is a multi-turn of a virtual member of the chart of U_h. Then, by definition of $\langle U_h \rangle_d$, we obtain $T \in \langle U_h \rangle_d$.

In this section, we show that $\langle U_1, \dots, U_k \rangle_d \cap \mathcal{I}$ is generated by supports of multi-turns of monomials from $\langle U_i \rangle_d$ for $i = 1, \dots, k$. This enables us to construct a linear basis of $\langle U_1, \dots, U_k \rangle_d / (\langle U_1, \dots, U_k \rangle_d \cap \mathcal{I})$ and a linear basis of $\mathbb{Z}_2\mathcal{F}/\mathcal{I}$ (and therefore show that $\mathbb{Z}_2\mathcal{F}/\mathcal{I}$ is non-trivial).

Our approach can be compared with the more standard one, that uses the Diamond Lemma to control the ring relations ([**1**]). Instead, we introduce a filtration (and the corresponding grading) on $\langle U_1, \dots, U_k \rangle_d$ and show its compatibility with the subspaces of linear dependencies (see Lemma 4.7 and Theorem 4.1). This allows us to deal with linear dependencies in each graded component independently. The structure of each graded component is described in Proposition 4.10 and Proposition 4.12.

Notice that the Diamond Lemma can also be reformulated in the language of grading.

4.1. The filtration on spaces $\langle U_1, \dots, U_k \rangle_d$. Let Z be a monomial. We introduce the following numerical characteristics of Z (f-characteristics of monomials):

$$f(Z) = (N(Z), K_\tau(Z)), \tag{19}$$

where $N(Z)$ is the number of elements in a minimal covering of Z, $K_\tau(Z)$ is the number of virtual members of the chart of Z. If Z_1 and Z_2 are monomials, we say that $f(Z_1) < f(Z_2)$ if and only if $N(Z_1) < N(Z_2)$ or $N(Z_1) = N(Z_2)$ and $K_\tau(Z_1) < K_\tau(Z_2)$.

The characteristics f satisfies the following property.

LEMMA 4.1. *Assume U and Z are monomials, where Z is a derived monomial of U. Then $f(Z) \leqslant f(U)$. Moreover, $f(Z) < f(U)$ if and only if in the corresponding sequence of replacements there exists at least one replacement of the form $La_hR \to La_jR$ such that a_h is a virtual member of the chart of La_hR and a_j is not a virtual member of the chart of La_jR.*

PROOF. It follows directly from Corollary 3.6 and Corollary 3.7. □

In this section, let us set

$$V = \langle U_1, \dots, U_k \rangle_d, U_1, \dots, U_k \in \mathcal{F}.$$

We will define a decreasing filtration on V.

First consider a space generated by one monomial and its derived monomials, namely, $W = \langle U \rangle_d$, $U \in \mathcal{F}$. Let us define a subspace $\mathrm{L}(W) \subseteq W$. We put

$$\mathrm{L}(W) = \langle Z \in \mathcal{F} \mid Z \textit{ is a derived monomial of } U \textit{ such that } f(Z) < f(U) \rangle. \tag{20}$$

If the set of monomials with strictly smaller f-characteristics than $f(U)$ is empty, by definition, we put $\mathrm{L}(W) = 0$.

Let a monomial $Z' \in \mathrm{L}(W)$, Z'' be a derived monomial of Z'. Then, by definition of $\mathrm{L}(W)$ and Lemma 4.1, $f(U) > f(Z') \geqslant f(Z'')$. Hence, Z'' also belongs to $\mathrm{L}(W)$, that is, $\mathrm{L}(W)$ is closed under taking derived monomials.

Now consider a space $Y = \langle Z_1, \dots, Z_k, \dots \rangle_d$, where $\{Z_i\}$ is either a finite or infinite set of monomials. By definition, we put

$$\mathrm{L}(Y) = \sum_i \mathrm{L}(\langle Z_i \rangle_d). \tag{21}$$

Since every $\mathrm{L}(\langle Z_i \rangle_d)$ is generated by monomials and closed under taking derived monomials, the space $\mathrm{L}(Y)$ is generated by monomials and closed under taking derived monomials as well. Hence,

$$\mathrm{L}(Y) = \langle Z'_1, \dots, Z'_k, \dots \rangle_d \textit{ for some } Z'_i \in \mathcal{F}. \tag{22}$$

Notice that one generalized fractional power may have an infinite number of incident monomials of Λ-measure less than $\tau - 2\varepsilon$, because they may contain different powers of w. Recall that such generalized fractional powers are never counted as a virtual members of the chart. Hence, even if the space Y is generated by derived monomials of finite number of monomials, the space $\mathrm{L}(Y)$ might be generated by derived monomials of countably many monomials.

In the sequel, we will widely use the following simple properties of derived monomials.

LEMMA 4.2. *Let Z_1 be a monomial, Z be a derived monomial of Z_1, $Y \subseteq \mathbb{Z}_2\mathcal{F}$ be a space generated by monomials and closed under taking derived monomials. Then the following statements hold:*

1. *If $Z \in \langle Z_1 \rangle_d \setminus \mathrm{L}(\langle Z_1 \rangle_d)$, then $\langle Z_1 \rangle_d = \langle Z \rangle_d$.*
2. *If $Z \in Y$ and $Z \in \langle Z_1 \rangle_d \setminus \mathrm{L}(\langle Z_1 \rangle_d)$, then $\langle Z_1 \rangle_d \subseteq Y$.*

PROOF. Suppose Z'_1 is a result of transformation (1) (Definition 4.1) such that Z'_1 is contained in $\langle Z_1 \rangle_d \setminus \mathrm{L}(\langle Z_1 \rangle_d)$. Then this transformation is invertible, hence Z_1 is also a derived monomial of Z'_1. Repeating this argument for every transformation in a sequence that connects Z_1 and Z, we obtain that Z_1 is a derived monomial of Z. Now assume U is a derived monomial of Z_1. Since Z_1 is a derived monomial of Z, U is also a derived monomial of Z. Hence $\langle Z_1 \rangle_d = \langle Z \rangle_d$. The first statement of the lemma is proved.

Now let us prove the second statement. Since $Z \in Y$ and Y is generated by monomials and closed under taking derived monomials, we obtain $\langle Z \rangle_d \subseteq Y$. But above we proved $\langle Z \rangle_d = \langle Z_1 \rangle_d$, hence, $\langle Z_1 \rangle_d \subseteq Y$. □

We defined the subspace $\mathrm{L}(Y)$ using a set of generators of Y (see formula (21)). Let us show that, in fact, $\mathrm{L}(Y)$ does not depend on the set of generators of Y.

PROPOSITION 4.3. *Assume*

$$Y = \langle Z_1, \dots, Z_k, \dots \rangle_d = \langle Z'_1, \dots, Z'_k, \dots \rangle_d,$$

where $\{Z_i\}$ and $\{Z'_i\}$ are either finite or infinite sets of monomials. Then

$$\sum_i \mathrm{L}(\langle Z_i \rangle_d) = \sum_i \mathrm{L}(\langle Z'_i \rangle_d).$$

PROOF. Let us enumerate all monomials from Y. Let $\{X_j\}$ be the set of all monomials of the space Y. Then, evidently, we obtain the following description of Y:

$$Y = \langle X_1, \ldots, X_k, \ldots\rangle = \langle X_1, \ldots, X_k, \ldots\rangle_d.$$

Clearly, it is sufficient to show that

$$\sum_i \mathrm{L}(\langle Z_i\rangle_d) = \sum_j \mathrm{L}(\langle X_j\rangle_d)$$

for an arbitrary set of monomials $\{Z_i\}$ such that $Y = \langle Z_1, \ldots, Z_k, \ldots\rangle_d$.

Since $\{Z_1, \ldots, Z_k, \ldots\} \subseteq \{X_1, \ldots, X_k, \ldots\}$, clearly,

$$\sum_i \mathrm{L}(\langle Z_i\rangle_d) \subseteq \sum_j \mathrm{L}(\langle X_j\rangle_d).$$

Assume $Z \in \sum_j \mathrm{L}(\langle X_j\rangle_d)$. Monomials form a basis of $\mathbb{Z}_2\mathcal{F}$ and every $\mathrm{L}(\langle X_j\rangle_d)$ is generated by monomials. Hence, Z is a derived monomial of some X_{j_0} such that we have $Z \in \mathrm{L}(\langle X_{j_0}\rangle_d)$ for some j_0. Therefore, there exists a sequence of transformations $\alpha_1, \ldots, \alpha_{s_1}$ of type (1), (2) such that

$$X_{j_0} \xmapsto{\alpha_1} \ldots \xmapsto{\alpha_{s_1}} Z. \tag{23}$$

Moreover, in the sequence, there exists at least one transformation $\alpha_l : Y_{l-1} \to Y_l$ such that $f(Y_l) < f(Y_{l-1})$.

Since $X_{j_0} \in Y = \langle Z_1, \ldots, Z_k, \ldots\rangle_d$, we have $X_{j_0} \in \langle Z_{i_0}\rangle_d$ for some i_0. Hence, there exists a sequence of transformations $\beta_1, \ldots, \beta_{s_2}$ of type (1), (2) such that

$$Z_{i_0} \xmapsto{\beta_1} \ldots \xmapsto{\beta_{s_2}} X_{j_0}. \tag{24}$$

Gluing (24) and (23), we obtain

$$Z_{i_0} \xmapsto{\beta_1} \ldots \xmapsto{\beta_{s_2}} X_{j_0} \xmapsto{\alpha_1} \ldots \xmapsto{\alpha_{s_1}} Z.$$

Consequently, since there exists α_l that decreases the value of the function f, we have $Z \in \mathrm{L}(\langle Z_{i_0}\rangle_d)$. Hence, $Z \in \sum_i \mathrm{L}(\langle Z_i\rangle_d)$ and

$$\sum_j \mathrm{L}(\langle X_j\rangle_d) \subseteq \sum_i \mathrm{L}(\langle Z_i\rangle_d).$$

This completes the proof. □

Now we define a decreasing filtration on $V = \langle U_1, \ldots, U_k\rangle_d$ in the following way. By definition, put

$$\begin{aligned} \mathrm{F}_0 V &= V, \\ \mathrm{F}_{n+1} V &= \mathrm{L}(\mathrm{F}_n V). \end{aligned} \tag{25}$$

Since for every $Y = \langle Z_1, \ldots, Z_k, \ldots\rangle_d$, where $\{Z_i\}$ is either a finite or infinite set of monomials, we have description (22), therefore formula (25) is applicable for every $n \geqslant 0$. Here we mean $\mathrm{L}(0) = 0$.

PROPOSITION 4.4. *The filtration defined above has finitely many levels, that is, there exists a number N such that $\mathrm{F}_N V = 0$. Moreover, we never have a situation $\mathrm{F}_{n+1} V = \mathrm{F}_n V$ for $\mathrm{F}_n V \neq 0$.*

PROOF. Recall that $V = \langle U_1, \ldots, U_k \rangle_d$, $U_1, \ldots, U_k \in \mathcal{F}$. We put

$$N_{\max} = \max_{i \in \{1,\ldots,k\}} (N(U_i)).$$

Consider f-characteristics of monomials in V. Let U be an arbitrary monomial from V. From Lemma 4.1 it follows that $N(U) \leqslant N_{\max}$. Since every virtual member of the chart is of Λ-measure not less than $\tau - 2\varepsilon$, it is contained in every covering of U. Hence, we evidently obtain $K_\tau(U) \leqslant N(U) \leqslant N_{\max}$. Therefore, there are finitely many different values of f-characteristics of monomials in V. Then

$$m^{(n)} = \max_{U \in \mathrm{F}_n V} f(U)$$

is finite for any n such that $\mathrm{F}_n V \neq 0$.

Recall that, by definition, $\mathrm{F}_{n+1} V = \mathrm{L}(\mathrm{F}_n V)$. Then from (20) and (21) it follows that $m^{(n)} > m^{(n+1)}$ if $\mathrm{F}_{n+1} V \neq 0$. So, we have a strictly decreasing sequence of f-characteristics

$$m^{(0)} > \ldots > m^{(n)} > \ldots \tag{26}$$

that corresponds to the decreasing sequence of non-trivial spaces

$$\mathrm{F}_0 V \supseteq \ldots \supseteq \mathrm{F}_n V \supseteq \ldots .$$

The sequence (26) can not be infinite, so, it ends up at some step n_0. This means that $\mathrm{F}_{n_0} V$ is the last non-trivial subspace of the filtration. That is, the filtration has finitely many non-zero levels.

Assume $\mathrm{F}_{n+1} V = \mathrm{F}_n V$ for $\mathrm{F}_n V \neq 0$. That is, $\mathrm{F}_n V = \mathrm{L}(\mathrm{F}_n V)$. Then by induction we have $\mathrm{F}_{n+k} V = \mathrm{F}_n V \neq 0$ for any $k \in \mathbb{N}$. But we already proved that the filtration has finitely many non-zero levels. This contradiction completes the proof. □

DEFINITION 4.3. Suppose Y is a subspace of $\mathbb{Z}_2\mathcal{F}$ linearly generated by an arbitrary set of monomials and closed under taking derived monomials. Every linear dependence from $\mathcal{T}'$ is, in fact, a linear dependence between a monomial U and its derived monomials. Hence, any multi-turn of a virtual member of the chart of a monomial from Y generates a linear dependence between the monomials from Y, because Y is closed under taking derived monomials. We consider the subspace of Y

$$\mathrm{Dp}(Y) = \Big\langle \sum_{j=1}^{k} U_j \mid U_h \textit{ is a monomial from } Y,$$
$$U_h = L a_h R, \textit{ where } a_h \textit{ is a virtual member of the chart of } U_h,$$
$$U_h \mapsto \sum_{\substack{j=1\\ j\neq h}}^{k} U_j \textit{ is a multi-turn that comes from}$$
$$\textit{an elementary multi-turn } a_h \mapsto \sum_{\substack{j=1\\ j\neq h}}^{k} a_j \Big\rangle$$

and call this subspace *the subspace of dependencies on* Y. Using this notion, $\mathcal{I} = \langle \mathcal{T}' \rangle = \mathrm{Dp}(\mathbb{Z}_2\mathcal{F})$.

Note that if a monomial $U \in W$ has the empty chart, then there are no multi-turns of U. Hence, when W consists only of monomials with the empty chart, by definition we put $\mathrm{Dp}(W) = 0$.

Since V is closed under taking derived monomials, we consider its subspace $\mathrm{Dp}(V)$. Evidently, for every monomial U the space $\mathrm{L}(\langle U\rangle_d)$ is linearly generated by monomials and closed under taking derived monomials. Therefore, by the definition, every F_nV is generated by monomials and closed under taking derived monomials. Hence, we can consider its subspace $\mathrm{Dp}(\mathrm{F}_nV)$ and define the filtration on $\mathrm{Dp}(V)$ in the following way

$$\mathrm{F}_n\mathrm{Dp}(V) = \mathrm{Dp}(\mathrm{F}_nV),$$

that is, $\mathrm{F}_n\mathrm{Dp}(V)$ is the vector space generated by linear dependencies coming from monomials of F_nV.

LEMMA 4.5. *Suppose U is a monomial, $U \in \mathrm{F}_nV$. If $U \notin \mathrm{F}_{n+1}V$, then*

$$\mathrm{F}_{n+1}V \cap \langle U\rangle_d = \mathrm{L}(\langle U\rangle_d).$$

PROOF. Let $\mathrm{F}_nV = \langle Z_1, \dots, Z_k, \dots\rangle_d$, where $\{Z_i\}$ is either a finite or infinite set of monomials. Since $U \in \mathrm{F}_nV$ and F_nV is closed under taking derived monomials, we can assume that $\mathrm{F}_nV = \langle U, Z_1, \dots, Z_k, \dots\rangle_d$. Since the definition of $\mathrm{F}_{n+1}V = \mathrm{L}(\mathrm{F}_nV)$ does not depend on the set of generators of F_nV (see Proposition 4.3), we have

$$\mathrm{F}_{n+1}V = \mathrm{L}(\mathrm{F}_nV) = \mathrm{L}(\langle U\rangle_d) + \sum_i \mathrm{L}(\langle Z_i\rangle_d).$$

Let $Z \in \mathrm{L}(\langle U\rangle_d)$. Then, using the last equality, we immediately obtain $Z \in \mathrm{F}_{n+1}V$. So, $\mathrm{L}(\langle U\rangle_d) \subseteq \mathrm{F}_{n+1}V \cap \langle U\rangle_d$.

Assume a monomial $Z \in \mathrm{F}_{n+1}V \cap \langle U\rangle_d$ and $Z \notin \mathrm{L}(\langle U\rangle_d)$. Since $U \notin \mathrm{F}_{n+1}V$, we have $U \notin \mathrm{L}(\langle Z_i\rangle_d)$ for all i. Monomials form a basis of $\mathbb{Z}_2\mathcal{F}$ and every $\mathrm{L}(\langle Z_i\rangle_d)$ is generated by monomials. So, since $Z \in \mathrm{F}_{n+1}V = \sum_i \mathrm{L}(\langle Z_i\rangle_d)$, we have $Z \in \mathrm{L}(\langle Z_{i_0}\rangle_d)$ for some i_0. We assumed that $Z \in \langle U\rangle_d \setminus \mathrm{L}(\langle U\rangle_d)$, hence, by Lemma 4.2, $\langle U\rangle_d \subseteq \mathrm{L}(\langle Z_{i_0}\rangle_d)$. But that contradicts the assumption that $U \notin \mathrm{F}_{n+1}V$. Therefore, $\mathrm{F}_{n+1}V \cap \langle U\rangle_d \subseteq \mathrm{L}(\langle U\rangle_d)$.

Thus, finally we obtain $\mathrm{F}_{n+1}V \cap \langle U\rangle_d = \mathrm{L}(\langle U\rangle_d)$. □

In Definition 4.1. we defined derived monomials with the use of a special set of transformations (1) and (2). However, in the next lemma we will use a slightly wider class of transformations. Let us prove that, using this wider class of transformations of a given monomial, we still will have its derived monomials.

Namely, let U be a monomial, a_h and b_h be virtual members of the chart of U. We consider two replacements $a_h \mapsto a_j$, $b_h \mapsto b_j$ (a_h and a_j are incident monomials, b_h and b_j are incident monomials). Assume that first we apply the transformation $a_h \mapsto a_j$, namely, $U = L_a a_h R_a \mapsto L_a a_j R_a$. To be precise, we suppose that the beginning of b_h lies from the right of the beginning of a_h. Denote by b'_h the intersection of b_h and R_a. If a_h and b_h are separated or touch at a point, $b'_h = b_h$.

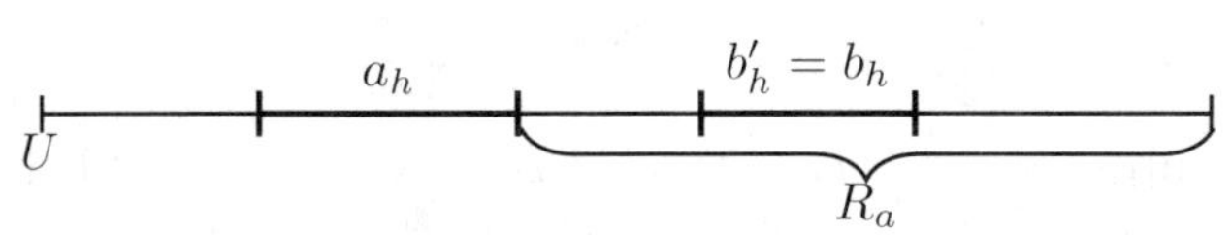

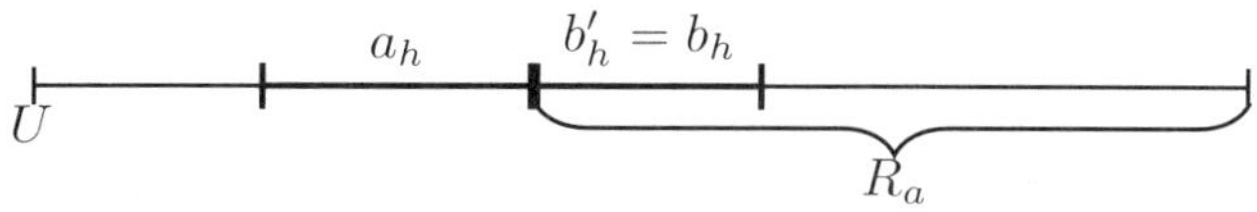

If a_h and b_h has an overlap c, then $b_h = cb'_h$.

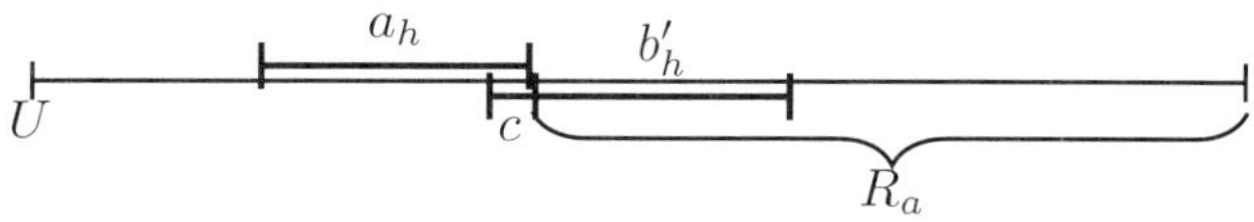

In any case, we can apply the transformation $b'_h \mapsto b'_j$ to $L_a a_j R_a$, where either $b'_j = b_j$ if a_h and b_h are separated or touch at a point, or $b'_j = c^{-1}b_j$ if a_h and b_h has an overlap c. Here we mean that if $a_j = 1$ and the monomial $L_a R_a$ has cancellations, we do not perform them. Instead, we perform a replacement $b'_h \mapsto b'_j$.

Assume $a_j \neq 1$ or $a_j = 1$ and $L_a R_a$ has no further cancellations. The replacement of the image of b'_h in $L_a a_j R_a$ (see Definition 3.1) can be represented as the replacement $b'_h \mapsto b'_j$ and the further cancellations if there are any (they may occur when b'_h is not a maximal occurrence of a generalized fractional power in $L_a a_j R_a$). So, if the image of b'_h is a virtual member of $L_a a_j R_a$, as a result of the transformation $b'_h \mapsto b'_j$ in $L_a a_j R_a$ and the further cancellations, we obtain a derived monomial of U.

If a_j is a virtual member of the chart of $L_a a_j R_a$, then, by Corollary 3.7, the image of b'_h is always a virtual member of $L_a a_j R_a$. If a_j is not virtual member of the chart of $L a_j R$, then the image of b'_h may not be a virtual member of $L_a a_j R_a$. The following lemma states that if we apply the replacement $b'_h \mapsto b'_j$ to $L_a a_j R_a$ in this case, we, nevertheless, obtain a derived monomial of U.

LEMMA 4.6. *Let U be a monomial, a_h and b_h be virtual members of the chart of U. We consider two replacements $a_h \mapsto a_j$, $b_h \mapsto b_j$ (a_h and a_j are incident monomials, b_h and b_j are incident monomials). Assume $U = L_a a_h R_a$, a_j is not a virtual member of the chart of $L_a a_j R_a$, b'_h is an intersection of R_a and b_h. Then the result of the replacement $b'_h \mapsto b'_j$ (corresponding to $b_h \mapsto b_j$) in $L_a a_j R_a$ is a derived monomial of U, possibly after the further cancellations. Moreover, its f-characteristics is strictly less than $f(U)$.*

PROOF. We will consider only the most interesting case when a_h and b_h has an overlap c. Two other cases are similar.

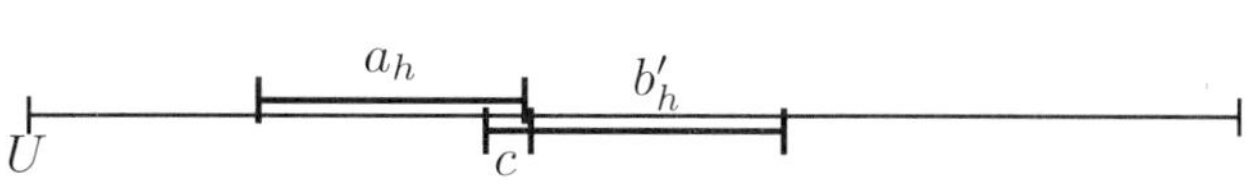

First we consider the case when $a_j \neq 1$. Then the resulting monomial $L_a a_j R_a$ is reduced. We have the transformation $b_h \mapsto b_j$. Since $b_h = cb'_h$, we also have the transformation $b'_h \mapsto c^{-1}b_j$. So, we have a sequence of replacements

$$U = L_a a_h R_a \mapsto L_a a_j R_a = L_a a_j b'_h R_b \mapsto L_a a_j c^{-1} b_j R_b. \tag{27}$$

If $\Lambda(b_h) \geqslant \tau + \varepsilon$, then $\Lambda(b'_h) \geqslant \tau$. The image of b'_h in $L_a a_j R_a$ prolongs b'_h. Therefore, its Λ-measure is not less than the Λ-measure of b'_h. So, the image of b'_h in $L_a a_j R_a$ is of Λ-measure $\geqslant \tau$. Therefore, it is a virtual member of the chart of $L_a a_j R_a$ and the result of (27) is a derived monomial of U, possibly after the further cancellations.

Suppose $\Lambda(b_h) < \tau + \varepsilon$. Consider an elementary multi-turn $b_h \mapsto \sum_{\substack{k=1\\k\neq h}}^{s} b_k$, where b_j is one of the resulting monomials. In Proposition 3.1, we proved that there exists b_{k_0} such that $\Lambda(b_{k_0}) \geqslant \tau$. However, using the same argument, one can easily prove a slightly stronger statement, namely, that there exists b_{k_0} such that $\Lambda(b_{k_0}) \geqslant \tau + \varepsilon$ (and, hence, $k_0 \neq h$).

Assume $U = L_b b_h R_b$. Consider the transformation

$$U = L_b b_h R_b \mapsto L_b b_{k_0} R_b.$$

Denote by a'_h the intersection of a_h and L_b in $L_b b_h R_b$. Then $a_h = a'_h c$. Let the image of a_h in $L_b b_{k_0} R_b$ be equal to $a'_h c'$, where c' is possibly empty. Also let us put $b_{k_0} = c' b'_{k_0}$. Since b_{k_0} is a virtual member of the chart of $L_b b_{k_0} R_b$, from Corollary 3.7 it follows that $a'_h c'$ is a virtual member of the chart of $L_b b_{k_0} R_b$.

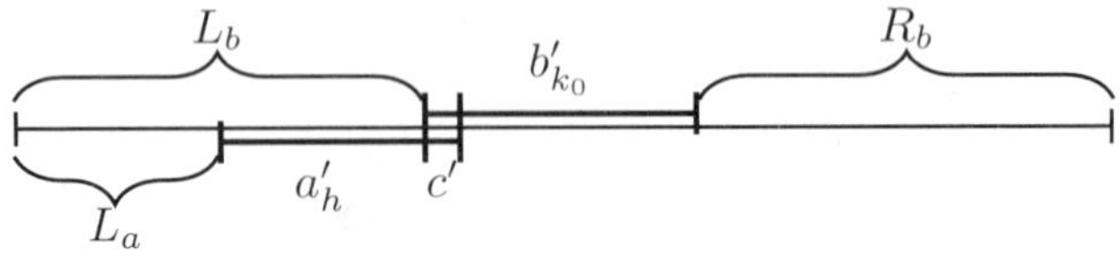

We have a transformation $a_h \mapsto a_j$. Multiplying it by $c^{-1}c'$ from the right, we obtain $a'_h c' \mapsto a'_j$, where $a'_j = a_j c^{-1} c'$. The monomials b_{k_0} and b_j are incident, so, we have a transformation $b_{k_0} \mapsto b_j$. Multiplying it by c'^{-1} from the left, we obtain $b'_{k_0} \mapsto b'_j$, where $b'_j = c'^{-1} b_j$. Since $\Lambda(b_{k_0}) \geqslant \tau + \varepsilon$, clearly, $\Lambda(b'_{k_0}) \geqslant \tau$. Therefore, a prolongation of b'_{k_0} is always a virtual member of the chart. Hence, we obtain a sequence of replacements of virtual members of the chart

$$\begin{aligned} L_b b_{k_0} R_b &= L_a a'_h c' b'_{k_0} R_b \mapsto L_a a'_j b'_{k_0} R_b \\ &\mapsto L_a a'_j b'_j R_b = L_a a_j c^{-1} c' c'^{-1} b_j R_b = L_a a_j c^{-1} b_j R_b. \end{aligned} \tag{28}$$

So, the results of (27) and of (28) are equal. But the result of sequence (28) is a derived monomial of U, by definition. Since a_j is not a virtual member of the chart of $L_a a_j R_a$, from Lemma 3.10 it follows that a'_j is not a virtual member of the chart of $L_a a'_j b'_{k_0} R_b$. Hence, $f(L_a a_j c^{-1} b_j R_b) < f(U)$.

Consider the case when $a_j = 1$. If there are no cancellations in $L_a R_a$, then we argue as above. Suppose there are cancellations in $L_a R_a$, namely, $b_h = c b_{h,m} b_{h,f}$, $L_a = L'_a b_{h,m}^{-1}$, where $b_{h,f}$ is possibly empty. For simplicity, assume that $L'_a b_{h,f} R_b$ has no further cancellations.

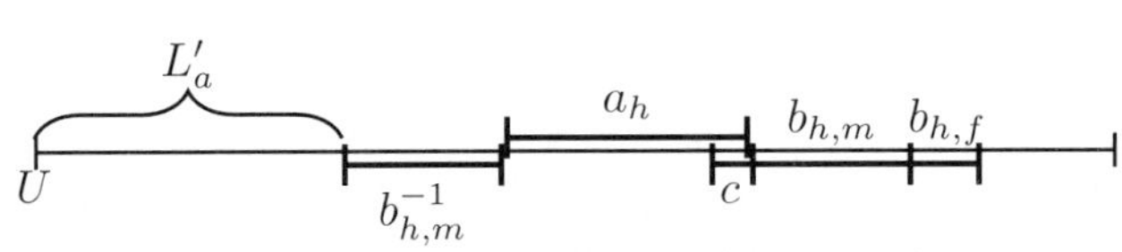

As above, let b'_h be an intersection of b_h and R_a, that is, $b_h = c b'_h$, $b'_h = b_{h,m} b_{h,f}$. We have a replacement $b_h \mapsto b_j$ and the corresponding replacement $b'_h = c^{-1} b_h \mapsto c^{-1} b_j$. After the replacement $L_a a_h R_a \mapsto L_a R_a$ there are cancellations of the resulting monomial. Let us not perform them, instead, we perform the replacement $b'_h \mapsto c^{-1} b_j$. So, we have a sequence of replacements

$$U = L_a a_h R_a \mapsto L_a R_a = L_a b'_h R_b = L_a c^{-1} b_j R_b. \tag{29}$$

Again consider an elementary multi-turn $b_h \mapsto \sum_{\substack{k=1\\k\neq h}}^{s} b_k$, where b_j is one of the resulting monomials. Multiplying it by $b_{h,m}^{-1}c^{-1}$ from the left side, we obtain an elementary multi-turn $b_{h,f} \mapsto \sum_{\substack{k=1\\k\neq h}}^{s} b_{h,m}^{-1}c^{-1}b_k$. If $\Lambda(b_{h,f}) \geqslant \tau$, then its image in $L'_a b_{h,f} R_b$ is a virtual member or the chart. Then we again can argue as in the beginning of the proof. Assume $\Lambda(b_{h,f}) < \tau$. Then, by Proposition 3.1, there exists b_{k_1} such that $\Lambda(b_{h,m}^{-1}c^{-1}b_{k_1}) \geqslant \tau$.

We have the replacement $a_h \mapsto 1$. As above, let a'_h be an intersection of a_h and L_b, that is, $a_h = a'_h c$. Let $a'_h c''$ be the image of a_h in $L_b b_{k_1} R_b$ (c'' is possibly empty). That is, c'' is an overlap of $a'_h c''$ and b_{k_1}, $b_{k_1} = c'' b'_{k_1}$. From Corollary 3.7 it follows that $a'_h c''$ is a virtual member of the chart of $L_b b_{k_1} R_b$.

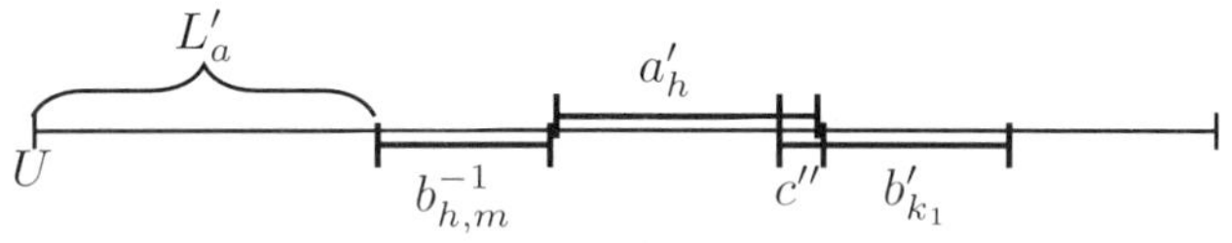

So, multiplying $a_h \mapsto 1$ by $c^{-1}c''$ from the right side, we obtain the replacement $a'_h c'' \mapsto c^{-1}c''$.

The monomials b_{k_1} and b_j are incident, hence, we have a transformation $b_{k_1} \mapsto b_j$. So, we also have $b_{h,m}^{-1}c^{-1}b_{k_1} \mapsto b_{h,m}^{-1}c^{-1}b_j$. Since $\Lambda(b_{h,m}^{-1}c^{-1}b_{k_1}) \geqslant \tau$ its prolongation is always a virtual member of the chart. Therefore, we obtain the sequence of replacements of virtual members of the chart:

$$\begin{aligned}(30)\quad U = L_b b_h R_b \mapsto L_b b_{k_1} R_b = L_a a'_h c'' b'_{k_1} R_b \mapsto \\ \mapsto L_a c^{-1} c'' b'_{k_1} R_b = L'_a b_{h,m}^{-1} c^{-1} b_{k_1} R_b \mapsto L'_a b_{h,m}^{-1} c^{-1} b_j R_b = L_a c^{-1} b_j R_b.\end{aligned}$$

So, the result of (29) is equal to the result of (30). But the result of sequence (30) is a derived monomial of U, by definition. In (30) we had the replacement $a'_h c'' \mapsto c^{-1}c''$, where $\Lambda(c^{-1}c'') \leqslant 2\varepsilon$. Hence, $c^{-1}c''$ can not be a virtual member of the chart. Therefore, $f(L_a c^{-1} b_j R_b) < f(U)$. □

LEMMA 4.7 (**Main Lemma**). *Let* $V = \langle U_1, \ldots, U_k\rangle_d, U_1, \ldots, U_k \in \mathcal{F}$. *Then we have*

$$\mathrm{Dp}(\mathrm{F}_n V) \cap \mathrm{F}_{n+1} V = \mathrm{Dp}(\mathrm{F}_{n+1} V).$$

PROOF. We have

$$\mathrm{Dp}(\mathrm{F}_n V) = \mathrm{Dp}(\mathrm{F}_{n+1} V) + \langle T_1^{(n)}, \ldots, T_k^{(n)}, \ldots\rangle,$$

where $T_i^{(n)}$ are linear dependencies coming from monomials of $\mathrm{F}_n V \setminus \mathrm{F}_{n+1} V$. Evidently, $\mathrm{Dp}(\mathrm{F}_n V) \cap \mathrm{F}_{n+1} V \supseteq \mathrm{Dp}(\mathrm{F}_{n+1} V)$. So, we need to show that $\mathrm{Dp}(\mathrm{F}_n V) \cap \mathrm{F}_{n+1} V \subseteq \mathrm{Dp}(\mathrm{F}_{n+1} V)$. Since $\mathrm{Dp}(\mathrm{F}_{n+1} V) \subseteq \mathrm{F}_{n+1} V$, it is sufficient to prove that

$$\langle T_1^{(n)}, \ldots, T_k^{(n)}, \ldots\rangle \cap \mathrm{F}_{n+1} V \subseteq \mathrm{Dp}(\mathrm{F}_{n+1} V).$$

First let us show that it is sufficient to prove Lemma 4.7 only for the case when $\mathrm{F}_n V$ is generated by one monomial and its derived monomials. Indeed, suppose Lemma 4.7 is proved for this case, then let us prove it for the general case. Assume $T_1, \ldots T_l$ are linear dependencies coming from monomials of $\mathrm{F}_n V \setminus \mathrm{F}_{n+1} V$ and $\sum_{i=1}^{l} T_i \in \mathrm{F}_{n+1} V$. So, all monomials from $\mathrm{F}_n V \setminus \mathrm{F}_{n+1} V$ have to cancel out in this sum.

Let T_i be generated by a multi-turn of a monomial $Z_i \in \mathrm{F}_n V \setminus \mathrm{F}_{n+1} V$, $i = 1, \ldots, l$. We choose a subset $\{Z_{i_1}, \ldots, Z_{i_k}\} \subseteq \{Z_1, \ldots, Z_l\}$ such that $\langle Z_{i_1}, \ldots, Z_{i_k} \rangle_d = \langle Z_1, \ldots, Z_l \rangle_d$ and $\langle Z_{i_j} \rangle_d \nsubseteq \langle Z_{i_{j'}} \rangle_d$ for $j \neq j'$. The dependencies $\{T_1, \ldots T_l\}$ can be split into groups $\{T_1^j, \ldots T_{l_j}^j\}$, $j = 1, \ldots k$, of dependencies coming from monomials of $\langle Z_{i_j} \rangle_d$. So, we have

$$\sum_{i=1}^{l} T_i = \sum_{j=1}^{k} \sum_{i=1}^{l_j} T_i^j \in \mathrm{F}_{n+1} V. \tag{31}$$

Every linear dependence T_i comes from a monomial that belongs to $\mathrm{F}_n V \setminus \mathrm{F}_{n+1} V$; hence, by the definition of $\mathrm{F}_{n+1} V$, every T_i^j comes from a monomial that belongs to $\langle Z_{i_j} \rangle_d \setminus \mathrm{L}(\langle Z_{i_j} \rangle_d)$. From Lemma 4.5 it follows that all monomials from $\langle Z_{i_j} \rangle_d \setminus \mathrm{L}(\langle Z_{i_j} \rangle_d)$ are contained in $\mathrm{F}_n V \setminus \mathrm{F}_{n+1} V$ and $\mathrm{L}(\langle Z_{i_j} \rangle_d) \subseteq \mathrm{F}_{n+1} V$. Hence, the monomials from every $\langle Z_{i_j} \rangle_d \setminus \mathrm{L}(\langle Z_{i_j} \rangle_d)$, $j = 1, \ldots, k$, have to cancel out in the sum (31).

Since $\langle Z_{i_j} \rangle_d \nsubseteq \langle Z_{i_{j'}} \rangle_d$ for $j \neq j'$, from Lemma 4.2 it follows that if a monomial Z belongs to $\langle Z_{i_j} \rangle_d \setminus \mathrm{L}(\langle Z_{i_j} \rangle_d)$, then $Z \notin \langle Z_{i_{j'}} \rangle_d$ for $j \neq j'$. Therefore, the monomials from $\mathrm{F}_n V \setminus \mathrm{F}_{n+1} V$ in the sum (31) have to cancel out in every sum $\sum_{i=1}^{l_j} T_i^j$ separately. Hence, we obtain $\sum_{i=1}^{l_j} T_i^j \in \mathrm{L}(\langle Z_{i_j} \rangle_d)$. Since we assumed that the statement of the lemma holds for the spaces $\langle Z_{i_j} \rangle_d$, we have

$$\sum_{i=1}^{l_j} T_i^j \in \mathrm{Dp}(\mathrm{L}(\langle Z_{i_j} \rangle_d)) \subseteq \mathrm{Dp}(\mathrm{F}_{n+1} V).$$

Thus,

$$\sum_{i=1}^{l} T_i = \sum_{j=1}^{k} \sum_{i=1}^{l_j} T_i^j \in \mathrm{Dp}(\mathrm{F}_{n+1} V).$$

So, it remains to prove the lemma only for the case when $\mathrm{F}_n V$ is generated by one monomial and its derived monomials.

We start with proving the following statement.

LEMMA 4.8. *Suppose $A_1, \ldots, A_k$ are vector spaces, $\mathrm{L}(A_i) \subseteq A_i$ is a subspace of A_i. Suppose D_i is a subspace of A_i such that $D_i \cap \mathrm{L}(A_i) = 0$. Consider $A_1 \otimes \ldots \otimes A_k$ and its subspaces $A_1 \otimes \ldots \otimes D_i \otimes \ldots \otimes A_k$. Define the subspaces*

$$\mathrm{L}(A_1 \otimes \ldots \otimes A_k) = \sum_{i=1}^{k} A_1 \otimes \ldots \otimes \mathrm{L}(A_i) \otimes \ldots \otimes A_k,$$

and

$$\mathrm{L}(A_1 \otimes \ldots \otimes D_i \otimes \ldots \otimes A_k) = \sum_{\substack{i'=1 \\ i \neq i'}}^{k} A_1 \otimes \ldots \otimes D_i \otimes \ldots \otimes \mathrm{L}(A_{i'}) \otimes \ldots \otimes A_k, i = 1, \ldots, k.$$

Then

$$\left(\sum_{i=1}^{k} A_1 \otimes \ldots \otimes D_i \otimes \ldots \otimes A_k \right) \bigcap \mathrm{L}(A_1 \otimes \ldots \otimes A_k) = \sum_{i=1}^{k} \mathrm{L}(A_1 \otimes \ldots \otimes D_i \otimes \ldots \otimes A_k).$$

PROOF. From the condition $D_i \cap \mathrm{L}(A_i) = 0$ it follows that the sum of this subspaces is a direct sum $D_i \oplus \mathrm{L}(A_i) \subseteq A_i$. Since A_i is a vector space, the subspace $D_i \oplus \mathrm{L}(A_i)$ has a direct complement J_i and we obtain $A_i = J_i \oplus D_i \oplus \mathrm{L}(A_i)$. Hence,

$$\begin{aligned}\mathrm{L}(A_1 \otimes \ldots \otimes A_k) &= \sum_{i=1}^{k} (J_1 \oplus D_1 \oplus \mathrm{L}(A_1)) \otimes \ldots \otimes \mathrm{L}(A_i) \otimes \ldots \otimes (J_k \oplus D_k \oplus \mathrm{L}(A_k)) \\ &= \sum_{i=1}^{k} (\bigoplus_j B_1^j \otimes \ldots \otimes B_{i-1}^j \otimes \mathrm{L}(A_i) \otimes B_{i+1}^j \otimes \ldots \otimes B_k^j),\end{aligned}$$

where every B_i^j is either J_i, or D_i, or $\mathrm{L}(A_i)$ and we encounter all possible combinations in the sum. Tensor products that contain more then one member $\mathrm{L}(A_i)$ repeat in the sum. If we take every repeating space only one time, we obtain the direct sum $\mathrm{L}(A_1 \otimes \ldots \otimes A_k) = \bigoplus_j B_1^j \otimes \ldots \otimes B_k^j$, where every B_i^j is either J_i, or D_i, or $\mathrm{L}(A_i)$, we encounter all combinations in the sum, such that at least one $B_i^j = \mathrm{L}(A_i)$. Similarly,

$$\begin{aligned}&\sum_{i=1}^{k} A_1 \otimes \ldots \otimes D_i \otimes \ldots \otimes A_k \\ &\qquad = \sum_{i=1}^{k} (J_1 \oplus D_1 \oplus \mathrm{L}(A_1)) \otimes \ldots \otimes D_i \otimes \ldots \otimes (J_k \oplus D_k \oplus \mathrm{L}(A_k)) \\ &\qquad = \sum_{i=1}^{k} (\bigoplus_j C_1^j \otimes \ldots \otimes C_{i-1}^j \otimes D_i \otimes C_{i+1}^j \otimes \ldots \otimes C_k^j),\end{aligned}$$

where every C_i^j is either J_i, or D_i, or $\mathrm{L}(A_i)$ and we encounter all possible combinations in the sum. Tensor products that contain more then one member D_i repeat in the sum. If we take every repeating space only one time, we obtain the direct sum $\sum_{i=1}^{k} A_1 \otimes \ldots \otimes D_i \otimes \ldots \otimes A_k = \bigoplus_j C_1^j \otimes \ldots \otimes C_k^j$, where every C_i^j is either J_i, or D_i, or $\mathrm{L}(A_i)$, we encounter all combinations in the sum, such that at least one $C_i^j = D_i$. Hence,

$$\left(\sum_{i=1}^{k} A_1 \otimes \ldots \otimes D_i \otimes \ldots \otimes A_k) \cap \mathrm{L}(A_1 \otimes \ldots \otimes A_k \right) = \bigoplus_j K_1^j \otimes \ldots \otimes K_k^j,$$

where every K_i^j is either J_i, or D_i, or $\mathrm{L}(A_i)$, we encounter all combinations in the sum, such that at least one $K_i^j = D_i$ and at least one $K_i^j = \mathrm{L}(A_i)$. Therefore,

$$\begin{aligned}&\left(\sum_{i=1}^{k} A_1 \otimes \ldots \otimes D_i \otimes \ldots \otimes A_k) \cap \mathrm{L}(A_1 \otimes \ldots \otimes A_k \right) \\ &\qquad = \sum_{i=1}^{k} \sum_{\substack{i'=1 \\ i \neq i'}}^{k} A_1 \otimes \ldots \otimes D_i \otimes \ldots \otimes \mathrm{L}(A_{i'}) \otimes \ldots \otimes A_k \\ &\qquad = \sum_{i=1}^{k} \mathrm{L}(A_1 \otimes \ldots \otimes D_i \otimes \ldots \otimes A_k).\end{aligned}$$

□

Using Lemma 4.8, we continue the proof of Lemma 4.7. Assume U is a monomial. Derived monomials of U are defined with the use of certain sequences of replacements of virtual members of the chart. When we perform replacements that preserve f-characteristics of monomials, they preserve, roughly speaking, the structure of the chart. Moreover, there is no interaction between the replaced occurrence and the separated virtual members of the chart and there is a very small interaction between the replaced occurrence and its neighbours. This kind of behaviour provides the idea to consider a tensor product of linear spaces that correspond to each place of the chart of U.

Assume the monomial U has k virtual members of the chart, that is, $K_\tau(U) = k$. Let $a^{(i)}$ be the virtual member of the chart of U placed on the i-th position from the beginning of U. Let $A_i[U]$ be a subspace of $\mathbb{Z}_2\mathcal{F}$ such that

$$A_i[U] = \left\langle a_j^{(i)} \mid a^{(i)} \textit{ and } a_j^{(i)} \textit{ are incident monomials}, j \in \mathbb{N} \right\rangle . \tag{32}$$

Suppose $U = L_i a^{(i)} R_i$. We define a subspace $\mathrm{L}_i[U] \subseteq A_i[U]$ by the following rule:

$$\mathrm{L}_i[U] = \Big\langle a_j^{(i)} \mid a^{(i)} \textit{ and } a_j^{(i)} \textit{ are incident monomials}, j \in \mathbb{N}, \tag{33}$$
$$L_i a_j^{(i)} R_i \in \mathrm{L}(\langle U \rangle_d) \Big\rangle .$$

REMARK 4.2. Suppose U' is a derived monomial of U such that $U' \in \langle U \rangle_d \setminus \mathrm{L}(\langle U \rangle_d)$. Let us construct the spaces $A_i[U']$ and $\mathrm{L}_i[U'] \subseteq A_i[U']$ corresponding to U' as above. Then, obviously, the spaces $A_i[U']$ are generated by the same sets of monomials as the spaces $A_i[U]$ up to shifts of the initial and the final points in the corresponding v-diagram by ε. Moreover, using Lemma 3.10, one can prove that the spaces $\mathrm{L}_i[U']$ are also generated by the same sets of monomials as the spaces $\mathrm{L}_i[U]$ up to shifts of the initial and the final points in the corresponding v-diagram by ε.

Although the precise forms of the spaces $A_i[U']$ and $\mathrm{L}_i[U'] \subseteq A_i[U']$ depend on the monomial U, in the sequel, we will omit $[U]$ in the denotation of this spaces when it does not lead to ambiguity.

We construct a linear mapping

$$\mu[U] : A_1 \otimes \ldots \otimes A_k \to \langle U \rangle_d .$$

Elements $b^{(1)} \otimes \ldots \otimes b^{(k)} \in A_1 \otimes \ldots \otimes A_k$, where $b^{(i)} \in A_i$ are generalized fractional powers incident to $a^{(i)}$, form a basis of $A_1 \otimes \ldots \otimes A_k$, because generalized fractional powers incident to $a^{(i)}$ form a basis of A_i. We will define $\mu[U]$ on these basis elements.

We distinguish between four possibilities:

Case 1 First we define $\mu[U]$ on elements $b^{(1)} \otimes \ldots \otimes b^{(k)}$ such that all $b^{(i)} \in A_i \setminus \mathrm{L}_i$. It encodes a sequence of replacements that starts from the monomial U. Recall that $a^{(1)}, \ldots, a^{(k)}$ are the virtual members of the chart of U enumerated from left to right. Let $U = L_1 a^{(1)} R_1$. Then we start with the transformation $L_1 a^{(1)} R_1 \mapsto L_1 b^{(1)} R_1$. Denote by $\widehat{a}^{(2)}$ the image of $\widehat{a}^{(2)}$ in $L_1 b^{(1)} R_1$. Since $\Lambda(b^{(1)}) > \varepsilon$, either $\widehat{a}^{(2)} = a^{(2)}$, or $\widehat{a}^{(2)} = c^{(2)} a^{(2)}$, or $\widehat{a}^{(2)} = {c^{(2)}}^{-1} a^{(2)}$, where $\Lambda(c^{(2)}) \leqslant \varepsilon$. The next step is the replacement of $\widehat{a}^{(2)}$ in $L_1 b^{(1)} R_1$ corresponding to the transformation $a^{(2)} \mapsto b^{(2)}$. That is, the replacement $\widehat{a}^{(2)} \mapsto \widehat{b}^{(2)}$ in $L_1 b^{(1)} R_1$, where either $\widehat{b}^{(2)} = b^{(2)}$ if

$\widehat{a}^{(2)} = a^{(2)}$, or $\widehat{b}^{(2)} = c^{(2)}b^{(2)}$ if $\widehat{a}^{(2)} = c^{(2)}a^{(2)}$, or $\widehat{b}^{(2)} = {c^{(2)}}^{-1}b^{(2)}$ if $\widehat{a}^{(2)} = {c^{(2)}}^{-1}a^{(2)}$. In the same way, we continue transformations for every position in the chart. Notice that, by Corollary 3.11, we actually can perform the replacements in any order.

Case 2 Assume $1 \leqslant i_0 \leqslant k$ is a place in the chart of U. We define $\mu[U]$ on elements $b^{(1)} \otimes \ldots \otimes b^{(k)}$ such that $b^{(i_0)} \in \mathrm{L}_{i_0}$ and $b^{(i)} \in A_i \setminus \mathrm{L}_i$ for $i \neq i_0$. First we replace virtual members of the chart of U in positions different from i_0 as we described above. Then we perform a corresponding replacement in the position i_0 and all the further cancellations if necessary.

Case 3 Assume $1 \leqslant i_0' < i_0 \leqslant n$ are two places in the chart of U. We define $\mu[U]$ on elements $b^{(1)} \otimes \ldots \otimes b^{(k)}$ such that $b^{(i)} \in A_i \setminus \mathrm{L}_i$ for $i \neq i_0', i_0$ and $b^{(i_0')} \in \mathrm{L}_{i_0'}$, $b^{(i_0)} \in \mathrm{L}_{i_0}$. First we replace virtual members of the chart of U in positions different from i_0', i_0 as we described above. Then we perform the corresponding replacement in the position i_0. Assume that as a result we obtain a monomial $L\widehat{b}^{(i_0)}R$. If there are any further cancellations in $L\widehat{b}^{(i_0)}R$ (when $\widehat{b}^{(i_0)} = 1$), we do not perform them right after the replacement. Instead we perform the corresponding replacement in the subword L or R in the place corresponding to i_0'. Then we perform the cancellations if there are any.

From Lemma 3.9 it follows that we can perform two last transformations starting from any position i_0' or i_0 and obtain the same final result. From Lemma 4.6 it follows that the resulting monomial is a derived monomial U.

Case 4 For elements $b^{(1)} \otimes \ldots \otimes b^{(k)}$ such that there are more than two $b^{(i)} \in \mathrm{L}_i$, we could continue in a similar way. But, in fact, we do not need to preserve full information about these elements. So, by definition, we put $\mu[U](b^{(1)} \otimes \ldots \otimes b^{(k)}) = 0$ in this case.

From Corollary 3.11, it follows that in Case 1 an element $\mu[U](b^{(1)}\otimes\ldots\otimes b^{(k)}) \in \langle U\rangle_d \setminus \mathrm{L}(\langle U\rangle_d)$. From Lemma 4.1 and Lemma 4.6, it follows that in Case 2 and Case 3 an element $\mu[U](b^{(1)} \otimes \ldots \otimes b^{(k)}) \in \mathrm{L}(\langle U\rangle_d)$.

Let Z be a derived monomial of U such that Z does not belong to $\mathrm{L}(\langle U\rangle_d)$. By Lemma 4.1, this means that there exists a sequence of replacements of virtual members of the chart such that every replacement preserves f-characteristics of monomials. From Corollary 3.11, it follows that the replacements can be performed in any order and after changing the order every replacement still preserves f-characteristics of monomials. Namely, we can first perform all replacements in the first position of the chart, then in the second position, etc, in the k-th position. Assume $\widehat{b}^{(1)}, \ldots, \widehat{b}^{(k)}$ are the virtual members of the chart of Z enumerated from left to right. Then the element $\widehat{b}^{(i)}$ may not be incident to $a^{(i)}$ (where $a^{(i)}$ is the corresponding member of the chart of U). But, clearly, its ends differ only by a shift by ε. Let $b^{(i)}$ be an incident monomial of $a^{(i)}$, corresponding to $\widehat{b}^{(i)}$. Then

$$Z = \mu[U](b^{(1)} \otimes \ldots \otimes b^{(k)}).$$

From Corollary 3.11 it follows that every replacement can be moved to the beginning of the sequence, which starts from U, and it still preserves the f-characteristic of the monomials. That is, every replacement $a^{(i)} \mapsto b^{(i)}$ in the monomial U does not decrease the f-characteristic of the resulting monomial. Hence, $b^{(i)} \in A_i \setminus \mathrm{L}_i$.

Suppose $b^{(1)} \otimes \ldots \otimes b^{(k)}$ and $d^{(1)} \otimes \ldots \otimes d^{(k)}$ are different elements such that $b^{(i)}, d^{(i)}$ are generalized fractional powers, $b^{(i)}, d^{(i)} \in A_i \setminus \mathrm{L}_i$. Since there exists $b^{(i_0)} \neq d^{(i_0)}$ and two different incident generalized fractional powers can not differ only by a piece equal to a possible overlap, we obtain

$$\mu[U](b^{(1)} \otimes \ldots \otimes b^{(k)}) \neq \mu[U](d^{(1)} \otimes \ldots \otimes d^{(k)}).$$

Thus, we have showed that $\mu[U]$ gives a bijective correspondence between elements $b^{(1)} \otimes \ldots \otimes b^{(k)}$ such that generalized fractional powers $b^{(i)} \in A_i \setminus \mathrm{L}_i$ and monomials from $\langle U \rangle_d \setminus \mathrm{L}(\langle U \rangle_d)$.

As above, we denote by $\mathrm{Dp}(A_i)$ the subspace of A_i generated by supports of corresponding elementary multi-turns (see Definition 4.3). Recall that A_i is generated by generalized fractional powers incident to one generalized fractional power $a^{(i)}$. Therefore, from the definition of an elementary multi-turn, it follows that a sum of two supports from $\mathrm{Dp}(A_i)$ is again the support of elementary multi-turn. So, actually $\mathrm{Dp}(A_i)$ consists precisely of supports of elementary multi-turns. Recall that a maximal occurrence of a generalized fractional power of Λ-measure $\geqslant \tau$ is always a virtual member of the chart. Hence, applying together (33) and Proposition 3.1, we obtain $\mathrm{Dp}(A_i) \cap \mathrm{L}_i = 0$.

In order to prove Lemma 4.7 it remains to show that

$$\mathrm{Dp}(\langle U \rangle_d) \cap \mathrm{L}(\langle U \rangle_d) = \mathrm{Dp}(\mathrm{L}(\langle U \rangle_d)). \tag{34}$$

We already proved that $\mathrm{Dp}(\langle U \rangle_d) \cap \mathrm{L}(\langle U \rangle_d) \supseteq \mathrm{Dp}(\mathrm{L}(\langle U \rangle_d))$. So, we will show that $\mathrm{Dp}(\langle U \rangle_d) \cap \mathrm{L}(\langle U \rangle_d) \subseteq \mathrm{Dp}(\mathrm{L}(\langle U \rangle_d))$.

Suppose T_s, $s = 1, \ldots, m$, are the supports of multi-turns coming from monomials of $\langle U \rangle_d \setminus \mathrm{L}(\langle U \rangle_d)$ and

$$\sum_{s=1}^{m} T_s \in \mathrm{L}(\langle U \rangle_d).$$

Assume T_s comes from a monomial $U_h^{(s)} \in \langle U \rangle_d \setminus \mathrm{L}(\langle U \rangle_d)$. Then there exists an element $b_h^{(1)} \otimes \ldots \otimes b_h^{(k)} \in A_1 \otimes \ldots \otimes A_n$ such that

$$\mu[U](b_{h_s}^{(1)} \otimes \ldots \otimes b_{h_s}^{(k)}) = U_h^{(s)},$$

where $b_{h_s}^{(i)}$ are generalized fractional powers such that $b_{h_s}^{(i)} \notin \mathrm{L}_i$. Assume this multi-turn comes from an elementary multi-turn of the virtual member of the chart of $U_h^{(s)}$ placed on the i_s-th position. The element $b_{h_s}^{(i_s)} \in A_{i_s}$ corresponds to this virtual member of the chart. By the definition of $\mu[U]$ in Case 1 and Case 2, we obtain

$$T_s = \mu[U](b_{h_s}^{(1)} \otimes \ldots \otimes t_s^{(i_s)} \otimes \ldots \otimes b_{h_s}^{(k)}),$$

where $t_s^{(i_s)} \in \mathrm{Dp}(A_{i_s})$ is the support of the corresponding elementary multi-turn of $b_{h_s}^{(i_s)}$.

So, we have

$$\sum_{s=1}^{m} T_s = \mu[U]\left(\sum_{s=1}^{m} b_{h_s}^{(1)} \otimes \ldots \otimes t_s^{(i_s)} \otimes \ldots \otimes b_{h_s}^{(k)}\right).$$

By the assumption, all monomials in the left-hand sum that do not belong to $\mathrm{L}(\langle U \rangle_d)$ cancel out. Since $\mu[U]$ gives a bijective correspondence between the elements $b^{(1)} \otimes \ldots \otimes b^{(k)}$ such that generalized fractional powers $b^{(i)} \in A_i \setminus \mathrm{L}_i$ and the monomials from $\langle U\rangle_d \setminus \mathrm{L}(\langle U\rangle_d)$, we obtain that all such elements cancel out in the right-hand sum as well. Then from Lemma 4.8 it follows that

$$\sum_{s=1}^{m} b_{h_s}^{(1)} \otimes \ldots \otimes t_s^{(i_s)} \otimes \ldots \otimes b_{h_s}^{(k)} \in \sum_{i=1}^{k} \sum_{\substack{i'=1 \\ i\neq i'}}^{k} A_1 \otimes \ldots \otimes \mathrm{Dp}(A_i) \otimes \ldots \otimes \mathrm{L}_{i'} \otimes \ldots \otimes A_k,$$

that is, the sum $\sum_{s=1}^{m} b_{h_s}^{(1)} \otimes \ldots \otimes t_s^{(i_s)} \otimes \ldots \otimes b_{h_s}^{(k)}$ is equal to a sum of elements of the form

$$d^{(1)} \otimes \ldots \otimes d^{(i'_0-1)} \otimes \widetilde{d}^{(i'_0)} \otimes d^{(i'_0+1)} \otimes \ldots \otimes d^{(i_0-1)} \otimes \widetilde{t}^{(i_0)} \otimes d^{(i_0+1)} \otimes \ldots \otimes d^{(k)},$$
$$d^{(1)} \otimes \ldots \otimes d^{(i_0-1)} \otimes \widetilde{d}^{(i_0)} \otimes d^{(i_0+1)} \otimes \ldots \otimes d^{(i'_0-1)} \otimes \widetilde{t}^{(i'_0)} \otimes d^{(i'_0+1)} \otimes \ldots \otimes d^{(k)}$$

for different $i_0 \neq i'_0$, where generalized fractional powers $d^{(i)} \in A_i \setminus \mathrm{L}_i$, a generalized fractional power $\widetilde{d}^{(i'_0)} \in \mathrm{L}_{i'_0}$, and the support of an elementary multi-turn $\widetilde{t}^{(i_0)} \in \mathrm{Dp}(A_{i_0})$. To be precise, assume $i'_0 < i_0$.

Assume $\widetilde{t}^{(i_0)} = \sum_{j=1}^{m_0} e_j^{(i_0)}$. Let us calculate

$$\mu[U](d^{(1)} \otimes \ldots \otimes \widetilde{d}^{(i'_0)} \otimes \ldots \otimes \sum_{j=1}^{m_0} e_j^{(i_0)} \otimes \ldots \otimes d^{(k)}).$$

From Lemma 3.9 and Corollary 3.11, it follows that in the definition of $\mu[U]$ first we can do replacements of all positions of the chart except i_0 and i'_0, then a replacement in the position i_0 and then a replacement in the position corresponding to i'_0 (the exact position may shift if $e_j^{(i_0)} \in \mathrm{L}_{i_0}$). Suppose Z is a result after the replacement in the position i_0. Then $Z = L\widehat{e}^{(i_0)}R$, where $\widehat{e}^{(i_0)}$ may differ from $e^{(i_0)}$ by shifts of its ends by ε. For simplicity, we illustrate the case when the virtual members of the chart on the positions i'_0 and i_0 are separated, but the other cases are analogous to this one.

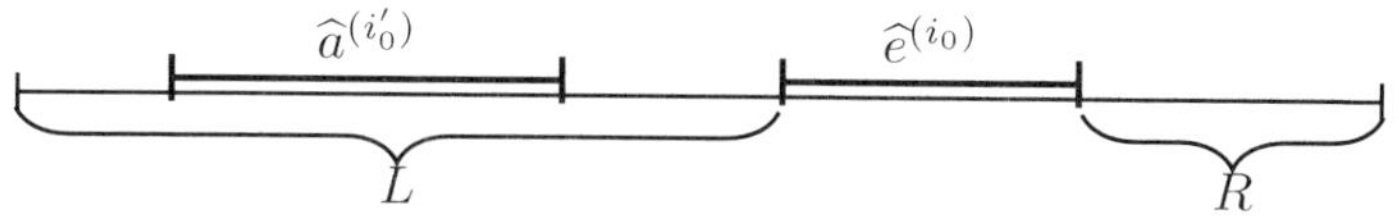

Then the last replacement in Z can be represented as a replacement of the corresponding maximal occurrence in L and the further cancellations if there are any. Denote the result of the transformation of L by L'.

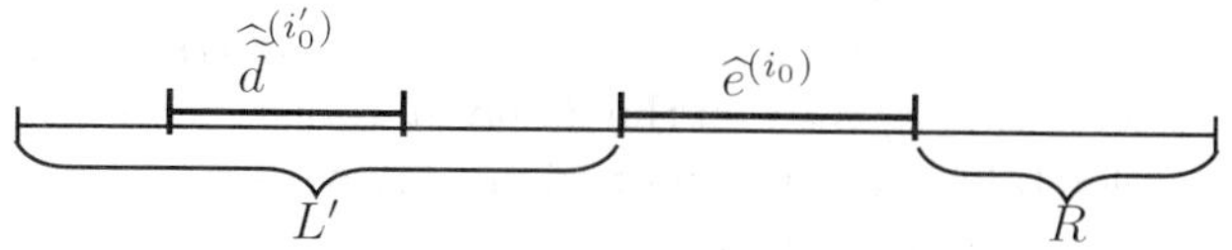

Hence, we obtain

$$\mu[U](d^{(1)} \otimes \ldots \otimes \widetilde{d}^{(i'_0)} \otimes \ldots \otimes e_j^{(i_0)} \otimes \ldots \otimes d^{(k)}) = L'\widehat{e_j}^{(i_0)}R,$$

where the word $L'\widehat{e_j}^{(i_0)}R$ is possibly non-reduced. So,

$$\mu[U](d^{(1)} \otimes \ldots \otimes \widetilde{d}^{(i_0')} \otimes \ldots \otimes \sum_{j=1}^{m_0} e_j^{(i_0)} \otimes \ldots \otimes d^{(k)}) = \sum_{j=1}^{m_0} L'\widehat{e}_j^{(i_0)}R$$
$$= L'\left(\sum_{j=1}^{m_0} \widehat{e}_j^{(i_0)}\right)R.$$

Since $\sum_{j=1}^{m_0} e_j^{(i_0)}$ is the support of an elementary multi-turn, the sum $\sum_{j=1}^{m_0} \widehat{e}_j^{(i_0)}$ is the support of an elementary multi-turn as well (possibly with shifts of the ends of the monomials by ε). Therefore, from Proposition 3.2 it follows that $\sum_{j=1}^{m_0} L'\widehat{e}_j^{(i_0)}R$ is the support of a multi-turn (after the cancellation in the monomials).

Since $\widetilde{d}^{(i_0')} \in \mathrm{L}(A_i)$, we have

$$\mu[U](d^{(1)} \otimes \ldots \otimes \widetilde{d}^{(i_0')} \otimes \ldots \otimes e_j^{(i_0)} \otimes \ldots \otimes d^{(k)}) \in \mathrm{L}(\langle U\rangle_d).$$

Then, since $\mu[U](d^{(1)} \otimes \ldots \otimes \widetilde{d}^{(i_0')} \otimes \ldots \otimes \sum_{j=1}^{m_0} e_j^{(i_0)} \otimes \ldots \otimes d^{(k)})$ is the support of a multi-turn, we have

$$\mu[U](d^{(1)} \otimes \ldots \otimes \widetilde{d}^{(i_0')} \otimes \ldots \otimes \sum_{j=1}^{m_0} e_j^{(i_0)} \otimes \ldots \otimes d^{(k)}) \in \mathrm{Dp}(\mathrm{L}(\langle U\rangle_d)).$$

Thus,

$$\sum_{s=1}^{m} T_s \in \mathrm{Dp}(\mathrm{L}(\langle U\rangle_d))$$

and the equality (34) holds. This concludes the proof of Lemma 4.7. □

Using Lemma 4.7, we obtain the following proposition.

PROPOSITION 4.9. *Suppose X, Y are subspaces of $\mathbb{Z}_2\mathcal{F}$ generated by monomials and closed under taking derived monomials, $Y \subseteq X$. Then $\mathrm{Dp}(X) \cap Y = \mathrm{Dp}(Y)$.*

PROOF. Clearly, $\mathrm{Dp}(Y) \subseteq \mathrm{Dp}(X) \cap Y$. Let us show that $\mathrm{Dp}(X) \cap Y \subseteq \mathrm{Dp}(Y)$. Suppose $T_1 + \ldots + T_m \in \mathrm{Dp}(X) \cap Y$, where every T_i, $i = 1, \ldots m$, belongs to the set of generating supports of multi-turns of $\mathrm{Dp}(X)$. Consider $T_{i_0} \in Y$. Since Y is generated by monomials, we obtain that every monomial of T_{i_0} belongs to Y. So, T_{i_0} is a linear dependence generated by a monomial from Y, that is, $T_{i_0} \in \mathrm{Dp}(Y)$.

So, we may suppose that every $T_i \notin Y$, $i = 1, \ldots, m$. Denote by X' the subspace of X generated by the monomials of T_i, $i = 1, \ldots, m$, and their derived monomials. Suppose the filtration on X' has N non-zero levels. Let us prove that $T_1 + \ldots + T_m \in \mathrm{Dp}(Y)$ by induction on N.

First let us do the step of induction. Assume T_i is generated by a multi-turn of a monomial Z_i. If $Z_i \in \mathrm{L}(X')$, then all the monomials of T_i belong to $\mathrm{L}(X')$ because $\mathrm{L}(X')$ is closed under taking derived monomials. Assume $Z_i \in X' \setminus \mathrm{L}(X')$. Let Z be an arbitrary monomial of T_i such that $Z \in X' \setminus \mathrm{L}(X')$. Since $Z \in X' \setminus \mathrm{L}(X')$, we have $Z \in \langle Z_i\rangle_d \setminus \mathrm{L}(\langle Z_i\rangle_d)$. Therefore, from Lemma 4.2 it follows that Z_i is a derived monomial of Z. Assume $Z \in Y$, then all its derived monomials belong to Y because Y is closed under taking derived monomials. So, we have $Z_i \in Y$; hence, all the monomials of T_i are contained in Y. This contradicts our assumption that

$T_i \notin Y$. Therefore, the monomials of T_i that are contained in $X' \setminus \mathrm{L}(X')$ are not contained in Y.

Since $T_1 + \ldots + T_m \in Y$, the monomials that are contained in $X' \setminus \mathrm{L}(X')$ cancel in the sum $T_1 + \ldots + T_m$; hence, $T_1 + \ldots + T_m \in \mathrm{L}(X')$. Then from Lemma 4.7 it follows that

$$T_1 + \ldots + T_m \in \mathrm{Dp}(\mathrm{L}(X')).$$

Therefore, $T_1 + \ldots + T_m = T'_1 + \ldots + T'_{m'}$, where $T'_i \in \mathrm{L}(U')$, $i = 1, \ldots, m'$, are supports of multi-turns that come from monomials of $\mathrm{L}(X')$. We have

$$T_1 + \ldots + T_m = T'_1 + \ldots + T'_{m'} = \sum_{T'_i \in Y} T'_i + \sum_{T'_i \notin Y} T'_i.$$

As above, every element of the first sum $\sum_{T'_i \in Y} T'_i$ belongs to $\mathrm{Dp}(Y)$. The space $\mathrm{L}(X')$ has $N-1$ non-zero levels of the filtration; hence the second sum $\sum_{T'_i \notin Y} T'_i$ belongs to $\mathrm{Dp}(Y)$ by the induction hypothesis, and therefore,

$$T_1 + \ldots + T_m \in \mathrm{Dp}(W).$$

Let us prove the basis of induction. Consider $N = 1$. As above, we obtain $T_1 + \ldots + T_m \in \mathrm{L}(X')$. But since $N = 1$, we have $\mathrm{L}(X') = 0$; therefore $T_1 + \ldots + T_m = 0$, and so, it belongs to $\mathrm{Dp}(Y)$.

Thus, we obtain $\mathrm{Dp}(X) \cap Y \subseteq \mathrm{Dp}(Y)$. This concludes the proof. □

Using the mapping $\mu[U]$ that was constructed in the second part of the proof of Lemma 4.7, we obtain the following statement.

PROPOSITION 4.10. *Let U be a monomial with k virtual members of the chart. Suppose A_i and $\mathrm{L}_i \subseteq A_i$, $i = 1, \ldots k$, are defined above by (32) and (33) subspaces of $\mathbb{Z}_2\mathcal{F}$ corresponding to U. Then we have*

$$\langle U\rangle_d / \mathrm{L}(\langle U\rangle_d) \cong A_1/\mathrm{L}_1 \otimes \ldots \otimes A_k/\mathrm{L}_k.$$

Moreover,

$$\langle U\rangle_d / (\mathrm{Dp}(\langle U\rangle_d) + \mathrm{L}(\langle U\rangle_d)) \cong A_1/(\mathrm{Dp}(A_1) + \mathrm{L}_1) \otimes \ldots \otimes A_k/(\mathrm{Dp}(A_k) + \mathrm{L}_k).$$

PROOF. Recall that in the proof of Lemma 4.7 we constructed a linear mapping

$$\mu[U] : A_1 \otimes \ldots \otimes A_k \to \langle U\rangle_d$$

(see page 46). We define a linear mapping

$$\overline{\mu}_1[U] : A_1/\mathrm{L}_1 \otimes \ldots \otimes A_k/\mathrm{L}_k \to \langle U\rangle_d / \mathrm{L}(\langle U\rangle_d)$$

by the following rule:

$$\overline{\mu}_1[U]((b^{(1)} + \mathrm{L}_1) \otimes \ldots \otimes (b^{(k)} + \mathrm{L}_k)) = \mu[U](b^{(1)} \otimes \ldots \otimes b^{(k)}) + \mathrm{L}(\langle U\rangle_d),$$

where $b^{(i)} \in A_i$ are generalized fractional powers. In Lemma 4.7, we proved (using Lemma 4.1 and Lemma 4.6) that

$$\mu[U]\left(\sum_{i=1}^{k} A_1 \otimes \ldots \otimes \mathrm{L}_i \otimes \ldots \otimes A_k\right) \subseteq \mathrm{L}(\langle U\rangle_d). \tag{35}$$

Hence, the mapping $\overline{\mu}_1[U]$ is well-defined. Since $\mu[U]$ gives a bijective correspondence between elements $b^{(1)} \otimes \ldots \otimes b^{(k)}$, where $b^{(i)}$ are generalized fractional powers such that $b^{(i)} \in A_i \setminus \mathrm{L}_i$, and the monomials from $\langle U\rangle_d \setminus \mathrm{L}(\langle U\rangle_d)$, the mapping $\overline{\mu}_1[U]$ is bijective. So, $\overline{\mu}_1[U]$ is an isomorphism of linear spaces.

In the same way, we define a linear mapping

$\overline{\mu}_2[U] : A_1/(\mathrm{Dp}(A_1)+\mathrm{L}_1)\otimes\ldots\otimes A_k/(\mathrm{Dp}(A_k)+\mathrm{L}_k) \to \langle U\rangle_d/(\mathrm{Dp}(\langle U\rangle_d)+\mathrm{L}(\langle U\rangle_d)$

by the following rule:

$$\begin{aligned}\overline{\mu}_2[U]((b^{(1)}+\mathrm{Dp}(A_1)+\mathrm{L}_1)\otimes\ldots\otimes(b^{(k)}+\mathrm{Dp}(A_k)+\mathrm{L}_k))\\ = \mu[U](b^{(1)}\otimes\ldots\otimes b^{(k)})+\mathrm{Dp}(\langle U\rangle_d)+\mathrm{L}(\langle U\rangle_d),\end{aligned} \tag{36}$$

where $b^{(i)}\in A_i$ are generalized fractional powers. Arguing in the same way as in Lemma 4.7, it is easy to show that

$$\mu[U]\left(\sum_{i=1}^{k} A_1\otimes\ldots\otimes\mathrm{Dp}(A_i)\otimes\ldots\otimes A_k\right)\subseteq \mathrm{Dp}(\langle U\rangle_d)+\mathrm{L}(\langle U\rangle_d).$$

Using this together with (35), we see that the mapping $\overline{\mu}_2[U]$ is well-defined. Since $\mu[U]$ gives a bijective correspondence between elements $b^{(1)}\otimes\ldots\otimes b^{(k)}$, where $b^{(i)}$ are generalized fractional powers such that $b^{(i)}\in A_i\setminus\mathrm{L}_i$, and the monomials from $\langle U\rangle_d\setminus\mathrm{L}(\langle U\rangle_d)$, one can show that the mapping $\overline{\mu}_2[U]$ is bijective. So, $\overline{\mu}_2[U]$ is an isomorphism of linear spaces. □

REMARK 4.3. In Proposition 4.10, in fact, we used the following construction. We consider a composition of linear mappings

$$A_1\otimes\ldots\otimes A_k \overset{\mu[U]}{\longrightarrow} \langle U\rangle_d \overset{\pi_1}{\longrightarrow} \langle U\rangle_d/\mathrm{L}(\langle U\rangle_d),$$
$$A_1\otimes\ldots\otimes A_k \overset{\mu[U]}{\longrightarrow} \langle U\rangle_d \overset{\pi_2}{\longrightarrow} \langle U\rangle_d/(\mathrm{Dp}(\langle U\rangle_d)+\mathrm{L}(\langle U\rangle_d)),$$

where π_1 and π_2 are the canonical homomorphisms. Then, by properties of $\mu[U]$ established in Lemma 4.7,

$$\ker(\pi_1\circ\mu[U]) = \sum_{i=1}^{k} A_1\otimes\ldots\otimes\mathrm{L}_i\otimes\ldots\otimes A_k,$$
$$\ker(\pi_2\circ\mu[U]) = \sum_{i=1}^{k} A_1\otimes\ldots\otimes\mathrm{L}_i\otimes\ldots\otimes A_k+\sum_{i=1}^{k} A_1\otimes\ldots\otimes\mathrm{Dp}(A_i)\otimes\ldots\otimes A_k.$$

Then, using the isomorphism theorem and the definition of a tensor product of vector spaces, we obtain the result of Proposition 4.10.

4.2. The structure of quotient spaces $\langle U_1,\ldots,U_k\rangle_d/\mathrm{Dp}(\langle U_1,\ldots,U_k\rangle_d)$. The following statement easily follows from Proposition 4.9.

COROLLARY 4.11. *Let $V=\langle U_1,\ldots,U_k\rangle_d$, where $U_1,\ldots,U_k$ are monomials. Let $\widehat{V}$ be the corresponding subspace in $\mathbb{Z}_2\mathcal{F}/\langle\mathcal{T}'\rangle$, that is, $\widehat{V}=(V+\langle\mathcal{T}'\rangle)/\langle\mathcal{T}'\rangle$. Then $\widehat{V}\cong V/\mathrm{Dp}(V)$.*

PROOF. From the Isomorphism Theorem, it follows that $\widehat{V}\cong V/(V\cap\langle\mathcal{T}'\rangle)$. According to Definition 4.3, we have $\langle\mathcal{T}'\rangle=\mathrm{Dp}(\mathbb{Z}_2\mathcal{F})$. Therefore, from Proposition 4.9 it follows that $V\cap\langle\mathcal{T}'\rangle=V\cap\mathrm{Dp}(\mathbb{Z}_2\mathcal{F})=\mathrm{Dp}(V)$; hence, $\widehat{V}\cong V/\mathrm{Dp}(V)$. □

The quotient space $V/\mathrm{Dp}(V)$ inherits the filtration from V,

$$\mathrm{F}_n(V/\mathrm{Dp}(V))=(\mathrm{F}_nV+\mathrm{Dp}(V))/\mathrm{Dp}(V).$$

Suppose V has N non-zero levels of the filtration

$$V = \mathrm{F}_0 V \supseteq \mathrm{F}_1 V \supseteq \ldots \supseteq \mathrm{F}_{N-1} V \supseteq \mathrm{F}_N V = 0.$$

We have the corresponding graded space

$$\mathrm{Gr}(V/\mathrm{Dp}(V)) = \bigoplus_{n=0}^{N-1} (\mathrm{F}_n(V/\mathrm{Dp}(V))/\mathrm{F}_{n+1}(V/\mathrm{Dp}(V))).$$

For any $n = 0, \ldots, N-1$, we have the mapping $\mathrm{gr}_n : \mathrm{F}_n V \to \mathrm{F}_n V/\mathrm{F}_{n+1} V$.

The following theorem establishes the compatibility of the filtration and the corresponding grading with the linear dependencies on the space V.

THEOREM 4.1. *Let $V = \langle U_1, \ldots, U_k \rangle_d$, $U_1, \ldots, U_k \in \mathcal{F}$. Then*

$$\mathrm{Gr}(V/\mathrm{Dp}(V)) \cong \bigoplus_{n=0}^{N-1} \mathrm{Gr}_n(V)/\mathrm{gr}_n(\mathrm{Dp}(\mathrm{F}_n V)),$$

where

$$\mathrm{Gr}_n(V) = \mathrm{F}_n V/\mathrm{F}_{n+1} V,$$
$$\mathrm{gr}_n(\mathrm{Dp}(\mathrm{F}_n V)) = (\mathrm{Dp}(\mathrm{F}_n V) + \mathrm{F}_{n+1} V)/\mathrm{F}_{n+1} V.$$

PROOF. Using isomorphism theorems, we obtain

$$\begin{aligned}\mathrm{F}_n(V/\mathrm{Dp}(V))/\mathrm{F}_{n+1}(V/\mathrm{Dp}(V)) &\cong (\mathrm{F}_n V + \mathrm{Dp}(V))/(\mathrm{F}_{n+1} V + \mathrm{Dp}(V)) \\ &= (\mathrm{F}_n V + \mathrm{F}_{n+1} V + \mathrm{Dp}(V))/(\mathrm{F}_{n+1} V + \mathrm{Dp}(V)) \\ &\cong \mathrm{F}_n V/(\mathrm{F}_n V \cap (\mathrm{F}_{n+1} V + \mathrm{Dp}(V))) \\ &= \mathrm{F}_n V/(\mathrm{F}_{n+1} V + \mathrm{F}_n V \cap \mathrm{Dp}(V)).\end{aligned}$$

Hence,

$$\mathrm{Gr}(V/\mathrm{Dp}(V)) = \bigoplus_{n=0}^{N-1} \mathrm{F}_n V/(\mathrm{F}_{n+1} V + \mathrm{F}_n V \cap \mathrm{Dp}(V)).$$

On the other hand, we have

$$\begin{aligned}\bigoplus_{n=0}^{N-1} \mathrm{Gr}_n V/\mathrm{gr}_n(\mathrm{Dp}(\mathrm{F}_n V)) &= \bigoplus_{n=0}^{N-1} (\mathrm{F}_n V/\mathrm{F}_{n+1} V)/((\mathrm{Dp}(\mathrm{F}_n V) + \mathrm{F}_{n+1} V)/\mathrm{F}_{n+1} V) \\ &\cong \bigoplus_{n=0}^{N-1} \mathrm{F}_n V/(\mathrm{Dp}(\mathrm{F}_n V) + \mathrm{F}_{n+1} V).\end{aligned}$$

From Corollary 4.9 it follows that $\mathrm{F}_n V \cap \mathrm{Dp}(V) = \mathrm{Dp}(\mathrm{F}_n V)$. Consequently,

$$\bigoplus_{n=0}^{N-1} \mathrm{Gr}_n V/\mathrm{gr}_n(\mathrm{Dp}(\mathrm{F}_n V)) \cong \mathrm{Gr}(V/\mathrm{Dp}(V)).$$

□

PROPOSITION 4.12. *Let $V = \langle U_1, \ldots, U_k \rangle_d$, $U_1, \ldots, U_k \in \mathcal{F}$. Assume $\mathrm{F}_n V$ is non-zero subspace of the filtration, $\mathrm{F}_n V = \langle Z_1, \ldots, Z_s, \ldots \rangle_d$, where $\{Z_i\}_{i \in I}$ is either a finite or infinite set of monomials, and let $V_i = \langle Z_i \rangle_d$. Then*

(1) *there exists a subset of indices $I' \subseteq I$ such that*

$$\mathrm{F}_n V/(\mathrm{Dp}(\mathrm{F}_n V) + \mathrm{F}_{n+1}V) \cong \bigoplus_{i\in I'} V_i/(\mathrm{Dp}(V_i) + \mathrm{L}(V_i)),$$

where $V_i \nsubseteq \mathrm{F}_{n+1}V$ and the spaces V_i, $i \in I'$ are pairwise different;

(2) *if $Z \in \mathrm{F}_n V$ is a monomial such that $Z \notin \mathrm{F}_{n+1}V$, then Z belongs to precisely one space V_i, $i \in I'$, and the isomorphism acts as follows*

$$Z + \mathrm{Dp}(\mathrm{F}_n V) + \mathrm{F}_{n+1}V \mapsto (0, \dots, 0, Z + \mathrm{Dp}(V_i) + \mathrm{L}(V_i), 0, \dots).$$

PROOF. Since $\mathrm{F}_n V = \langle Z_1, \dots, Z_s, \dots\rangle_d$, $\mathrm{F}_n V = \sum_{i=1}^{\infty} V_i$ if there are infinitely many V_i and $\mathrm{F}_n V = \sum_{i=1}^{m} V_i$ otherwise. In what following we only discuss the case of infinitely many V_i, $i \in I$. The other case is similar. So, we have

$$\mathrm{F}_n V/\mathrm{F}_{n+1}V = \sum_{i=1}^{\infty} ((V_i + \mathrm{F}_{n+1}V)/\mathrm{F}_{n+1}V).$$

Let us choose a special subset of $\{V_i\}_{i\in I}$. We take all the different spaces from $\{V_i\}_{i\in I}$ and remove V_i such that $V_i \nsubseteq \mathrm{F}_{n+1}V$. Since $\mathrm{F}_n V \neq 0$, from Proposition 4.4 it follows that $\mathrm{F}_n V \neq \mathrm{F}_{n+1}V$. Hence, we do not remove all V_i in this process and obtain a non-empty set $\{V_i\}_{i\in I'}$ such that

$$\mathrm{F}_n V/\mathrm{F}_{n+1}V = \sum_{i\in I'} ((V_i + \mathrm{F}_{n+1}V)/\mathrm{F}_{n+1}V).$$

Assume V_j is one of the spaces $\{V_i\}_{i\in I'}$, that is, $j \in I'$, and

$$((V_j + \mathrm{F}_{n+1}V)/\mathrm{F}_{n+1}V) \bigcap \sum_{\substack{i\neq j\\ i\in I'}} ((V_i + \mathrm{F}_{n+1}V)/\mathrm{F}_{n+1}V) \neq 0. \tag{37}$$

Since every V_i, $i \in I'$, and $\mathrm{F}_{n+1}V$ are generated by monomials, it follows from (37) that there exist monomials $Z'_j \in V_j$ such that $Z'_j \notin \mathrm{F}_{n+1}V$ and a space V_{i_0}, $i_0 \in I'$, $i_0 \neq j$, such that $Z'_j \in V_{i_0}$. Since $Z'_j \notin \mathrm{F}_{n+1}V$, we have $Z'_j \notin \mathrm{L}(V_j)$ and $Z'_j \notin \mathrm{L}(V_{i_0})$. Then from Lemma 4.2 it follows that $V_j = V_{i_0}$. But this contradicts the assumption that all spaces V_i, $i \in I'$, are pairwise different. Hence,

$$((V_j + \mathrm{F}_{n+1}V)/\mathrm{F}_{n+1}V) \bigcap \sum_{\substack{i\neq j\\ i\in I'}} ((V_i + \mathrm{F}_{n+1}V)/\mathrm{F}_{n+1}V) = 0.$$

Thus, we obtain a direct sum

$$\mathrm{F}_n V/\mathrm{F}_{n+1}V = \bigoplus_{i\in I'} ((V_i + \mathrm{F}_{n+1}V)/\mathrm{F}_{n+1}V). \tag{38}$$

Let $Z \in \mathrm{F}_n V$ be a monomial, $Z \notin \mathrm{F}_{n+1}V$. Notice that, by the same argument as above, we obtain that Z belongs to precisely one of the spaces V_i, $i \in I'$.

Let T be the support of a multi-turn $U_h \mapsto \sum_{\substack{j=1\\ j\neq h}}^{k} U_j$, where $U_h \in \mathrm{F}_n V \setminus \mathrm{F}_{n+1}V$. Hence, U_h belongs to a space V_{i_0}, $i_0 \in I'$. Then every U_j also belongs to V_{i_0} and $T \in \mathrm{Dp}(V_{i_0})$. Therefore, we have

$$(\mathrm{Dp}(\mathrm{F}_n V) + \mathrm{F}_{n+1}V)/\mathrm{F}_{n+1}V = \bigoplus_{i\in I'} ((\mathrm{Dp}(V_i) + \mathrm{F}_{n+1}V)/\mathrm{F}_{n+1}V).$$

Hence,

(39)

$$
\begin{aligned}
\mathrm{F}_nV/(\mathrm{Dp}(\mathrm{F}_nV)+\mathrm{F}_{n+1}V) &\cong (\mathrm{F}_nV/\mathrm{F}_{n+1}V)/\left((\mathrm{Dp}(\mathrm{F}_nV)+\mathrm{F}_{n+1}V)/\mathrm{F}_{n+1}V\right)\\
&= \left(\bigoplus_{i\in I'}((V_i+\mathrm{F}_{n+1}V)/\mathrm{F}_{n+1}V)\right)/\left(\bigoplus_{i\in I'}((\mathrm{Dp}(V_i)+\mathrm{F}_{n+1}V)/\mathrm{F}_{n+1}V)\right)\\
&\cong \bigoplus_{i\in I'}((V_i+\mathrm{F}_{n+1}V)/\mathrm{F}_{n+1}V)/((\mathrm{Dp}(V_i)+\mathrm{F}_{n+1}V)/\mathrm{F}_{n+1}V)\\
&\cong \bigoplus_{i\in I'}(V_i+\mathrm{F}_{n+1}V)/(\mathrm{Dp}(V_i)+\mathrm{F}_{n+1}V).
\end{aligned}
$$

By the Isomorphism Theorem, we obtain

(40)

$$
\begin{aligned}
(V_i+\mathrm{F}_{n+1}V)/(\mathrm{Dp}(V_i)+\mathrm{F}_{n+1}V) &= (V_i+\mathrm{Dp}(V_i)+\mathrm{F}_{n+1}V)/(\mathrm{Dp}(V_i)+\mathrm{F}_{n+1}V)\\
&\cong V_i/((\mathrm{Dp}(V_i)+\mathrm{F}_{n+1}V)\cap V_i) = V_i/(\mathrm{Dp}(V_i)+\mathrm{F}_{n+1}V\cap V_i)
\end{aligned}
$$

Since V_i, $i\in I'$, is not contained in $\mathrm{F}_{n+1}V$, from Lemma 4.5 it follows that $\mathrm{F}_{n+1}V\cap V_i=\mathrm{L}(V_i)$. Thus, from (39) and (40) it follows that

$$\mathrm{F}_nV/(\mathrm{Dp}(\mathrm{F}_nV)+\mathrm{F}_{n+1}V)\cong\bigoplus_{i\in I'}V_i/(\mathrm{Dp}(V_i)+\mathrm{L}(V_i)).$$

So, the first statement is proved.

Assume $Z\in\mathrm{F}_nV$ is a monomial, $Z\notin\mathrm{F}_{n+1}V$. We noticed above that Z belongs to one space V_{i_0}, $i_0\in I'$. Then from the definition of the canonical isomorphisms in (39) and (40), it easily follows that the final isomorphism maps $Z+\mathrm{Dp}(\mathrm{F}_nV)+\mathrm{F}_{n+1}V$ to the corresponding quotient space $V_{i_0}/(\mathrm{Dp}(V_{i_0})+\mathrm{L}(V_{i_0}))$, namely,

$$Z+\mathrm{Dp}(\mathrm{F}_nV)+\mathrm{F}_{n+1}V\mapsto(0,\dots,0,Z+\mathrm{Dp}(V_{i_0})+\mathrm{L}(V_{i_0}),0,\dots).$$

So, the second statement is proved. □

5. Description of a basis in $\mathbb{Z}_2\mathcal{F}/\mathcal{I}$

Let us enumerate all the monomials from $\mathcal{F}$, namely, $\mathcal{F}=\{U_1,U_2,\dots,U_k,\dots\}$. Consider an increasing sequence of subspaces of $\mathbb{Z}_2\mathcal{F}$

$$\langle U_1\rangle_d\subseteq\langle U_1,U_2\rangle_d\subseteq\dots\subseteq\langle U_1,\dots,U_k\rangle_d\subseteq\dots$$

Clearly, the union of all these subspaces gives the whole ring $\mathbb{Z}_2\mathcal{F}$. Consider the corresponding increasing sequence of subspaces of $\mathbb{Z}_2\mathcal{F}/\langle\mathcal{T}'\rangle=\mathbb{Z}_2\mathcal{F}/\mathcal{I}$

$$
\begin{aligned}
(41)\qquad (\langle U_1\rangle_d+\langle\mathcal{T}'\rangle/\langle\mathcal{T}'\rangle) &\subseteq (\langle U_1,U_2\rangle_d+\langle\mathcal{T}'\rangle)/\langle\mathcal{T}'\rangle\subseteq\dots\\
&\subseteq(\langle U_1,\dots,U_k\rangle_d+\langle\mathcal{T}'\rangle)/\langle\mathcal{T}'\rangle\subseteq\dots.
\end{aligned}
$$

The union of all these subspaces gives the whole ring $\mathbb{Z}_2\mathcal{F}/\langle\mathcal{T}'\rangle$. By Corollary 4.11, we obtain

$$(\langle U_1,\dots,U_k\rangle_d+\langle\mathcal{T}'\rangle)/\langle\mathcal{T}'\rangle\cong\langle U_1,\dots,U_k\rangle_d/\mathrm{Dp}(\langle U_1,\dots,U_k\rangle_d).$$

In Section 5.2, we will construct a basis $\mathcal{B}_k$ in every subspace $(\langle U_1,\dots,U_k\rangle+\langle\mathcal{T}'\rangle)/\langle\mathcal{T}'\rangle$ such that we obtain the increasing sequence

$$\mathcal{B}_1\subseteq\mathcal{B}_2\subseteq\dots\subseteq\mathcal{B}_k\subseteq\dots.$$

Since the union of the subspaces (41) gives the whole ring $\mathbb{Z}_2\mathcal{F}/\langle\mathcal{T}'\rangle$, the union $\bigcup_k \mathcal{B}_k$ is a basis of $\mathbb{Z}_2\mathcal{F}/\langle\mathcal{T}'\rangle = \mathbb{Z}_2\mathcal{F}/\mathcal{I}$.

As above, assume $V = \langle U_1, \ldots, U_k\rangle_d$, $U_1, \ldots, U_k \in \mathcal{F}$. Let the filtration on the space V have N non-zero levels. Putting together the results from Section 4.2, we obtain

$$(V + \langle\mathcal{T}'\rangle)/\langle\mathcal{T}'\rangle \cong \bigoplus_{n=0}^{N-1} \bigoplus_{i\in I'_n} V_i^{(n)}/(\mathrm{Dp}(V_i^{(n)}) + \mathrm{L}(V_i^{(n)})),$$

where $V_i^{(n)} = \langle Z_i^{(n)}\rangle_d$, $Z_i^{(n)}$ is a monomial from $\mathrm{F}_n V$. In Section 5.1 we will characterize non-trivial quotient spaces $V_i^{(n)}/(\mathrm{Dp}(V_i^{(n)}) + \mathrm{L}(V_i^{(n)})$. In Section 5.2 we will construct a basis in every space $V_i^{(n)}/(\mathrm{Dp}(V_i^{(n)}) + \mathrm{L}(V_i^{(n)}))$ and, using this, we will explicitly describe a basis in the whole ring $\mathbb{Z}_2\mathcal{F}/\mathcal{I}$.

5.1. λ-semicanonical words. Fix a constant $\frac{1}{2} \ll \lambda \ll 1$ ($\ll$ on the ε-scale). For example, one can use $\lambda \geqslant \frac{2}{3}$. We introduce *$\lambda$-forbidden words*. Recall our notation $v = v_i v_m v_f$, where v_i is some initial part, v_m is some middle part and v_f is some final part of v (any part is allowed to be empty). A word of the form v_m^{-1} is called *λ-forbidden* if $\Lambda(v_m) > \lambda$.

$$v_m^{-1}, \Lambda(v_m) > \lambda. \tag{42}$$

DEFINITION 5.1. A word is called *λ-semicanonical* if it does not contain occurrences of λ-forbidden words.

We are motivated by an informal analogue for small cancellation groups. Let $G = \langle X \mid \mathcal{R}\rangle$, where $\mathcal{R}$ satisfies small cancellation conditions [**7**] with a certain measure Λ on the relators R_i. That is $\Lambda(R_i) = 1$ and $\Lambda(P) \leqslant \varepsilon$ for any small piece P. We call a subword S of $R_i^{\pm 1}$ *λ-forbidden* if $\Lambda(S) > \lambda$, where λ as above is between $\frac{1}{2}$ and 1. A word U is *λ-semicanonical* if it does not contain λ-forbidden words. In the Cayley graph of G any λ-semicanonical word is a quasigeodesic [**3**], [**4**].

A small cancellation group (with appropriate constants) is hyperbolic. Recall that if the elements of a hyperbolic group are represented by quasigeodesics, then their product takes the form of a thin triangle ([**5**]). In our case, first we will construct a special basis of $\mathbb{Z}_2\mathcal{F}/\mathcal{I}$ with the use of λ-semicanonical words. Then in Section 6, we will construct a special set of linear generators of $\mathbb{Z}_2\mathcal{F}/\mathcal{I}$ (not linearly independent) such that it contains the basis and we can express the product of two elements of this set as a sum of elements of this set that form thin triangles with the factors.

PROPOSITION 5.1. *Let a_h be a generalized fractional power, $W = \langle a_h\rangle_d$. Then a_h is either λ-semicanonical, or is equal modulo* $\mathrm{Dp}(W)$ *to a sum of λ-semicanonical generalized fractional powers from W. Namely, either a_h is λ-semicanonical, or there exists an elementary multi-turn $a_h \mapsto \sum_{\substack{j=1\\ j\neq h}}^{k} a_j$ such that a_j, $j = 1, \ldots, k$, $j \neq h$, are λ-semicanonical generalized fractional powers and they have shorter word length in $\mathcal{F}$ and less Λ-measure than a_h.*

PROOF. Let a_h contain a λ-forbidden subword b_h, $a_h = Lb_hR$. Returning to forbidden words (42), we consider the following relation in $\mathbb{Z}_2\mathcal{F}/\mathcal{I}$:

$$v_m^{-1} = v_f w v_i + v_f v_i. \tag{43}$$

Multiplying the relation $v^{-1} = 1 + w$ by v_f on the left side and by v_i on the right side, we get (43). By definition, the words $v_f w v_i$ and $v_f v_i$ are non λ-forbidden. If v_m^{-1} is a λ-forbidden word, that is, $\Lambda(v_m^{-1}) > \lambda$, then we obtain

$$\Lambda(v_f v_i) = \Lambda(v_f w v_i) = \Lambda(v_f) + \Lambda(v_i) < 1 - \lambda < \lambda < \Lambda(v_m^{-1}).$$

Moreover, since $|w| \ll |v|$, we obtain $|v_f v_i| < |v_f w v_i| < |v_m^{-1}|$, where $|\cdot|$ is word length in $\mathcal{F}$.

By Definition 2.4, the transformation $v_m^{-1} \mapsto v_f w v_i + v_f v_i$ is an elementary multi-turn and according to the above, it gives the reduction of λ-forbidden words to sums of shorter (in word length in $\mathcal{F}$ and in Λ-measure) λ-semicanonical words.

So, b_h can be substituted by a sum of λ-semicanonical generalized fractional powers $\sum_{\substack{j=1\\ j\neq h}}^n b_j = v_f w v_i + v_f v_i$. Since $\Lambda(b_h) > \varepsilon$, it appears in v^{-1} only once and, therefore, has fixed initial and fixed final point in the corresponding v-diagram. Since $a_h = Lb_hR$ is itself a generalized fractional power, we obviously obtain that every word Lb_jR is a generalized fractional power, possibly after cancellations, with the same initial point and the same final point in the corresponding v-diagram as the word a_h has. So, since $v_m^{-1} \mapsto v_f w v_i + v_f v_i$ is an elementary multi-turn, $a_h \mapsto \sum_{\substack{j=1\\ j\neq h}}^n Lb_jR$ is an elementary multi-turn as well. Hence, Lb_jR are derived monomials of a_h and $a_h = \sum_{\substack{j=1\\ j\neq h}}^n Lb_jR$ modulo $\mathrm{Dp}(W)$.

If the obtained word Lb_jR contains a λ-forbidden subword, we repeat the process and reduce it using the transformation (43). Successive elementary multi-turns match to addition of the corresponding expressions (9)–(14). Thus, applying several elementary multi-turns consecutively, we obtain an elementary multi-turn as a result. So, we obtain an elementary multi-turn of a_h after each step of the reduction process.

From (43) it follows that after every transformation that reduces λ-forbidden subword of a monomial, the word length in $\mathcal{F}$ of the resulting monomials is strictly smaller than of the initial monomial. Thus, the reduction process finishes and we obtain the result as an elementary multi-turn $a_h \mapsto \sum_{\substack{j=1\\ j\neq h}}^k a_j$, where $j = 1, \dots, k$, $j \neq h$, are λ-semicanonical generalized fractional powers with shorter word length in $\mathcal{F}$ and less Λ-measure than a_h. □

COROLLARY 5.2. *Assume $V = \langle U_1, \dots, U_k\rangle_d$, $U_1, \dots, U_k \in \mathcal{F}$. Let $U \in V$ be a monomial. Then U is equal to a sum of λ-semicanonical words from V modulo* $\mathrm{Dp}(V)$. *In particular, U is equal to a sum of λ-semicanonical words from $\langle U\rangle_d$ modulo* $\mathrm{Dp}(\langle U\rangle_d)$.

PROOF. Suppose $U \in \mathcal{F}$ and a_h is a maximal occurrence of a generalized fractional power in U, $U = La_hR$. Suppose a_h contains a λ-forbidden word. Then, clearly, a_h is a virtual member of the chart of U. From Proposition 5.1, it follows that there exists an elementary multi-turn $a_h \mapsto \sum_{\substack{j=1\\ j\neq h}}^k a_i$, where $j = 1, \dots, k$,

$j \neq h$, are λ-semicanonical generalized fractional powers with the word length in $\mathcal{F}$ strictly smaller than the word length of a_h. Then La_jR are derived monomials of U and

$$U = La_hR = \sum_{\substack{j=1 \\ j \neq h}}^{k} La_jR \mod \mathrm{Dp}(\langle U \rangle_d),$$

where the word length in $\mathcal{F}$ of all monomials appearing in the right-hand side is strictly smaller than the word length of U. We continue the reduction process for the obtained monomials if necessary. Since word length of the monomials strictly decreases after each step of the reduction, the process finishes and we obtain the result as a sum of λ-semicanonical words.

Assume $U \in V = \langle U_1, \dots, U_k \rangle_d$. Since V is generated by monomials and closed under taking derived monomials, $\mathrm{Dp}(\langle U \rangle_d) \subseteq \mathrm{Dp}(V)$. Hence, U is equal to the same resulting sum of λ-semicanonical monomials as above modulo $\mathrm{Dp}(V)$. $\square$

The most important property of λ-semicanonical word for us is the following.

PROPOSITION 5.3 (Non-degeneracy). *Suppose a_h is a λ-semicanonical generalized fractional power, $a_h \mapsto \sum_{\substack{j=1 \\ j \neq h}}^{l} a_j$ is an elementary multi-turn. Then there exists a_j, $j \neq h$, such that $\Lambda(a_j) \geqslant \tau$.*

PROOF. Let I and F be the initial and the final points of the paths corresponding to a_j, $j = 1, \dots, l$, in the v-diagram. There are the following possible positions of I and F:

Case 1 The points I and F lie on the v-arc.
Case 2 The points I and F lie on a w-arc.
Case 3 The point I lies on the v-arc, the point F lies on a w-arc.
Case 4 The point I lies on a w-arc, the point F lies on the v-arc.

Consider Case 1 for $a_1, \dots, a_l$. Assume the contrary, that is, $\Lambda(a_j) < \tau$, $j = 1, \dots, l$, $j \neq h$. Recall possible forms of monomials of Λ-measure less than τ (we enumerated them in the proof of Proposition 3.1). We denote the smallest path from I to F by b', the smallest path from I to O by b_1, and the smallest path from O to F by b_2. Since $\Lambda(a_j) < \tau$ for $j \neq h$, the smallest path between I and F is necessarily of Λ-measure less than τ. Then there are the following possibilities (for a graphical illustration see Proposition 3.1).

1.1 b' contains the point O;
1.2 b' does not contain the point O and $\Lambda(b_1) + \Lambda(b_2) < \tau$;
1.3 b' does not contain the point O and $\Lambda(b_1) + \Lambda(b_2) \geqslant \tau$.

We will study the following particular configuration in 1.1 in detail. Recall that we use the notation $v = v_i v_m v_f$.

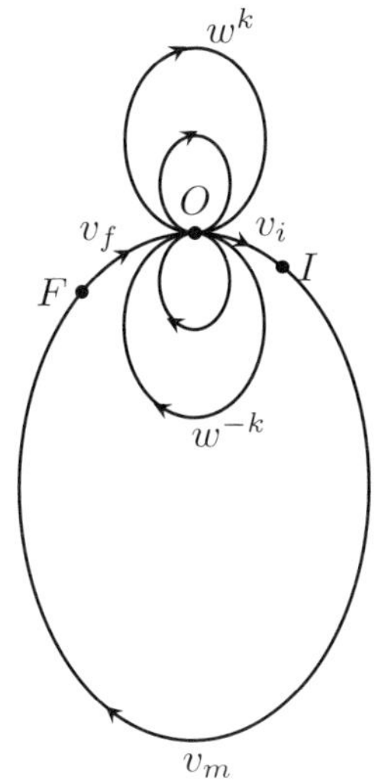

$$b_1 = v_i^{-1}, b_2 = v_f^{-1},$$
$$b' = v_i^{-1}v_f^{-1};$$

Returning to pictures (5), we obtain that the following list of generalized fractional powers corresponds to the above picture:

$$v_m v_f M(v,w) v_f^{-1},$$
$$v_m v_f M(v,w) v_i v_m,$$
$$v_i^{-1} M(v,w) v_i v_m,$$
$$v_i^{-1} M(v,w) v_f^{-1}.$$

So, every a_j, $j = 1, \ldots, l$, is of the above form. Denote possible forms of a_h by

$$v_m v_f M_h(v,w) v_f^{-1},$$
$$v_m v_f M_h(v,w) v_i v_m,$$
$$v_i^{-1} M_h(v,w) v_i v_m,$$
$$v_i^{-1} M_h(v,w) v_f^{-1}.$$

Since a_h is λ-semicanonical, the monomial $M_h(v,w)$ does not contain negative powers of v. By the assumption, $\Lambda(a_j) < \tau$ for $j \neq h$; hence,

$$a_j = v_i^{-1} w^{k_j} v_f^{-1}.$$

So, there are the following possible forms of the elementary multi-turn $a_h \mapsto \sum_{\substack{j=1 \\ j\neq h}}^{l} a_j$ depending on the form of a_h:

$$v_m v_f M_h(v,w) v_f^{-1} \mapsto \sum_{j=1}^{l} v_i^{-1} w^{k_j} v_f^{-1},$$

$$v_m v_f M_h(v,w) v_i v_m \mapsto \sum_{j=1}^{l} v_i^{-1} w^{k_j} v_f^{-1},$$

$$v_i^{-1} M_h(v,w) v_i v_m \mapsto \sum_{j=1}^{l} v_i^{-1} w^{k_j} v_f^{-1},$$

$$v_i^{-1} M_h(v,w) v_f^{-1} \mapsto \sum_{j=1}^{l} v_i^{-1} w^{k_j} v_f^{-1}.$$

We transform every elementary multi-turn above to an elementary multi-turn of a monomial over v, w, multiplying every expression above by corresponding generalized fractional powers:

$$vM_h(v,w) \mapsto \sum_{j=1}^{l} w^{k_j},$$

$$vM_h(v,w)v \mapsto \sum_{j=1}^{l} w^{k_j},$$

$$M_h(v,w)v \mapsto \sum_{j=1}^{l} w^{k_j},$$

$$M_h(v,w) \mapsto \sum_{j=1}^{l} w^{k_j}.$$

That is, we obtain an elementary multi-turn of the form

$$\widetilde{M}_h(v,w) \mapsto \sum_{j=1}^{l} w^{k_j}, \tag{44}$$

where the monomial $\widetilde{M}_h(v,w)$ does not contain negative powers of v. Since all monomials on the right-hand side are of zero Λ-measure, by Proposition 3.1, $\Lambda(\widetilde{M}_h(v,w)) \geqslant \tau$. Hence, the monomial $\widetilde{M}_h(v,w)$ contains at least one positive power of v. Then, clearly, the polynomial

$$P(v,w) = \widetilde{M}(v,w) + \sum_{j=1}^{l} w^{k_j}$$

does not satisfy condition (8), so, (44) can not be a multi-turn. We obtain a contradiction.

All the rest of the configurations in Case 1–Case 4 are processed in the same way. □

Using Corollary 5.2, Proposition 4.10 and Proposition 5.3, we obtain the following important statement.

COROLLARY 5.4. *Assume $W = \langle Z \rangle_d$, where Z is a monomial. Then the space $W/(\mathrm{Dp}(W) + \mathrm{L}(W))$ is non-trivial if and only if there exists a λ-semicanonical monomial $\widetilde{Z} \in W$ such that $W = \langle \widetilde{Z} \rangle_d$. Moreover, if X is a λ-semicanonical monomial such that $X \in W \setminus \mathrm{L}(W)$ (and then $W = \langle X \rangle_d$), then $X + \mathrm{Dp}(W) + \mathrm{L}(W) \neq 0$ as an element of $W/(\mathrm{Dp}(W) + \mathrm{L}(W))$.*

PROOF. Assume the quotient space $W/(\mathrm{Dp}(W) + \mathrm{L}(W))$ is non-trivial. Consider the set of all monomials from $W \setminus \mathrm{L}(W)$. If every monomial from this set belongs to $\mathrm{Dp}(W) + \mathrm{L}(W)$, then the quotient space $W/(\mathrm{Dp}(W) + \mathrm{L}(W))$ is trivial. Hence, there exists a monomial $Z' \in W \setminus \mathrm{L}(W)$ such that $Z' \notin \mathrm{Dp}(W) + \mathrm{L}(W)$. From Corollary 5.2 it follows that

$$Z' = \sum_{i=1}^{l} Z_i \mod \mathrm{Dp}(W),$$

where $Z_i \in W$ are λ-semicanonical monomials. Since $Z' \notin \mathrm{Dp}(W) + \mathrm{L}(W)$, there exists a monomial $Z_{i_0} \in W \setminus \mathrm{L}(W)$ in this sum. Hence, from Lemma 4.2 it follows that $W = \langle Z_{i_0} \rangle_d$.

Assume $W = \langle \widetilde{Z} \rangle_d$, where $\widetilde{Z}$ is λ-semicanonical monomial. Then, by Proposition 4.10, we have

$$W/(\mathrm{Dp}(W) + \mathrm{L}(W)) \cong A_1/(\mathrm{Dp}(A_1) + \mathrm{L}_1) \otimes \ldots \otimes A_k/(\mathrm{Dp}(A_k) + \mathrm{L}_k),$$

where $A_1, \ldots, A_k$ are spaces generated by generalized fractional powers and corresponding to each place in the chart of $\widetilde{Z}$ by formula (32). In particular, if $a^{(1)}, \ldots, a^{(k)}$ are all virtual members of the chart of $\widetilde{Z}$ enumerated from left to right, then $a^{(1)} \in A_1, \ldots, a^{(k)} \in A_k$.

Recall that $\mathrm{Dp}(A_i)$ consists of supports of elementary multi-turns. Therefore, by Proposition 5.3, we obtain $a^{(i)} \notin \mathrm{Dp}(A_i) + \mathrm{L}_i, i = 1, \ldots, k$. Hence,

$$(a^{(1)} + \mathrm{Dp}(A_1) + \mathrm{L}_1) \otimes \ldots \otimes (a^{(k)} + \mathrm{Dp}(A_k) + \mathrm{L}_k) \neq 0. \tag{45}$$

Then, by definition of the isomorphism

$$\overline{\mu}_2[\widetilde{Z}] : A_1/(\mathrm{Dp}(A_1) + \mathrm{L}_1) \otimes \ldots \otimes A_k/(\mathrm{Dp}(A_k) + \mathrm{L}_k) \to W/(\mathrm{Dp}(W) + \mathrm{L}(W)),$$

we have

$$\begin{aligned}\overline{\mu}_2[\widetilde{Z}]((a^{(1)} &+ \mathrm{Dp}(A_1) + \mathrm{L}_1) \otimes \ldots \otimes (a^{(k)} + \mathrm{Dp}(A_k) + \mathrm{L}_k)) \\ &= \mu[\widetilde{Z}](a^{(1)} \otimes \ldots \otimes a^{(k)}) + \mathrm{Dp}(W) + \mathrm{L}(W) = \widetilde{Z} + \mathrm{Dp}(W) + \mathrm{L}(W).\end{aligned}$$

Then, by (45), we obtain that $\widetilde{Z} + \mathrm{Dp}(W) + \mathrm{L}(W) \neq 0$. Thus, the quotient space $W/(\mathrm{Dp}(W) + \mathrm{L}(W))$ is non-trivial.

Let X be a λ-semicanonical monomial from $W \setminus \mathrm{L}(W)$. Then, by Lemma 4.2, we have $W = \langle X \rangle_d$. Therefore, by the same argument as above, we obtain that $X + \mathrm{Dp}(W) + \mathrm{L}(W) \neq 0$. This completes the proof. □

5.2. How to calculate a basis in $\mathbb{Z}_2\mathcal{F}/\mathcal{I}$ using λ-semicanonical words. As before, we denote the space $\langle U_1,\dots,U_k\rangle_d$, where $U_1,\dots,U_k$ are monomials, by V. Putting together the results of Corollary 4.11, Theorem 4.1 and Proposition 4.12, we obtain

$$(V+\langle\mathcal{T}'\rangle)/\langle\mathcal{T}'\rangle \cong \bigoplus_{n=0}^{N-1}\bigoplus_{i\in I'_n} V_i^{(n)}/(\mathrm{Dp}(V_i^{(n)})+\mathrm{L}(V_i^{(n)})), \tag{46}$$

where $V_i^{(n)}=\langle Z_i^{(n)}\rangle_d$, $Z_i^{(n)}$ is a monomial from F_nV, $V_i^{(n)}\not\subseteq \mathrm{F}_{n+1}V$ and the spaces $V_i^{(n)}$ are pairwise different.

PROPOSITION 5.5. *Assume*

$$\left\{\overline{W}_j^{(i,n)} \mid \overline{W}_j^{(i,n)} \in V_i^{(n)}/(\mathrm{Dp}(V_i^{(n)})+\mathrm{L}(V_i^{(n)})), j\in\mathbb{N}\right\},$$

is a basis of $V_i^{(n)}/(\mathrm{Dp}(V_i^{(n)})+\mathrm{L}(V_i^{(n)}))$, *where* $n=0,\dots,N-1$ *and* $i\in I'_n$. *Let* $W_j^{(i,n)}\in V_i^{(n)}$ *be an arbitrary representative of the coset* $\overline{W}_j^{(i,n)}$. *Then*

$$\bigcup_{n=0}^{N-1}\bigcup_{i\in I'_n}\left\{W_j^{(i,n)}+\langle\mathcal{T}'\rangle \mid j\in\mathbb{N}\right\}$$

is a basis of $(V+\langle\mathcal{T}'\rangle)/\langle\mathcal{T}'\rangle$.

PROOF. While the statement is pretty obvious, we prefer to give a proof to recollect the previously stated facts.

In Proposition 4.12 we proved that there exists an isomorphism

$$\varphi:\mathrm{F}_nV/(\mathrm{Dp}(\mathrm{F}_nV)+\mathrm{F}_{n+1}V)\to\bigoplus_{i\in I'_n}V_i^{(n)}/(\mathrm{Dp}(V_i^{(n)})+\mathrm{L}(V_i^{(n)}))$$

that acts as follows

$$\varphi(U+\mathrm{Dp}(\mathrm{F}_nV)+\mathrm{F}_{n+1}V)=(0,\dots,0,U+\mathrm{Dp}(V_i^{(n)})+\mathrm{L}(V_i^{(n)}),0,\dots), \tag{47}$$

where U is a monomial, $U\in\mathrm{F}_nV\setminus\mathrm{F}_{n+1}V$. Suppose

$$\left\{\overline{W}_j^{(i,n)} \mid \overline{W}_j^{(i,n)} \in V_i^{(n)}/(\mathrm{Dp}(V_i^{(n)})+\mathrm{L}(V_i^{(n)})), j\in\mathbb{N}\right\}$$

is a basis of $V_i^{(n)}/(\mathrm{Dp}(V_i^{(n)})+\mathrm{L}(V_i^{(n)}))$. Then from (47) it easily follows that

$$\bigcup_{i\in I'}\left\{W_j^{(i,n)}+\mathrm{Dp}(\mathrm{F}_nV)+\mathrm{F}_{n+1}V \mid j\in\mathbb{N}\right\},$$

where $W_j^{(i,n)}\in\mathrm{F}_nV$ is an arbitrary representative of the coset $\overline{W}_j^{(i,n)}$, is a basis of $\mathrm{F}_nV/(\mathrm{Dp}(\mathrm{F}_nV)+\mathrm{F}_{n+1}V)$.

Recall that

$$\mathrm{F}_n(V/\mathrm{Dp}(V))=(\mathrm{F}_nV+\mathrm{Dp}(V))/\mathrm{Dp}(V)$$

and, by the isomorphism theorems, we have the following canonical isomorphisms:

$$\pi_1:\mathrm{F}_n(V/\mathrm{Dp}(V))/\mathrm{F}_{n+1}(V/\mathrm{Dp}(V))\to(\mathrm{F}_nV+\mathrm{Dp}(V))/(\mathrm{F}_{n+1}V+\mathrm{Dp}(V)),$$
$$\pi_2:(\mathrm{F}_nV+\mathrm{Dp}(V))/(\mathrm{F}_{n+1}V+\mathrm{Dp}(V))\to\mathrm{F}_nV/(\mathrm{F}_{n+1}V+\mathrm{Dp}(V)\cap\mathrm{F}_nV).$$

Let us recall that from Proposition 4.9 it follows that $\mathrm{Dp}(V) \cap \mathrm{F}_n V = \mathrm{Dp}(\mathrm{F}_n V)$. Therefore, we have

$$\begin{aligned}\mathrm{F}_n(V/\mathrm{Dp}(V))/\mathrm{F}_{n+1}(V/\mathrm{Dp}(V)) &\xrightarrow{\pi_1} (\mathrm{F}_n V + \mathrm{Dp}(V))/(\mathrm{F}_{n+1} V + \mathrm{Dp}(V)) \\ &\xrightarrow{\pi_2} \mathrm{F}_n V/(\mathrm{F}_{n+1} V + \mathrm{Dp}(\mathrm{F}_n V)).\end{aligned}$$

Let W be an arbitrary element of $\mathrm{F}_n V$ (not necessarily a monomial). Then, by the well known construction of the canonical isomorphisms in the isomorphism theorems, we have

$$\begin{aligned}&\pi_2 \circ \pi_1(W + \mathrm{Dp}(V) + \mathrm{F}_{n+1}(V/\mathrm{Dp}(V))) \\ &\qquad = \pi_2(W + \mathrm{Dp}(V) + \mathrm{F}_{n+1} V) = W + \mathrm{F}_{n+1} V + \mathrm{Dp}(\mathrm{F}_n V).\end{aligned}$$

So, since

$$\bigcup_{i \in I'} \left\{ W_j^{(i,n)} + \mathrm{Dp}(\mathrm{F}_n V) + \mathrm{F}_{n+1} V \mid j \in \mathbb{N} \right\}$$

is a basis of $\mathrm{F}_n V/(\mathrm{F}_{n+1} V + \mathrm{Dp}(\mathrm{F}_n V))$, we obtain that

$$\bigcup_{i \in I'} \left\{ W_j^{(i,n)} + \mathrm{Dp}(V) + \mathrm{F}_{n+1}(V/\mathrm{Dp}(V)) \mid j \in \mathbb{N} \right\} \tag{48}$$

is a basis of $\mathrm{F}_n(V/\mathrm{Dp}(V))/\mathrm{F}_{n+1}(V/\mathrm{Dp}(V))$.

We have the graded space

$$\mathrm{Gr}(V/\mathrm{Dp}(V)) = \bigoplus_{n=0}^{N-1} \mathrm{F}_n(V/\mathrm{Dp}(V))/\mathrm{F}_{n+1}(V/\mathrm{Dp}(V)),$$

where $\mathrm{F}_n(V/\mathrm{Dp}(V)) = (\mathrm{F}_n V + \mathrm{Dp}(V))/\mathrm{Dp}(V)$. Assume that

$$\left\{ \overline{Z}_j^{(n)} \mid \overline{Z}_j^{(n)} \in \mathrm{F}_n(V/\mathrm{Dp}(V))/\mathrm{F}_{n+1}(V/\mathrm{Dp}(V)), j \in \mathbb{N} \right\}$$

is a basis of $\mathrm{F}_n(V/\mathrm{Dp}(V))/\mathrm{F}_{n+1}(V/\mathrm{Dp}(V))$, $n = 0, 1, \ldots, N-1$. Then one can easily show that

$$\bigcup_{n=0}^{N-1} \left\{ Z_j^{(n)} \mid Z_j^{(n)} \in \mathrm{F}_n(V/\mathrm{Dp}(V)), j \in \mathbb{N} \right\},$$

where $Z_j^{(n)}$ is an arbitrary representative of the coset $\overline{Z}_j^{(n)}$, forms a basis of $V/\mathrm{Dp}(V)$. Hence, using this observation together with (48) and Corollary 4.11, we obtain that

$$\bigcup_{n=0}^{N-1} \bigcup_{i \in I'_n} \left\{ W_j^{(i,n)} + \langle \mathcal{T}' \rangle \mid j \in \mathbb{N} \right\}$$

is a basis of $(V + \langle \mathcal{T}' \rangle)/\langle \mathcal{T}' \rangle$. □

PROPOSITION 5.6. *Let $\{X_j\}$ be all the λ-semicanonical monomials of V. Then there exists one-to-one correspondence between all the different spaces $\langle X_j \rangle_d$ and all the spaces $\mathrm{V}_i^{(n)}$ from (46) such that $\mathrm{V}_i^{(n)}/(\mathrm{Dp}(\mathrm{V}_i^{(n)}) + \mathrm{L}(\mathrm{V}_i^{(n)}))$ is non-trivial. Namely,*

1. *every space $\mathrm{V}_i^{(n)}$ such that $\mathrm{V}_i^{(n)}/(\mathrm{Dp}(\mathrm{V}_i^{(n)}) + \mathrm{L}(\mathrm{V}_i^{(n)})) \neq 0$ is equal to some space $\langle X_j \rangle_d$;*
2. *every space $\langle X_j \rangle_d$ is equal to some space $\mathrm{V}_i^{(n)}$ such that $\mathrm{V}_i^{(n)}/(\mathrm{Dp}(\mathrm{V}_i^{(n)}) + \mathrm{L}(\mathrm{V}_i^{(n)})) \neq 0$.*

PROOF. Consider a space $V_i^{(n)}$ such that $V_i^{(n)}/(\mathrm{Dp}(V_i^{(n)})+\mathrm{L}(V_i^{(n)})) \neq 0$. Then from Corollary 5.4 it follows that there exists a λ-semicanonical monomial $X_{j_0} \in V_i^{(n)}$ such that $V_i^{(n)} = \langle X_{j_0}\rangle_d$.

Let $X_j \in V$ be a λ-semicanonical monomial. Since $\mathrm{F}_N V = 0$, there exists a number n_0 such that $X_j \in \mathrm{F}_{n_0} V$ and $X_j \notin \mathrm{F}_{n_0+1} V$. Then, by Proposition 4.12, $X_j \in V_i^{(n_0)}$ for some $i \in I'_{n_0}$. Since $X_j \notin \mathrm{F}_{n_0+1}V$, we have $X_j \notin \mathrm{L}(V_i^{(n_0)})$. Then, by Lemma 4.2, $V_i^{(n_0)} = \langle X_j\rangle_d$. It remains to notice that, by Corollary 5.4, we have $V_i^{(n_0)}/(\mathrm{Dp}(V_i^{(n_0)}) + \mathrm{L}(V_i^{(n_0)})) \neq 0$. □

The next theorem follows from Proposition 5.5 and Proposition 5.6.

THEOREM 5.1. *Let $\{X_j\}_{j\in\mathbb{N}}$ be all λ-semicanonical monomials from $\mathcal{F}$. Let $\{V_i\}_{i\in\mathbb{N}}$ be all the different spaces $\{\langle X_j\rangle_d\}_{j\in\mathbb{N}}$ ($V_{i_1} \neq V_{i_2}$ for $i_1 \neq i_2$). If*

$$\left\{\overline{W}_j^{(i)} \mid \overline{W}_j^{(i)} \in V_i/(\mathrm{Dp}(V_i) + \mathrm{L}(V_i)), j \in \mathbb{N}\right\}$$

is a basis of $V_i/(\mathrm{Dp}(V_i) + \mathrm{L}(V_i))$, then

$$\bigcup_{i\in\mathbb{N}} \left\{W_j^{(i)} + \mathcal{I} \mid j \in \mathbb{N}\right\}$$

is a basis of $\mathbb{Z}_2\mathcal{F}/\mathcal{I}$, where $W_j^{(i)}$ is an arbitrary representative of the coset $\overline{W}_j^{(i)}$. In particular,

$$\mathbb{Z}_2\mathcal{F}/\mathcal{I} \cong \bigoplus_{i\in\mathbb{N}} V_i/(\mathrm{Dp}(V_i) + \mathrm{L}(V_i))$$

as vector spaces, and the right-hand side is explicitly described in Proposition 4.10.

PROOF. Let us enumerate all the monomials from $\mathcal{F}$, namely, $\mathcal{F} = \{U_1, U_2, \dots, U_k, \dots\}$. Then, evidently,

$$\mathbb{Z}_2\mathcal{F}/\mathcal{I} = \bigcup_{k=1}^{\infty} \left((\langle U_1, \dots, U_k\rangle_d + \langle \mathcal{T}'\rangle)/\langle \mathcal{T}'\rangle\right).$$

Consider spaces $\langle U_1, \dots, U_k\rangle_d \subseteq \langle U_1, \dots, U_k, U_{k+1}\rangle_d$. For the quotient space $(\langle U_1, \dots, U_k\rangle_d + \langle \mathcal{T}'\rangle)/\langle \mathcal{T}'\rangle$ we have decomposition (46), that is,

$$(\langle U_1, \dots, U_k\rangle_d + \langle \mathcal{T}'\rangle)/\langle \mathcal{T}'\rangle \cong \bigoplus_{n=0}^{N_k-1} \bigoplus_{i\in I'_n} V_i^{(n)}/(\mathrm{Dp}(V_i^{(n)}) + \mathrm{L}(V_i^{(n)})),$$

where N_k is the number of non-zero levels of the filtration on $\langle U_1, \dots, U_k\rangle_d$.

Let $\{X_j\}$ be all the λ-semicanonical monomials from $\langle U_1, \dots, U_k\rangle_d$. Then, obviously, the set of all λ-semicanonical monomials from $\langle U_1, \dots, U_{k+1}\rangle_d$ can be represented as

$$\{X_j\} \cup \{\widetilde{X}_j\},$$

where $\{\widetilde{X}_j\}$ collects the additional λ-semicanonical monomials that contain in $\langle U_1, \dots, U_k, U_{k+1}\rangle_d$ but do not contain in $\langle U_1, \dots, U_k\rangle_d$ (in particular, the set $\{\widetilde{X}_j\}$ may be empty). Then from Proposition 5.6, it follows that decomposition (46) for

$(\langle U_1, \dots, U_{k+1}\rangle_d + \langle \mathcal{T}'\rangle)/\langle \mathcal{T}'\rangle$ can be written as

$$(\langle U_1, \dots, U_{k+1}\rangle_d + \langle \mathcal{T}'\rangle)/\langle \mathcal{T}'\rangle \cong \left(\bigoplus_{n=0}^{N_k-1} \bigoplus_{i\in I'_n} V_i^{(n)}/(\mathrm{Dp}(V_i^{(n)}) + \mathrm{L}(V_i^{(n)}))\right) \bigoplus \left(\bigoplus_{n=0}^{N_{k+1}-1} \bigoplus_{h\in I''_n} \widetilde{V}_h^{(n)}/(\mathrm{Dp}(\widetilde{V}_h^{(n)}) + \mathrm{L}(\widetilde{V}_h^{(n)}))\right),$$

where N_{k+1} is the number of non-zero levels of the filtration on $\langle U_1, \dots, U_k, U_{k+1}\rangle_d$. The set I''_n collects the additional spaces $\widetilde{V}_h^{(n)}$, not appearing in I'_n, that correspond to additional monomials $\{\widetilde{X}_j\}$.

From Corollary 5.4, it follows that the spaces $V_i^{(n)}/(\mathrm{Dp}(V_i^{(n)}) + \mathrm{L}(V_i^{(n)}))$ and $\widetilde{V}_h^{(n)}/(\mathrm{Dp}(\widetilde{V}_h^{(n)}) + \mathrm{L}(\widetilde{V}_h^{(n)}))$ are non-trivial. We will choose a basis in every direct summand $V_i^{(n)}/(\mathrm{Dp}(V_i^{(n)}) + \mathrm{L}(V_i^{(n)}))$ and $\widetilde{V}_h^{(n)}/(\mathrm{Dp}(\widetilde{V}_h^{(n)}) + \mathrm{L}(\widetilde{V}_h^{(n)}))$ independently. Then from Proposition 5.5, it easily follows that there exists a basis $\mathcal{B}_{k+1}$ of $(\langle U_1, \dots, U_{k+1}\rangle_d + \langle \mathcal{T}'\rangle)/\langle \mathcal{T}'\rangle$ such that it expands the previously chosen basis $\mathcal{B}_k$ of $(\langle U_1, \dots, U_k\rangle_d + \langle \mathcal{T}'\rangle)/\langle \mathcal{T}'\rangle$. Therefore, finally, we will construct the increasing sequence of basis

$$\mathcal{B}_1 \subseteq \mathcal{B}_2 \subseteq \dots \subseteq \mathcal{B}_k \subseteq \dots .$$

Since the ring $\mathbb{Z}_2\mathcal{F}/\langle \mathcal{T}'\rangle$ is a union of the subspaces $(\langle U_1, \dots, U_k\rangle_d + \langle \mathcal{T}'\rangle)/\langle \mathcal{T}'\rangle$, the union $\bigcup_{k=1}^{\infty} \mathcal{B}_k$ is a basis of $\mathbb{Z}_2\mathcal{F}/\langle \mathcal{T}'\rangle = \mathbb{Z}_2\mathcal{F}/\mathcal{I}$. □

From Theorem 5.1 we obtain the claim that served as the main motivation.

COROLLARY 5.7. *The quotient ring $\mathbb{Z}_2\mathcal{F}/\mathcal{I}$ is non-trivial.*

REMARK 5.1. It is possible to choose a basis in every direct summand $V_i/(\mathrm{Dp}(V_i) + \mathrm{L}(V_i))$, where $V_i = \langle Z_i\rangle_d$, Z_i is a λ-semicanonical monomial, in the following way. In Proposition 4.10, we have proved that $V_i/(\mathrm{Dp}(V_i) + \mathrm{L}(V_i))$ is isomorphic to a tensor product of spaces that corresponds to each virtual member of the chart of Z_i. Namely, there exist spaces $A_1, \dots, A_{k_i}$ defined by (32) and the isomorphism

$$\begin{aligned}\overline{\mu}_2[Z_i] :& A_1/(\mathrm{Dp}(A_1) + \mathrm{L}(A_1)) \otimes \dots \otimes A_{k_i}/(\mathrm{Dp}(A_{k_i}) + \mathrm{L}(A_{k_i})) \\ &\to V_i/(\mathrm{Dp}(V_i) + \mathrm{L}(V_i))\end{aligned}$$

defined by (36). So, if we choose a basis in all the different spaces $A_j/(\mathrm{Dp}(A_j) + \mathrm{L}(A_j))$, this enables us to construct the corresponding basis in every direct summand $V_i/(\mathrm{Dp}(V_i) + \mathrm{L}(V_i))$.

From Proposition 5.2 it follows that $A_j/(\mathrm{Dp}(A_j)+\mathrm{L}(A_j))$ is generated by cosets of λ-semicanonical generalized fractional powers that belong to A_j. We choose a basis of $A_j/(\mathrm{Dp}(A_j) + \mathrm{L}(A_j))$ that consists of such cosets. Let

$$(b_l^{(1)} + \mathrm{Dp}(A_1) + \mathrm{L}(A_1)) \otimes \dots \otimes (b_l^{(k_i)} + \mathrm{Dp}(A_{k_i}) + \mathrm{L}(A_{k_i})), l \in \mathbb{N},$$

be the basis elements, where $b_l^{(1)}, \dots, b_l^{(k_i)}$ are λ-semicanonical generalized fractional powers. Then their images

$$\begin{aligned}\overline{\mu}_2[Z_i](b_l^{(1)} &+ \mathrm{Dp}(A_1) + \mathrm{L}(A_1)) \otimes \dots \otimes (b_l^{(k_i)} + \mathrm{Dp}(A_{k_i}) + \mathrm{L}(A_{k_i})) \\ &= \mu[Z_i](b_l^{(1)} \otimes \dots \otimes b_l^{(k_i)}) + \mathrm{Dp}(V_i) + \mathrm{L}(V_i)\end{aligned}$$

form a basis of $V_i/(\mathrm{Dp}(V_i + \mathrm{L}(V_i))$.

We put

$$W_l^{(i)} = \mu[Z_i](b_l^{(1)} \otimes \ldots \otimes b_l^{(k_i)}). \tag{49}$$

Then from Theorem 5.1 it follows that

$$\bigcup_{i=1}^{\infty} \{W_l^{(i)} + \mathcal{I} \mid l \in \mathbb{N}\} \tag{50}$$

is a basis of $\mathbb{Z}_2\mathcal{F}/\mathcal{I}$. Notice that, by definition of $\mu[Z_i]$, $W_l^{(i)}$ are monomials. From the description given in Section 3.1, it follows that the monomials $W_l^{(i)}$ are $(\lambda+2\varepsilon)$-semicanonical.

Notice that the constructed basis satisfies the following property. Assume U is a monomial and there is the base decomposition

$$U + \mathcal{I} = \sum_{i,l} W_l^{(i)} + \mathcal{I}.$$

Then every $W_l^{(i)}$ in this decomposition is a derived monomial of U, in particular, $f(W_l^{(i)}) \leqslant f(U)$.

6. Geometry of the multiplication

The linear span of all λ-semicanonical words is denoted by S_λ. However, it turns out that that it is more convenient to define a subspace $\widetilde{S}_\lambda$ for which $S_\lambda \subseteq \widetilde{S}_\lambda \subseteq S_{\lambda+2\varepsilon}$ that contains all the elements of the basis of $\mathbb{Z}_2\mathcal{F}/\mathcal{I}$. The cosets of elements of $\widetilde{S}_\lambda$ linearly generate $\mathbb{Z}_2\mathcal{F}/\mathcal{I}$, and they display nice multiplication properties. Namely, the result of the multiplication can be represented as a sum of monomials from $\widetilde{S}_\lambda$ that form thin triangles with the factors.

6.1. Definition of the space $\widetilde{S}_\lambda$. Let U be a $(\lambda + 2\varepsilon)$-semicanonical word and a be a virtual member of its chart. Assume l and r are virtual members of the chart of U such that l and r are the left and the right neighbour of a in the chart, respectively. Let a_l be an overlap between a and l (possibly empty), a_r be an overlap between a and r (possibly empty). Then $a = a_l a_m a_r$.

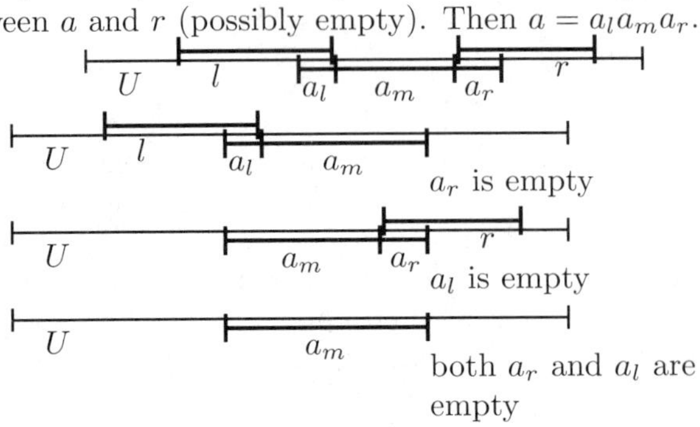

Definition 6.1 ($\widetilde{S}_\lambda$-condition). We say that a virtual member of the chart a *satisfies $\widetilde{S}_\lambda$-condition* if the subword a_m of a defined above is λ-semicanonical.

We denote by $\widetilde{S}_\lambda$ a linear span of words such that each virtual member of the chart of a word satisfies the $\widetilde{S}_\lambda$-condition. Clearly, every monomial that belongs to $\widetilde{S}_\lambda$ is $(\lambda + 2\varepsilon)$-semicanonical.

Notice that unlike words from S_λ, words from $\widetilde{S}_\lambda$ may contain subwords that do not belong to $\widetilde{S}_\lambda$.

Recall that in the previous section we constructed a linear basis in $\mathbb{Z}_2\mathcal{F}/\mathcal{I}$, see (50). By formula (49) (since $b_l^{(1)}, \dots, b_l^{(k_i)}$ are λ-semicanonical), we obtain that the monomials $W_l^{(i)}$ (the fixed representatives of the basis elements) belong to $\widetilde{S}_\lambda$.

Assume a monomial $U_h = La_hR \in \widetilde{S}_\lambda$ and let a_h be a virtual member of its chart. Let $U_h \mapsto \sum_{\substack{j=1\\ j\neq h}}^m U_j$ be a multi-turn that comes from an elementary multi-turn $a_h \mapsto \sum_{\substack{j=1\\ j\neq h}}^m a_j$ such that a_j are λ-semicanonical generalized fractional powers. Consider monomials $U_j = La_jR$ such that a_j is a virtual member of the chart of U_j. Then, by Corollary 3.7, we obtain that the images of all virtual members of the chart of U are virtual members of the chart of U_j. Hence, by the definition of the space $\widetilde{S}_\lambda$, we immediately obtain that these monomials belong to $\widetilde{S}_\lambda$ even if the neighbours of a_h are prolonged in U_j.

Assume a monomial $U_h = La_hR$ is a λ-semicanonical word and we perform a multi-turn of a_h as above. If again we consider monomials $U_j = La_jR$ such that a_j is a virtual member of the chart of U_j, then U_j is not necessarily λ-semicanonical, because the neighbours of a_h may prolong in U_j and become λ-forbidden. That is why the space $\widetilde{S}_\lambda$ is more convenient in our further proofs.

6.2. Behaviour of $\widetilde{S}_\lambda$ with respect to multi-turns. As above, assume $U_h = La_hR \in \widetilde{S}_\lambda$, a_h is a virtual member of its chart, $U_h \mapsto \sum_{\substack{j=1\\ j\neq h}}^m U_j$ is a multi-turn that comes from an elementary multi-turn $a_h \mapsto \sum_{\substack{j=1\\ j\neq h}}^m a_j$ such that a_j are λ-semicanonical generalized fractional powers. Let us describe in detail monomials U_j such that a_j is not a virtual member of the chart of U_j. We split these monomials into three groups in the same way as we did in Section 3.1.

(1) La_jR, where $\Lambda(a_j) > \varepsilon$;
(2) La_jR, where $a_j = 1$;
(3) La_jR, where $\Lambda(a_j) \leqslant \varepsilon$, $a_j \neq 1$.

Being a λ-semicanonical word depends only on the Λ-measure of every virtual member of the chart. Hence, a resulting monomial of a multi-turn of a λ-semicanonical word may become non-λ-semicanonical only if the Λ-measure of some virtual members of the chart increases and λ-forbidden subwords appear in this monomial. One can see that the $\widetilde{S}_\lambda$-condition depends on a virtual member of the chart itself and on its neighbours and the notion of virtual member of the chart depends on the whole monomial. Hence, the $\widetilde{S}_\lambda$-condition on a virtual member of the chart may not be hold in a resulting monomial of a multi-turn not only when the virtual member of the chart changes itself, but when the virtual member of the chart stay unchanged but some other maximal occurrence becomes shorter. Let us study the possibilities in detail.

Consider monomials of type 1. Let us use notations from Section 3.1. We assume that b_h is the left neighbour of a_h in the chart of U_h, b_j is a maximal occurrence of a generalized fractional power corresponding to b_h in the chart of U_j.

(1) If b_h and a_h are separated or touch at a point and $\Lambda(b_h) < \tau$, then b_j may become not a virtual member of the chart and the $\widetilde{S}_\lambda$-condition may not hold for its left neighbour.

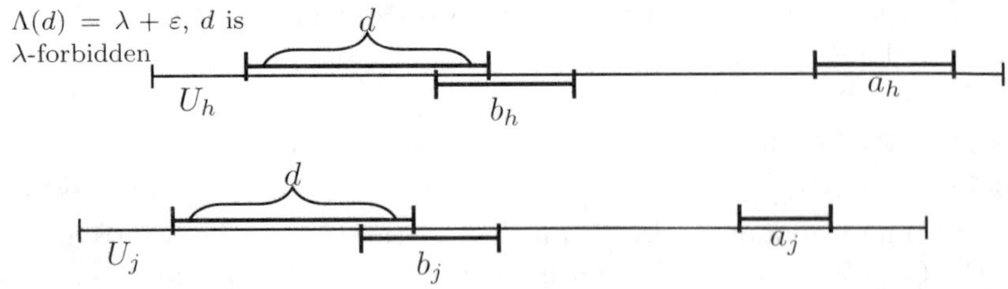

(2) If we have configurations 1.3 or 1.4 (page 13) for b_j, then b_j may become not a virtual member of the chart and the $\widetilde{S}_\lambda$-condition may not hold for its left neighbour.

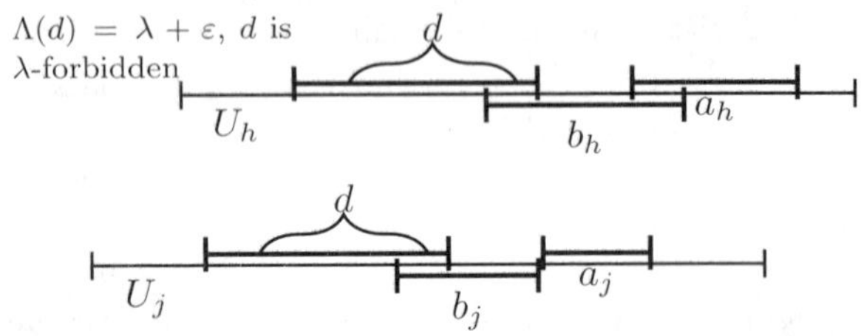

(3) Suppose we have configurations 1.1, 1.2 (page 12) or $b_j = b_h$. If b_j contains a λ-forbidden subword d that has an overlap with a_j, then the $\widetilde{S}_\lambda$-condition may not hold for b_j (because a_j is not a virtual member of the chart).

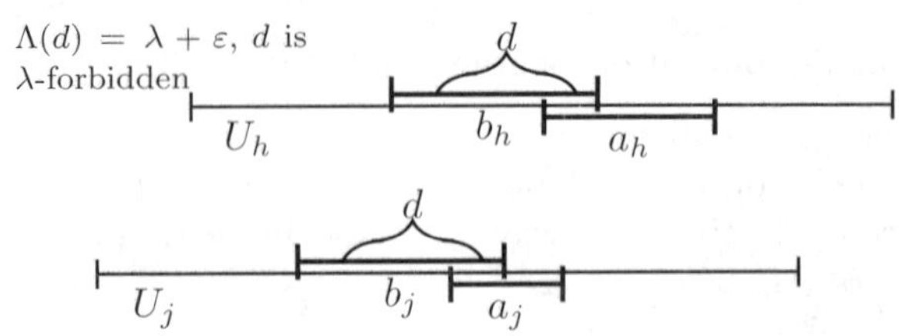

Obviously, the effects for the right neighbour of a_h in the chart are the same.

Consider monomials of type 2. First we do cancellations in $U_j = LR$ if possible, after that we obtain $U'_j = L'R'$. Recall that we suppose that a is a maximal occurrence of a generalized fractional power that is a terminal subword of L', b is a maximal occurrence of a generalized fractional power that is an initial subword of R', a' and b' are generalized fractional powers that prolong a and b in $L'R'$ respectively.

(1) In any configuration except 2.1, there may be a maximal occurrence b_h, $\Lambda(b_h) < \tau$, fully contained in the subword L' that becomes not a virtual member of the chart of $L'R'$. Then the $\widetilde{S}_\lambda$-condition may not hold for its left neighbour.

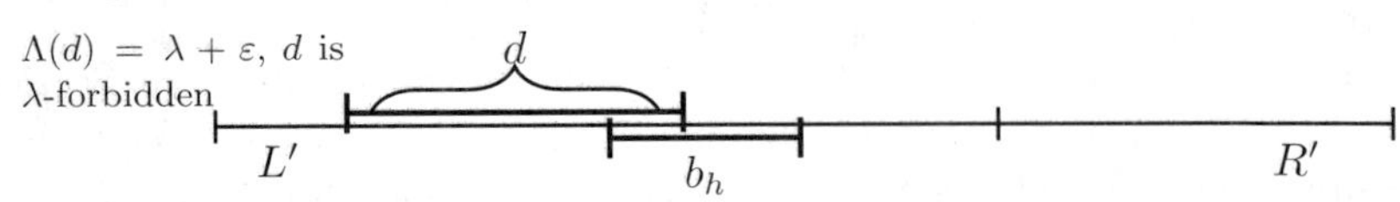

The same may happen in the subword R'.

(2) Suppose we have configuration 2.2 (page 14). Then a' or b' may contain a λ-forbidden subword and do not satisfy the $\widetilde{S}_\lambda$-condition. We may obtain

this situation if the right neighbour of a' in the chart of L cancels out up to the end of a' or if the left neighbour of b' in the chart of R cancels out up to the beginning of b'.

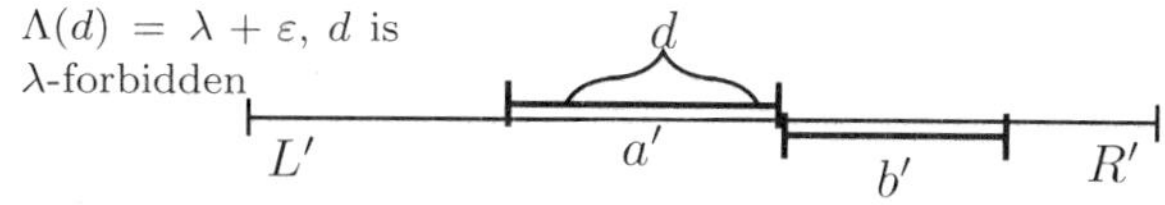

(3) Suppose we have configuration 2.2 or 2.3 (page 14). In this case, a' may become too short after cancellations and may not be counted as a virtual member of the chart. If the left neighbour of a_j contains a λ-forbidden subword d that has an overlap with a', then the word $L'R'$ may not belong to $\widetilde{S}_\lambda$.

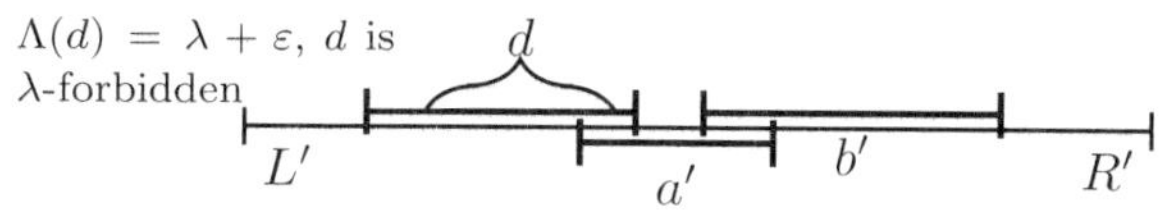

We may obtain a similar effect for the right neighbour of b'.

(4) If we have configuration 2.3 (page 14), then a' or b' may contain a λ-forbidden subword that does not satisfy the $\widetilde{S}_\lambda$-condition. For example, we may obtain this case when b' is not a virtual member of the chart and a' contains a λ-forbidden subword d that has an overlap with b'.

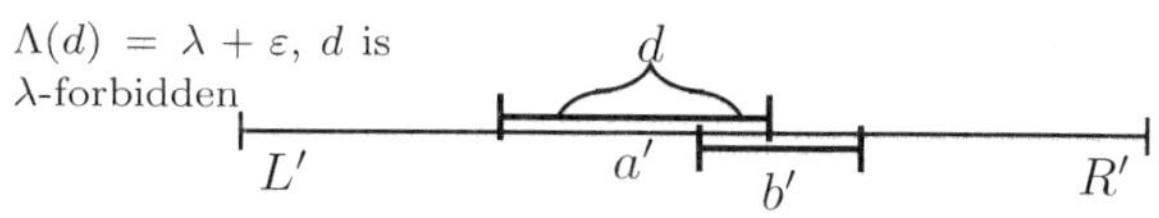

(5) Suppose we have configuration 2.4 (page 14) and ab is not a virtual member of the chart. If the left or the right neighbour of ab contains a λ-forbidden subword d that has an overlap with ab, then the word $L'R'$ may not belong to $\widetilde{S}_\lambda$ (similar to 3).

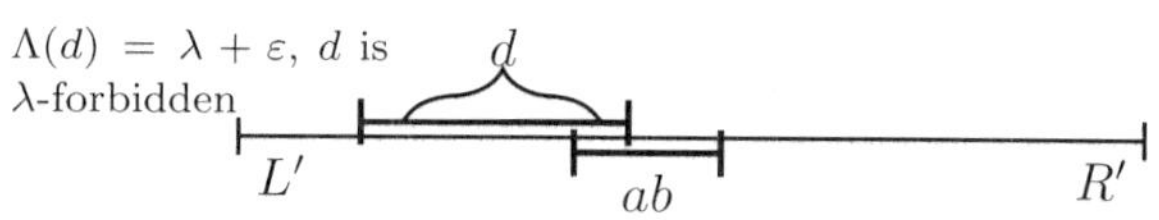

(6) If we have configuration 2.4 (page 14), the $\widetilde{S}_\lambda$-condition may not hold for ab.

Consider monomials of type 3. Recall that we suppose that a is a maximal occurrence of a generalized fractional power that is a terminal subword of L, b is a maximal occurrence of a generalized fractional power that is an initial subword of R, a' and b' are generalized fractional powers that prolong a and b in La_jR, respectively.

(1) In any configuration except 3.1 there may be a maximal occurrence b_h, $\Lambda(b_h) < \tau$, fully contained in the subword L that becomes not a virtual member of the chart of U_j. Then the $\widetilde{S}_\lambda$-condition may not hold for its left neighbour.

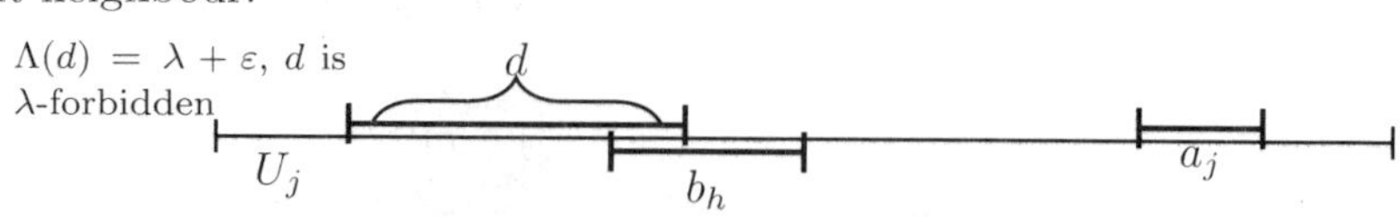

The same may happen in the subword R.

(2) If we obtain configuration 3.2 (page 15), then a' may be too short to be counted as a virtual member of the chart. In this case, the $\widetilde{S}_\lambda$-condition may not hold for its left neighbour.

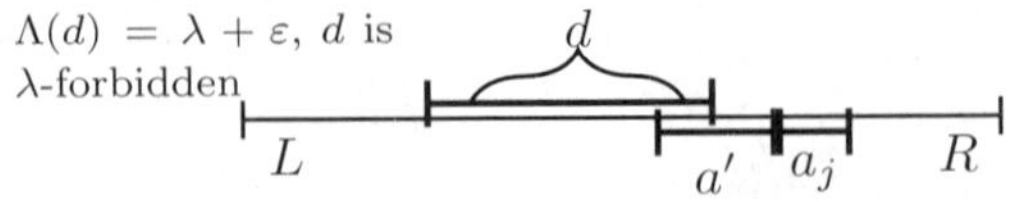

Clearly, we may obtain a similar effect for the right neighbour of b'.

(3) Suppose we have configuration 3.3 or 3.4 (page 15). Then a' or b' may contain a λ-forbidden subword d that does not satisfy the $\widetilde{S}_\lambda$-condition.

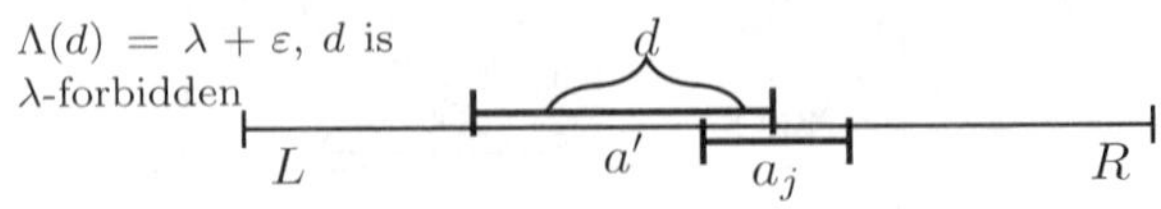

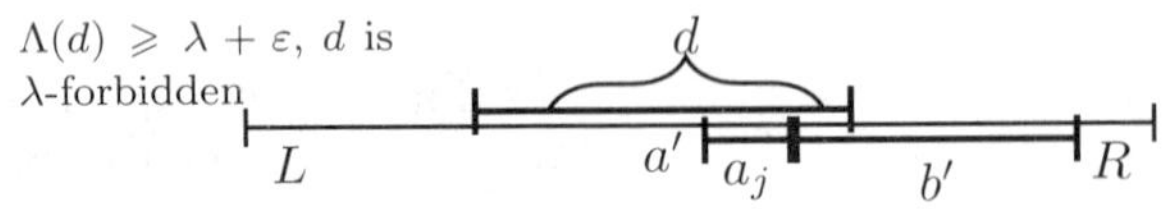

(4) If we obtain configuration 3.5 (page 16), the $\widetilde{S}_\lambda$-condition may not hold for the virtual member aa_jb.

6.3. Multiplication of words from $\widetilde{S}_\lambda$. In Proposition 5.1 we showed one possible way of transformation of an arbitrary generalized fractional power into a sum of λ-semicanonical generalized fractional powers. But some of the obtained elements may have Λ-measure less than the given threshold τ. In the further argument in Section 6.3, it is more convenient to deal with multi-turns such that all the resulting λ-semicanonical generalized fractional powers are of Λ-measure not less than τ. In the following proposition, we construct an elementary multi-turn with this property.

PROPOSITION 6.1. *Suppose a_h is a generalized fractional power. Then there exists an elementary multi-turn $a_h \mapsto \sum_{\substack{j=1 \\ j\neq h}}^{k} a_j$ such that a_j are λ-semicanonical generalized fractional powers and $\Lambda(a_j) \geqslant \tau$, $j \neq h$, i.e., a_j is always counted as a virtual member of the chart (possibly with greater word length and Λ-measure than a_h).*

Moreover, one can choose this multi-turn such that every a_j, $j \neq h$, does not contain subwords of v^{-1} of Λ-measure greater than ε.

PROOF. Let us construct this elementary multi-turn directly using Definition 2.4. If we consider v as an element of the field $\mathbb{Z}_2(w)$ such that $v^{-1} = 1 + w$, then in $\mathbb{Z}_2(w)$ we have $1 = vw + v$ and

$$v^{-1} = 1 + w = (vw + v) + (vw + v)w = v + vw^2. \tag{51}$$

Therefore, in $\mathbb{Z}_2(w)$ we have $v^{-m} = (v+vw^2)^m$. Hence, the non-commutative Laurent polynomial $x_1^{-m}+(x_1+x_1x_2^2)^m$, $m > 0$, satisfies (8). Hence, the transformation in $\mathbb{Z}_2\mathcal{F}$

$$v^{-m} \mapsto (v + vw^2)^m, m > 0, \tag{52}$$

is an elementary multi-turn.

Suppose $M_h(v,w)$ is a non-commutative monomial in v,w, that is,

$$M_h(v,w) = w^{k_1}v^{l_1}\cdots w^{k_n}v^{l_n},$$

where $k_i, l_i \in \mathbb{Z}$. Again consider $M_h(v,w)$ as an element of the field $\mathbb{Z}_2(w)$ such that $v^{-1} = 1+w$. Then, since v and w commute in $\mathbb{Z}_2(w)$, we have $w^{k_1}v^{l_1}\cdots w^{k_n}v^{l_n} = w^kv^l$, where $k = \sum_{i=1}^n k_i$, $l = \sum_{i=1}^n l_i$. Hence, the non-commutative Laurent polynomial $x_2^{k_1}x_1^{l_1}\cdots x_2^{k_n}x_1^{l_n} + x_2^kx_1^l$ satisfies (8). Therefore, the transformation in $\mathbb{Z}_2\mathcal{F}$

$$w^{k_1}v^{l_1}\cdots w^{k_n}v^{l_n} \mapsto w^kv^l, k = \sum_{i=1}^n k_i, l = \sum_{i=1}^n l_i, \tag{53}$$

is an elementary multi-turn. If $l = \sum_{i=1}^n l_i < 0$, from (52) it follows that

$$w^{k_1}v^{l_1}\cdots w^{k_n}v^{l_n} \mapsto w^k(v+vw^2)^{-l} \tag{54}$$

is an elementary multi-turn. Since $1 \mapsto v + vw$ is an elementary multi-turn, if $l = \sum_{i=1}^n l_i = 0$, we obtain that

$$w^{k_1}v^{l_1}\cdots w^{k_n}v^{l_n} \mapsto w^k(v+vw) \tag{55}$$

is an elementary multi-turn. Using (53)–(55), for every $l = \sum_{i=1}^n l_i$ we obtain a multi-turn

$$M_h(v,w) \mapsto \sum_{\substack{j=1\\ j\neq h}}^{s} M_j(v,w), \tag{56}$$

where every $M_j(v,w)$, $j \neq h$, does not contain negative powers of v and contains at least one positive power of v.

By definition, a generalized fractional power is a subword in a non-commutative monomial over the words v and w. Recall that the types of generalized fractional powers are enumerated in (5) – (7). Clearly, if we consider generalized fractional powers with the possibility of cancellations, we obtain that a_h before cancellations has one of the following forms:

$$\begin{aligned}
&v_fM_h(v,w)v_i,\\
&v_mv_fM_h(v,w)v_iv_m,\\
&w_fM_h(v,w)v_i,\\
&v_fM_h(v,w)w_i,\\
&w_fM_h(v,w)w_i,\\
&w_mw_fM_h(v,w)w_iw_m,
\end{aligned}$$

where $M_h(v,w)$ is a monomial over the words v and w.

Suppose $a_h = v_fM_h(v,w)v_i$, where $v_fM(v,w)_hv_i$ may have cancellations. From (56) it follows that the transformation $v_fM(v,w)_hv_i \mapsto \sum_{\substack{j=1\\ j\neq h}}^{s} v_fM_j(v,w)v_i$ is an elementary multi-turn after all possible cancellations in monomials are done. That is,

$$a_h \mapsto \sum_{\substack{j=1\\ j\neq h}}^{s} v_fM_j(v,w)v_i \tag{57}$$

is an elementary multi-turn.

Let us check that (57) satisfies the conditions of Proposition 6.1. Since every $M_j(v,w)$, $j \neq h$, contains only positive powers of v and v has no cancellations with w and w^{-1}, every monomial in the right hand side of this multi-turn is of Λ-measure not less than 1, that is, it is especially of Λ-measure not less than τ. Since w^k can not contain subwords of v and v^{-1} of Λ-measure greater than ε, every monomial in the right hand side does not contain subwords of v^{-1} of Λ-measure greater than ε. Thus, multi-turn (57) satisfies the conditions of Proposition 6.1.

In a similar way we deal with a_h of the other form and we obtain multi-turns

$$
\begin{aligned}
&a_h \mapsto \sum_{\substack{j=1\\ j\neq h}}^{s} v_m v_f M_j(v,w) v_i v_m, \; a_h \mapsto \sum_{\substack{j=1\\ j\neq h}}^{s} w_f M_j(v,w) v_i,\\
&a_h \mapsto \sum_{\substack{j=1\\ j\neq h}}^{s} v_f M_j(v,w) w_i, \; a_h \mapsto \sum_{\substack{j=1\\ j\neq h}}^{s} w_f M_j(v,w) w_i,\\
&a_h \mapsto \sum_{\substack{j=1\\ j\neq h}}^{s} w_m w_f M_j(v,w) w_i w_m,
\end{aligned} \tag{58}
$$

possibly after cancellations in the right hand sides. Notice that only subwords of w may be involved in cancellations. By the same argument as (57), it is proved that multi-turns (58) satisfy the conditions of Proposition 6.1. □

Let U_1 and U_2 be monomials from $\widetilde{S}_\lambda$. Then their product U_1U_2 may not belong to $\widetilde{S}_\lambda$. According to Corollary 5.2, every monomial is equal to a sum of words from S_λ modulo $\mathcal{I}$. Moreover, every monomial is equal to a sum of words from $\widetilde{S}_\lambda$ modulo $\mathcal{I}$, since $S_\lambda \subseteq \widetilde{S}_\lambda$. We will represent U_1U_2 as a sum of monomials from $\widetilde{S}_\lambda$. In this section, we analyse the process of reduction of the word U_1U_2 to a sum of words from $\widetilde{S}_\lambda$ modulo $\mathcal{I}$ in detail. In parallel we will describe the multiplication diagram that encrypts the history of this process.

The process of the reduction of U_1U_2 is as follows.

Step 1 Cancel U_1U_2 if possible. The multiplication diagram that encrypts cancellations is of the form

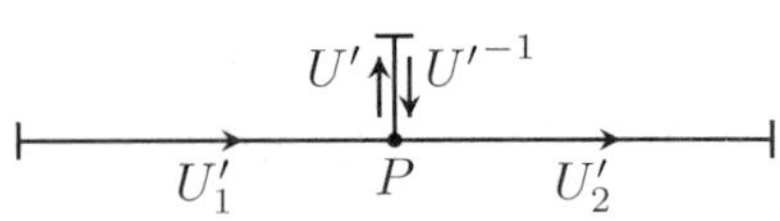

$U_1 = U_1'U'$

$U_2 = U'^{-1}U_2'$

$U = U_1'U_2'$ has no further cancellations

Let $U = U_1'U_2'$. Now we consider the chart of U. As we described in Section 3.1, we have one of configurations 2.1–2.4 (pages 14–14). If we have configuration 2.4 and the generalized fractional power that is obtained as a result of merging does not satisfy $\widetilde{S}_\lambda$-condition, then we go to Step 2 of the reduction process. So, let us first describe this case.

Denote the generalized fractional power that is obtained as a result of merging by a_h, $U = La_hR$.

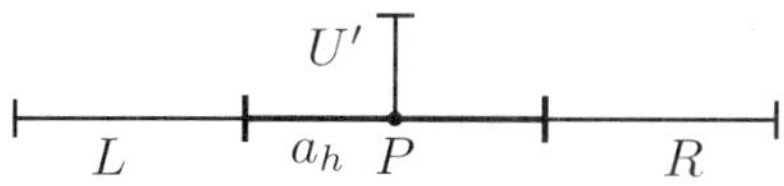

Suppose a_h does not satisfy the $\widetilde{S}_\lambda$-condition. Then the next step of the multiplication process is as follows.

Step 2 According to Proposition 6.1, a λ-forbidden virtual member a_h can be reduced to a sum of λ-semicanonical generalized fractional powers using an appropriate elementary multi-turn $a_h \mapsto \sum_{\substack{j=1\\ j\neq h}}^k a_j$, $\Lambda(a_j) \geqslant \tau$. So, perform the multi-turn $U = La_hR \mapsto \sum_{\substack{j=1\\ j\neq h}}^k La_jR$, which comes from this elementary multi-turn.

Since $\Lambda(a_j) \geqslant \tau$, the resulting monomials La_jR belong to $\widetilde{S}_\lambda$.

The monomial a_h corresponds to a unique path in the v-diagram (see Definition 2.3 is Section 2) with the initial point I and the final point F. The monomials a_j, $j = 1, \dots, k$, correspond to certain paths in this v-diagram with the same initial point I and the same final point F. Then we add this v-diagram to the bottom of the multiplication diagram, gluing the point I with the end of L and point F with the beginning of R. The previous part of the multiplication diagram is glued to the added v-diagram at the same point on a_h as before.

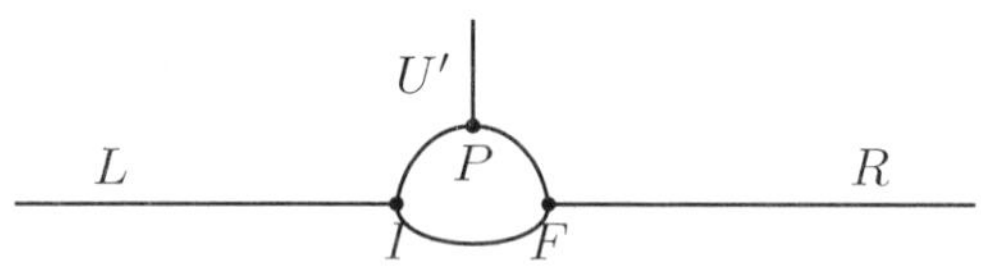

So, one can read the monomials La_jR, $j = 1, \dots, k$, in the low level of the obtained diagram. So far for Step 2.

Now consider configurations 2.1–2.3 in $U = U_1'U_2'$ after Step 1. Then the monomial U also may not belong to $\widetilde{S}_\lambda$. According to the classification given in Section 6.2, U contains at most two subwords that do not satisfy the $\widetilde{S}_\lambda$-condition, one comes from the chart of U_1 and the other comes from the chart of U_2 (see analysis of monomials of type 2, page 68; although here we deal with a product of two words from $\widetilde{S}_\lambda$, this analysis is still applicable).

First assume that U contains only one virtual member of the chart that does not satisfy the $\widetilde{S}_\lambda$-condition. To be precise, assume that it comes from the chart of U_1, denote it by b_h. Then there are the following possible configurations depending on the position b_h with respect to the point P:

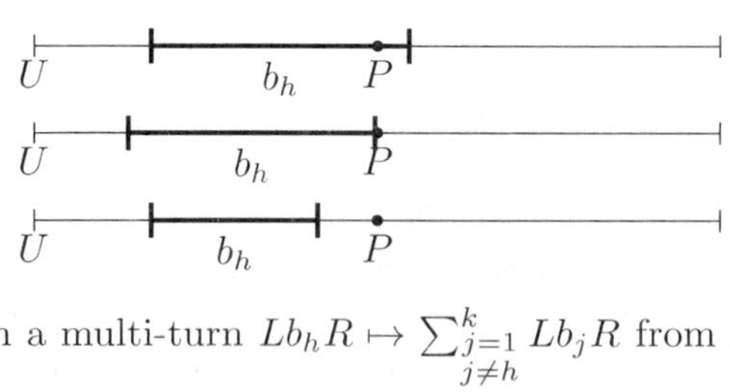

As above we perform a multi-turn $Lb_hR \mapsto \sum_{\substack{j=1\\ j\neq h}}^k Lb_jR$ from Proposition 6.1 that reduces b_h. Since $\Lambda(b_j) \geqslant \tau$, the resulting monomials Lb_jR belong to $\widetilde{S}_\lambda$.

We add a new v-diagram to the multiplication diagram as we described above and obtain the final multiplication diagram.

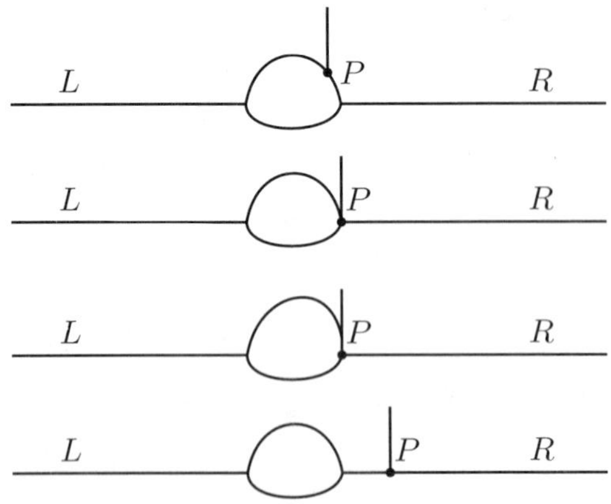

Now assume that U contains two virtual members of the chart that do not satisfy the $\widetilde{S}_\lambda$-condition, denote them by b_h and c_h. Then there are the following possible configurations depending on the position of b_h and c_h with respect to the point P.

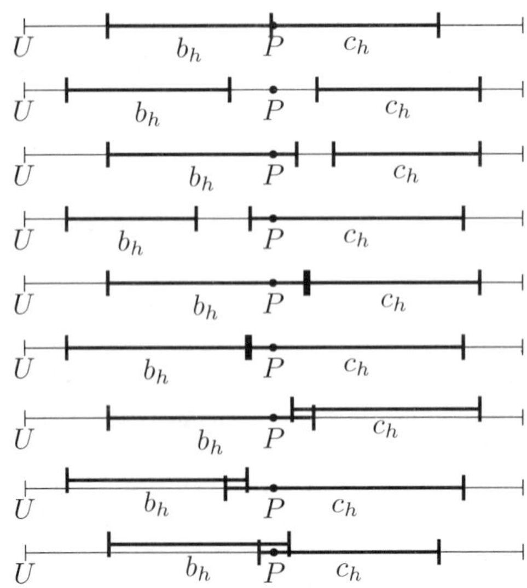

Then first, as above, we perform a multi-turn $Lb_hR \mapsto \sum_{\substack{j=1 \\ j\neq h}}^{k} Lb_jR$ from Proposition 6.1 that reduces b_h. Since $\Lambda(b_j) \geqslant \tau$, in every monomial Lb_jR the image of c_h is the only virtual member of the chart that does not satisfy the $\widetilde{S}_\lambda$-condition. Denote the image of c_h in the monomial Lb_jR by $c_h^{(j)}$. So, in every monomial Lb_jR we perform a multi-turn from Proposition 6.1 that reduces $c_h^{(j)}$. As a result, we obtain a sum of monomials from $\widetilde{S}_\lambda$.

After the reduction of $c_h^{(j)}$, we glue a new v-diagram to the multiplication diagram obtained after the reduction of b_h and obtain the final multiplication diagram. Let us illustrate the first three configurations of b_h and c_h. The multiplication diagrams for the remain configurations are constructed similarly.

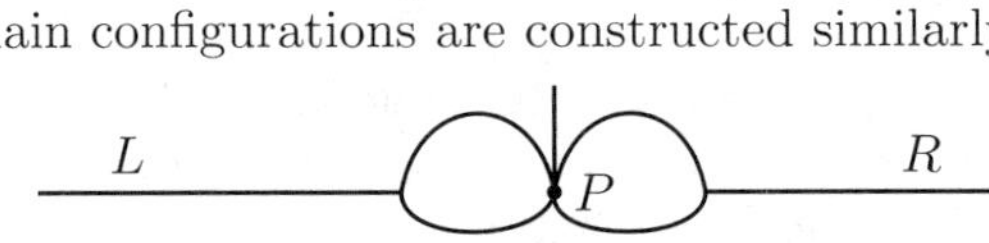

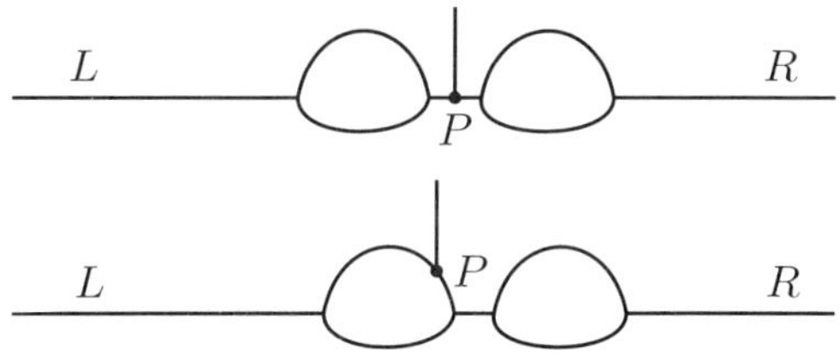

So, finally we obtain the multiplication diagram as a thin triangle. The result of the multiplication is a sum of words from $\widetilde{S}_\lambda$ that correspond to certain paths in the low level of the obtained thin triangle.

Thereby, we proved the following theorem.

THEOREM 6.1. *Assume $U_1+\mathcal{I}, U_2+\mathcal{I} \in \mathbb{Z}_2\mathcal{F}/\mathcal{I}$, where U_1 and U_2 are monomials from $\widetilde{S}_\lambda$. Then there exist monomials $Z_1, \dots, Z_k$ from $\widetilde{S}_\lambda$ such that $U_1U_2+\mathcal{I} = \sum_{i=1}^k Z_i + \mathcal{I}$, and U_1, U_2 and Z_i form a thin triangle consists of v-diagrams for every $i = 1, \dots, k$.*

REMARK 6.1. If we use not an elementary multi-turn from Proposition 6.1, but rather an arbitrary elementary multi-turn $a_h \mapsto \sum_{\substack{j=1 \\ j\neq h}}^k a_j$ in order to reduce a_h at Step 2, some a_j may be of Λ-measure $\leqslant \varepsilon$. In this case, we may obtain further merging of virtual members of the chart of La_jR. Namely, the cases 2.4 and 3.5 (pages 14, 16), in which the neighbours of a_j in the chart merge, are possible. For example, for 3.5 we have

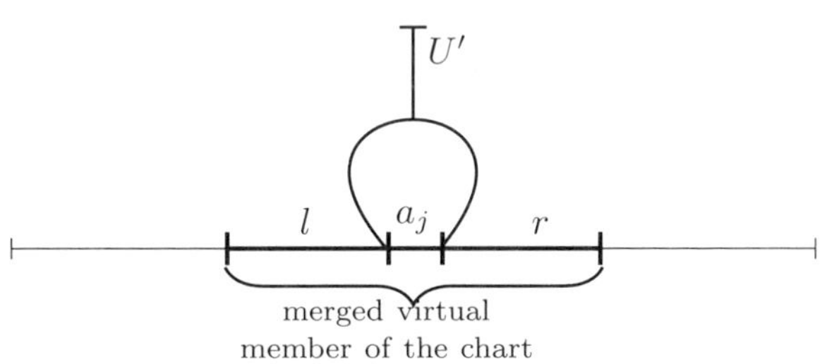

If there exists $a_j = 1$, we may obtain cancellations in the resulting monomial that contain $a_j = 1$.

So, when we reduce U_1U_2, we obtain a sequence of Steps 1 and 2. We can think of it as of a branching process, starting from the monomial U_1U_2. After each execution of Step 1 we obtain one monomial, after each execution of Step 2 we obtain a sum of monomials. Each monomial in the sum is treated independently and potentially gives us a new branch of the process. Since we may obtain merging of virtual members of the chart or cancellations at most in one resulting monomial, the process stops in all branches except at most one after each step. From Proposition 4.4, it follows that finally the process stops in all branches.

For every branch of the process, we can construct a multiplication diagram in the same way as above. If we do an arbitrary multi-turn in order to reduce the monomials obtained after the end of the sequence of Steps 1 and 2 in a branch, we may obtain a multiplication diagram that is not a thin triangle, but has a form of tree. Nevertheless, the reduction of these monomials can be done using multi-turns described in Proposition 6.1, and in this case the multiplication diagram for every branch of the process again is a thin triangle but of more complicated form

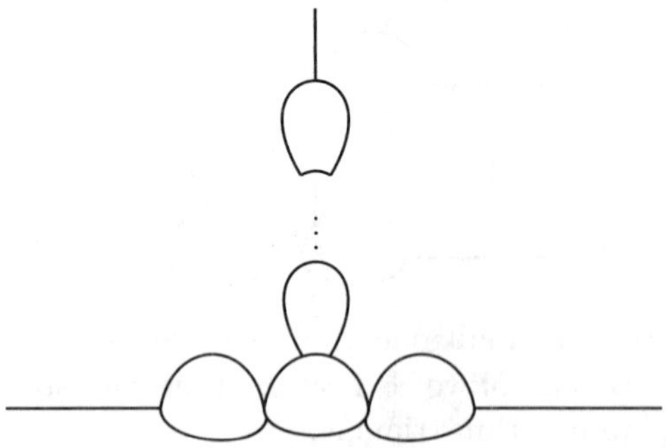

So, we obtain the multiplication diagram of the whole multiplication process as a set of thin triangles. The result of the multiplication is a sum of words from $\widetilde{S}_\lambda$ that correspond to certain paths in the low levels of the obtained thin triangles.

References

[1] G. M. Bergman, *The diamond lemma for ring theory*, Adv. in Math. **29** (1978), no. 2, 178–218, DOI 10.1016/0001-8708(78)90010-5. MR506890

[2] M. R. Bridson and A. Haefliger, *Metric spaces of non-positive curvature*, Grundlehren der Mathematischen Wissenschaften [Fundamental Principles of Mathematical Sciences], vol. 319, Springer-Verlag, Berlin, 1999. MR1744486

[3] M. Coornaert, T. Delzant, and A. Papadopoulos, *Géométrie et théorie des groupes: Les groupes hyperboliques de Gromov. [Gromov hyperbolic groups]* (French), Lecture Notes in Mathematics, vol. 1441, Springer-Verlag, Berlin, 1990. MR1075994

[4] É. Ghys and P. de la Harpe, *Espaces métriques hyperboliques* (French), Sur les groupes hyperboliques d'après Mikhael Gromov (Bern, 1988), Progr. Math., vol. 83, Birkhäuser Boston, Boston, MA, 1990, pp. 27–45, DOI 10.1007/978-1-4684-9167-8_2. MR1086650

[5] M. Gromov, *Hyperbolic groups*, Essays in group theory, Math. Sci. Res. Inst. Publ., vol. 8, Springer, New York, 1987, pp. 75–263, DOI 10.1007/978-1-4613-9586-7_3. MR919829

[6] S. Lang, *Algebra*, Springer-Verlag New York, 2002.

[7] R. C. Lyndon and P. E. Schupp, *Combinatorial group theory*, Springer-Verlag, Berlin-New York, 1977. MR0577064

[8] A. Yu. Ol′shanskiĭ, *Geometry of defining relations in groups*, Mathematics and its Applications (Soviet Series), vol. 70, Kluwer Academic Publishers Group, Dordrecht, 1991. Translated from the 1989 Russian original by Yu. A. Bakhturin. MR1191619

[9] E. Rips and Z. Sela, *Canonical representatives and equations in hyperbolic groups*, Invent. Math. **120** (1995), no. 3, 489–512, DOI 10.1007/BF01241140. MR1334482

Department of Mathematics, Bar-Ilan University, 5290002 Ramat Gan, Israel
Email address: atkarskaya.agatha@gmail.com

Department of Mathematics, Bar-Ilan University, 5290002 Ramat Gan, Israel
Email address: beloval@macs.biu.ac.il

Department of Mathematics, Bar-Ilan University, 5290002 Ramat Gan, Israel
Email address: plotkin.evgeny@gmail.com

Department of Mathematics, The Hebrew University of Jerusalem, Givat Ram, 9190401 Jerusalem, Israel
Email address: eliyahu.rips@mail.huji.ac.il

Contemporary Mathematics
Volume **726**, 2019
https://doi.org/10.1090/conm/726/14606

Geometric view on homogeneous groups

E. Aladova

ABSTRACT. In our work, we study algebraic structures within the framework of Logical Geometry. The main aim of Logical Geometry is to consider the interaction between the algebraic structure of an algebra and its geometrical and logical properties. This theory uses the ideas and considers the problems related to Universal Algebra, Model Theory and Algebraic Geometry.

The present paper is focused on homogeneous groups. We will deal with three notions of homogeneous groups: *algebraically homogeneous groups, logically homogeneous groups and weakly algebraically homogeneous groups.* We survey recent results in this area and consider some open problems.

1. Introduction

In our work, we study algebraic structures within the framework of Logical Geometry. This theory arose in the papers of B.I. Plotkin (see, for instance, [**21**], [**22**], [**23**], [**24**]). The main aim of the theory is to consider the interaction between algebraic structure of an algebra and its geometrical and logical properties. This theory uses the ideas and treats the problems related to Universal Algebra, Model Theory and Algebraic Geometry.

Homogeneity is one of the working notions in Universal Algebra and Model Theory. The present paper is focused on various views on homogeneity in groups. We will deal with the following three notions of homogeneous groups: *algebraically homogeneous groups, logically homogeneous groups and weakly algebraically homogeneous groups.*

Recall that a group G is *algebraically homogeneous* if every isomorphism between two of its finitely generated subgroups can be extended up to an automorphism of G. The definition of logical homogeneity is based on the same idea as the notion of algebraic homogeneity; the precise definition will be given in Section 4. Even though these notions are similar in nature, they also have many differences.

Weakly algebraically homogeneous groups (for the definition see Section 5) lie between algebraically homogeneous and logically homogeneous groups. For example, finitely generated free abelian groups are not algebraically homogeneous, but they are weakly algebraically homogeneous and logically homogeneous [**36**].

2010 *Mathematics Subject Classification.* 20A15, 03C50.

Research partially supported by the Israel Science Foundation grants No. 533/14, No. 1207/12, by the Emmy Noether Institute and by the Gelbart Institute for Mathematical Sciences, Department of Mathematics, Bar-Ilan University.

The paper is organized as follows. In Section 2, we give a short explanation of the term "Logical Geometry". Further, in the subsequent sections, having the necessary structures and machinery of Logical Geometry, we will show how they work in the case of groups. Section 3 is devoted to algebraically homogeneous groups. We recall here the algebraic definition of a homogeneous group. Section 4 deals with logically homogeneous groups. We recall the model theoretical notion of a type and give the definition of a logically-geometrical type, which is a more appropriate tool to work with logically homogeneous groups within the framework of Logical Geometry. Besides, we recall known results about logically homogeneous groups and consider some open problems. In Section 5, we discuss the notion of a weakly homogeneous group and prove that *free solvable groups of derived length* $n > 1$ *and* rank 2 *are not weakly algebraically homogeneous.*

2. What is Logical Geometry?

Let us start with the basic setting of Classical Algebraic Geometry (cf. [**8**], [**34**]). In some sense, Classical Algebraic Geometry studies solutions of equations of the form

$$f(x_1, \dots, x_n) = 0,$$

where $f(x_1, \dots, x_n)$ is a commutative and associative polynomial over a field K, i.e., an element of the algebra $K[x_1, \dots, x_n]$. The solution of $f(x_1, \dots, x_n) = 0$ is a tuple $\bar{a} = (a_1, \dots, a_n)$ of elements from the field K.

If T is a system of polynomial equations

$$\begin{cases} f_1(x_1, \dots, x_n) = 0, \\ \quad \dots \\ f_k(x_1, \dots, x_n) = 0, \end{cases}$$

then the set A_T of all common zeros of these polynomials is a subset in the n-th Cartesian power of K, which is regarded as the affine space K^n, i.e.,

$$A_T = \{\bar{a} \mid f_i(\bar{a}) = 0, \text{ for all } f_i \in T\} \subseteq K^n.$$

On the other hand, if A is a subset in K^n, one can ask to find a set T_A of all polynomials vanishing at all points from A, i.e.,

$$T_A = \{f_i \mid f_i(\bar{a}) = 0, \quad \text{for all } \bar{a} \in A\} \subseteq K[x_1, \dots, x_n].$$

From an algebraic point of view, the system T corresponds to the ideal generated by the polynomials $f_1, \dots, f_k$. Moreover, the correspondence between ideals in the algebra $K[x_1, \dots, x_n]$ and subsets in the affine space K^n is the Galois correspondence (cf. [**11**]), that is, $T \subseteq T_{A_T}$ and $A \subseteq A_{T_A}$.

Suppose we would like to generalize this situation and consider equations in groups. First of all, we need to understand what is the role of polynomials, the algebra $K[x_1, \dots, x_n]$ and the field K.

If we work with polynomial equations, we deal with the variety Com-K of associative and commutative algebras over a field K. By variety, we mean a class of algebras satisfying a set of identities.

In the variety Com-K the algebra $K[x_1, \dots, x_n]$ is a finitely generated free algebra, polynomials f_i are elements of this algebra and the field K is an algebra.

If we deal with groups, we should take a variety of groups *Grp* and a finitely generated free group in this variety denoted by $F(x_1, \dots, x_n)$. The system of equations in this situation can be written in the form

$$\begin{cases} g_1(x_1, \dots, x_n) = 1, \\ \quad \dots \\ g_n(x_1, \dots, x_n) = 1 \\ \quad \dots \end{cases}$$

where $g_i(x_1, \dots, x_n)$ are elements of the free group $F(x_1, \dots, x_n)$. One can solve this system of equations in a group G. Note that in this case we do not have an analogue of Hilbert's Basis Theorem (cf. [**10**]) and we cannot assume that every system of equations is equivalent to a finite system.

There is also the Galois correspondence between normal subgroups in $F(x_1, \dots, x_n)$ (which corresponds to the system of equations in groups) and subsets in the affine space G^n.

Now we are ready to describe what is Logical Geometry. In the sense of the consideration above, Logical Geometry studies solutions of arbitrary systems of first order logic formulas "over" equalities, i.e., all atomic formulas are of the form $g = 1$, $g \in F(x_1, \dots, x_n)$. All other formulas are built over "equalities" $g = 1$ using logical connectives and quantifiers. For example, one can consider logical equations

$$\neg(x_1x_2^3 = 1) \wedge (x_1x_2x_3^{-1} = 1),$$

$$\big(\exists x_1(x_1 = x_2^3)\big) \vee \big(x_1x_2 = x_3\big).$$

REMARK 2.1. We should note that we use the term "a logical equation" instead of the model-theoretical term "a formula" to emphasize the geometrical trend of our consideration.

REMARK 2.2. We will not speak here about systems of formulas "over" arbitrary predicates (or relations). This is also part of Logical Geometry and it has applications in the databases and knowledge bases theories (see, for instance, [**2**], [**3**], [**28**], [**29**], [**30**]).

One can define *a solution of a logical equation* by induction (cf. [**13**]), or using a more algebraic approach (cf. [**25**]). In simple words, a tuple $(a_1, \dots, a_n) \in G^n$ is a solution of a logical equation $u(x_1, \dots, x_n)$ if $(a_1, \dots, a_n)$ satisfies the formula $u(x_1, \dots, x_n)$.

Recall that in Algebraic Geometry, systems of equations over groups correspond to normal subgroups of $F(x_1, \dots, x_n)$; thus it is enough to work with the finitely generated free group $F(x_1, \dots, x_n)$ and its normal subgroups. In the case of the system of logical equations it is not enough. We need the place where the formulas are situated.

Logical equations with variables $x_1, \dots, x_n$ lie in the algebra of formulas $\Phi(X)$, $X = \{x_1, \dots, x_n\}$. We will not give a precise definition of this algebra, for details see [**1**], [**22**], [**25**]. We only note that $\Phi(X)$ is a boolean algebra with operations $\wedge, \vee, \neg$. Moreover, it is a quantifier algebra, that is, the existential quantifiers $\exists x_i$, for $x_i \in X$, are defined as unary operations (cf. [**7**], [**20**]). The signature of $\Phi(X)$ also includes infinitely many unary operations, denote them by one symbol s_*. Each operation $s_* : \Phi(X_i) \to \Phi(X_j)$ is induced by a homomorphism of free groups $s : F(X_i) \to F(X_j)$.

The reason to introduce these operations is as follows. In Logic and Model Theory we are working with an infinite set of variables, but for geometrical considerations we need a finite set of variables in order to work with finite dimensional affine spaces. To bypass this difficulty, we deal with the collection of all algebras $\Phi(X_j)$ for all finite subset X_j of an infinite set X^0. The collection of all such algebras one can consider as a multi-sorted algebra or as a category (for details see [**1**], [**22**], [**25**]). Such an approach allows us to define infinitely many unary operations s_* (morphisms in terms of categories). Note that from the logical viewpoint, these operations correspond to substitutions of variables.

Now, having in mind the necessary structures and machinery of Logical Geometry we will show how they work in the case of groups. The three subsequent sections are devoted to algebraically homogeneous, logically homogeneous and weakly algebraically homogeneous groups viewed from the perspective of Logic and Geometry.

3. Algebraically homogeneous groups

We start with the algebraic approach to the notion of homogeneity.

DEFINITION 3.1. A group G is called *algebraically homogeneous* if every isomorphism between two of its finitely generated subgroups can be extended up to an automorphism of G.

Such groups were constructed by P. Hall [**6**]. We reformulate this definition in the other terms.

Let $F(X)$ be a free group with the free generating set $X = \{x_1, \dots, x_n\}$, G be a group, $\mu_1, \mu_2 : F(X) \to G$ be homomorphisms.

DEFINITION 3.2. A group G is *algebraically homogeneous* if for all homomorphisms $\mu_1, \mu_2 : F(X) \to G$ whenever $\operatorname{Ker} \mu_1 = \operatorname{Ker} \mu_2$ then there is an automorphism σ of G such that $\sigma \circ \mu_1 = \mu_2$.

The homogeneity property is considered in Model Theory with respect to an arbitrary algebraic structure. In the next section, we present model-theoretical and logically-geometrical approaches.

4. Logically homogeneous groups

To define logically homogeneous groups, we need to recall the notion of *a type* from Model Theory (cf. [**13**]).

Let G be a group, $\mu = (g_1, \dots g_n)$ be a point in G^n.

DEFINITION 4.1. *A complete model-theoretical type (MT-type) of a point* $\mu = (g_1, \dots, g_n)$ is the set of all first order logic formulas (in a language $\mathbb{L}$) in free variables $x_1, \dots, x_n$ which hold true on the point μ:

$$tp^G(\mu) = \{u(x_1, \dots, x_n, y_1, \dots) \in \mathbb{L} \mid G \models u(g_1, \dots, g_n)\}.$$

In informal words, a language $\mathbb{L}$ is a set of symbols, which we use to compose the formulas. It includes logical connectives, symbol of operations (functional symbols), constant symbols and predicate symbols. Thus, in our case

$$\mathbb{L} = \{\wedge, \vee, \neg, \exists, =, \cdot, \ ^{-1}, 1\}.$$

REMARK 4.2. Each formula from the model-theoretical type $tp^G(\mu)$ has free variables from the set $\{x_1, \dots, x_n\}$ and bounded variables from the set $\{y_1, y_2, \dots\}$.

Model-theoretical types are not very convenient from a geometrical point of view. Indeed, take a point $\mu = (g_1, \dots, g_n) \in G^n$ and consider all formulas satisfying this point, that is, set $tp^G(\mu)$. Take all points satisfying all formulas from $tp^G(\mu)$. We get points from $G^\infty = \prod G$, i.e., from the Cartesian power of G (see Figure 1). This is because each formula from $tp^G(\mu)$ has free variables from the set $\{x_1, \dots, x_n\}$ and some bounded variables from $\{y_1, y_2, \dots\}$.

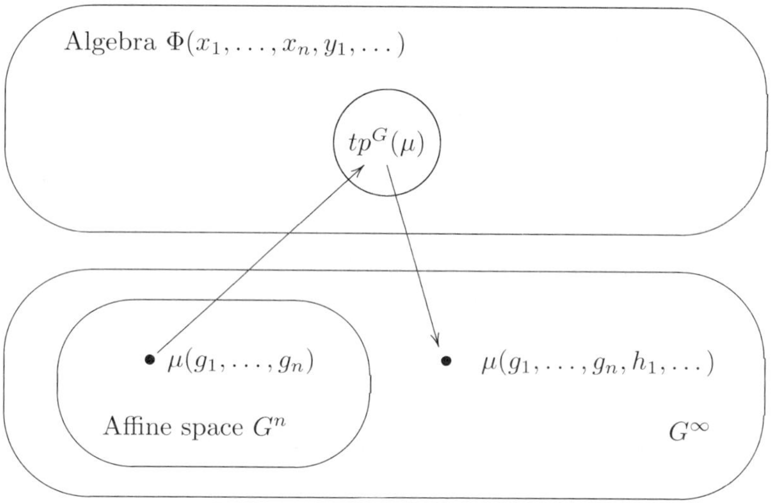

FIGURE 1. Geometric view on model-theoretical type

The notion of a logically-geometrical type is associated with finite dimensional spaces and thus oriented towards geometry related to sets of formulas (see [**22**], [**25**], [**31**]).

DEFINITION 4.3. *A logically-geometrical type (LG-type) of a point* $\mu = (g_1, \dots, g_n)$ is the set of all first order logic formulas in variables $x_1, \dots, x_n$ (not necessarily free) written in a language $\mathbb{L}^*$ which hold true on the point μ:

$$\mathrm{LKer}(\mu) = \{u(x_1, \dots, x_n) \in \mathbb{L}^* \mid G \models u(g_1, \dots, g_n)\}.$$

Note that the language $\mathbb{L}^*$ in the case of LG-types is different from $\mathbb{L}$. It includes the symbol of operation s_* which we mentioned above, speaking about the algebra $\Phi(X)$ (see Section 2 or [**25**]). Thus,

$$\mathbb{L}^* = \{\wedge, \vee, \neg, \exists, =, \cdot, \ ^{-1}, 1, s_*\}.$$

The following important result of G. Zhitomirski [**35**] (see also [**27**], [**36**]) gives a bridge between model-theoretical and logically-geometrical types. The detailed comparison of a model-theoretical type and a logically-geometrical type is given in Section 4 of [**26**].

THEOREM 4.4. *Model-theoretical types of two points coincide if and only if their logically-geometrical types coincide.*

This fact allows us to reformulate the definition of a homogeneous structure (cf. [**13**]) in terms of LG-types.

DEFINITION 4.5. A group G is called *logically homogeneous* if for every two points $\mu_1, \mu_2 \in G^n$ whenever $\mathrm{LKer}(\mu_1) = \mathrm{LKer}(\mu_2)$, then there exists an automorphism σ of G such that $\sigma \circ \mu_1 = \mu_2$.

REMARK 4.6. We use the term "logically homogeneous group" in order to distinguish algebraic and model-theoretical notions of homogeneous structures. Some authors use the notion "strongly homogeneous" (see [**19**]) or "logically perfect" (see [**35**], [**36**]).

The following results about logical homogeneity of free groups are very important.

THEOREM 4.7 ([**35**],[**36**]). *Finitely generated free abelian groups are logically homogeneous.*

THEOREM 4.8 ([**18**], [**19**]). *Finitely generated free non-abelian groups are logically homogeneous.*

The result of Theorem 4.8 was proved by A. Ould Houcine and, independently, by R. Sklinos and C. Perin. They used a model-theoretical notion of a type and model-theoretical technics. In turn, the result of G. Zhitomorski (Theorem 4.7) is based on methods and notions of Logical Geometry.

The following two problems still remain open.

PROBEM 4.9. Is it true that finitely generated free nilpotent groups are logically homogeneous?

PROBEM 4.10. Is it true that finitely generated free solvable groups are logically homogeneous?

The recent result of G. Zhitomirski gives an approximation to the solution of Problem 4.9.

THEOREM 4.11 ([**36**]). *Let G be a finitely generated free nilpotent group of class $c > 1$ and rank n. Let $\mu, \nu : F(x_1, \dots, x_m) \to G$ be two points in G^m such that $\mathrm{LKer}(\mu) = \mathrm{LKer}\,\nu$. Then there exists an automorphism σ of G which transports μ to ν modulo commutant, i.e, $\sigma \circ \mu(x_i) = \nu(x_i)q_i$, where $q_i \in G'$, $i = 1, \dots, m$. Moreover, if the subgroup A generated by $\mu(x_1), \dots, \mu(x_m)$ satisfies the condition $A \cap G' = A'$, then the automorphism σ satisfies the condition $\sigma \circ \mu(x_i) = \nu(x_i)$.*

As for Problem 4.10, there are numerous papers concerning logical properties of solvable groups. A. Malcev [**12**] proved that a noncyclic free solvable group of derived length $n \geq 2$ has an undecidable theory. The famous Tarski problem, *whether (absolutely) free groups of finite rank are elementarily equivalent or not*, was solved in the case of solvable groups by P. Rogers, H. Smith and D. Solitar [**32**]. They proved that free solvable groups of finite rank are elementarily equivalent if and only if they are isomorphic. In contrast, the original question of Tarski has been open for many years and only recently was solved in the affirmative: free non-abelian groups of finite rank are elementarily equivalent (see [**9**], [**33**]).

A.G. Myasnikov and N.S. Romanovskii [**14**], [**15**], [**16**], [**17**] investigate various logical properties of free solvable groups; in particular, they proved that every free solvable group of finite rank is strongly defined by its types. Note that a finitely generated group G is strongly defined by types if and only if any elementary embedding of a group to itself is an isomorphism [**17**]. This property turns out to be weaker than homogeneity but most likely stronger than the isotypic setting. In [**17**] the following result is proved.

THEOREM 4.12. *Every free solvable group of finite rank is strongly defined by its types.*

The recent interesting papers of G. Baumslag, R. Mikhailov and K.E. Orr [**4**], [**5**] give a new look at solvable, in particular, at metabelian groups. They begin to develop how geometrical group theory and algebraic geometry approaches might help to study finitely generated metabelian groups.

5. Weakly algebraically homogeneous groups

Studying logically homogeneous groups and semigroups, G. Zhitomirski introduces the notion of a weakly (algebraically) homogeneous algebra (see [**35**], [**36**]). We formulate this definition in the case of groups.

DEFINITION 5.1. A group G is called *weakly algebraically homogeneous* if for every isomorphism $\tilde{\sigma} : H_1 \to H_2$ between two of its finitely generated subgroups H_1 and H_2 the following condition takes place: if $\tilde{\sigma}$ itself and its inverse map $\tilde{\sigma}^{-1} : H_2 \to H_1$ can be extended to endomorphisms of G, then $\tilde{\sigma}$ can be extended to an automorphism of G.

In view of the next theorem of G. Zhitomirski, the notion of a weakly algebraically homogeneous group (algebra) is a useful instrument to study logically homogeneous structures.

THEOREM 5.2 ([**36**]). *Every weakly algebraically homogeneous finitely generated free algebra is logically homogeneous.*

In particular, the statement of Theorem 4.7 is based of the fact that finitely generated free abelian groups are weakly homogeneous.

Theorem 5.2 gives only necessary and not sufficient conditions for an algebra to be logically homogeneous. As was shown in [**36**], a finitely generated free non-abelian group is not weakly homogeneous; nevertheless, such a group is logically homogeneous [**18**], [**19**].

We present one more example of a class of groups which are not weakly homogeneous (cf. [**35**]).

THEOREM 5.3. *Free solvable groups of derived length* $n > 1$ *and rank* 2 *are not weakly homogeneous.*

PROOF. Let $\mathbb{S}_2$ be a free solvable group of derived length $n > 1$ and rank 2 with free generators x_1, x_2. Take in $\mathbb{S}_2$ two elements $a = x_1^2 x_2 x_1^{-1} x_2$ and $b = x_1 x_2$. Define endomorphisms α and β setting $\alpha(x_1) = x_1 x_2$, $\alpha(x_2) = 1$ and $\beta(x_1) = x_1^2 x_2$, $\beta(x_2) = x_1^{-1} x_2$. One can see that $\alpha(a) = b$ and $\beta(b) = a$. Thus, α induces an isomorphism $\tilde{\sigma} : \langle a \rangle \to \langle b \rangle$ of two finitely generated subgroups of $\mathbb{S}_2$ and β induces the inverse isomorphism $\tilde{\sigma}^{-1}$.

Suppose that $\tilde{\sigma}$ can be extended up to an automorphism σ of $\mathbb{S}_2$. Let $\sigma(x_1) = w_1$, $\sigma(x_2) = w_2$, where w_1, w_2 are elements of $\mathbb{S}_2$. Since $\sigma(a) = b$, in the free solvable group $\mathbb{S}_2$ there is the relation

$$x_1 x_2 = w_1^2 w_2 w_1^{-1} w_2. \tag{1}$$

Thus, it should be an identity in the variety of solvable groups of derived length $\leq n$. We will show that it is not true, so not every isomorphism of two finitely generated subgroups with the property from Definition 5.1 can be extended up to an automorphism of $\mathbb{S}_2$.

Let l_1, l_2 be the sums of all exponents of x_1, x_2 incoming in w_1, and m_1, m_2 the sums of all exponents of x_1, x_2 incoming in w_2, respectively. Then $l_1 + 2m_1 = 1$ and $l_2 + 2m_2 = 1$. Thus, l_1 and l_2 must be odd numbers.

Consider the group S_3 of all permutations on the set $\{1, 2, 3\}$. This group is a homomorphic image of $\mathbb{S}_2$ under the map γ which sends x_1 to the transposition (12) and x_2 to (23). According to (1), in the group S_3 the equality $\gamma(x_1x_2) = \gamma(w_1^2w_2w_1^{-1}w_2)$ holds true. We will see that it is not true.

Let $y_1 = (12)$, $y_2 = (23)$. For an arbitrary element $w \in \mathbb{S}_2$ the image of w under the homomorphism γ can be one of the following elements of the group S_3: 1, y_1, y_2, y_1y_2, y_2y_1, $y_1y_2y_1$. Moreover, w_1 and w_2 are images of generators of $\mathbb{S}_2$ under the automorphism σ; thus, they are generators of $\mathbb{S}_2$. Since the sums of all exponents of x_1, x_2 incoming in w_1 should be odd numbers, for $\gamma(w_1)$ there are only two possibilities: y_1y_2 or y_2y_1. Thus, for $\gamma(w_2)$ there are three possibilities: y_1, y_2, $y_1y_2y_1$.

Direct calculation shows that in the mentioned cases $\gamma(w_1^2w_2w_1^{-1}w_2) = 1$, but $\gamma(x_1x_2) = (132)$.

So, (1) is not an identity in the solvable (metabelian) group S_3 and, moreover, in the free solvable group $\mathbb{S}_2$. This means that the isomorphism $\tilde{\sigma} : \langle a \rangle \to \langle b \rangle$ can not be extended up to an automorphism of $\mathbb{S}_2$. Thus, $\mathbb{S}_2$ is not weakly homogeneous. □

However, we cannot say that $\mathbb{S}_2$ is not logically homogeneous. But we intend to consider the question about logically homogeneity of finitely generated free metabelian groups using new techniques from [**4**], [**5**] and [**17**].

The research was partially supported by the Israel Science Foundation grants No. 533/14, No. 1207/12, by the Emmy Noether Institute and by Gelbart Institute for Mathematical Sciences, Department of Mathematics, Bar-Ilan University.

References

[1] E. Aladova, *Syntax versus semantics in knowledge bases I*, Internat. J. Algebra and Comput., 2018, to appear.

[2] E. Aladova, A. Gvaramiya, B. Plotkin, and T. Plotkin, *Multisorted logic, models and logical geometry* (Russian, with English and Russian summaries), Fundam. Prikl. Mat. **19** (2014), no. 3, 5–22; English transl., J. Math. Sci. (N.Y.) **214** (2016), no. 6, 742–754. MR3431877

[3] E. Aladova, E. Plotkin, and T. Plotkin, *Isotypeness of models and knowledge bases equivalence*, Math. Comput. Sci. **7** (2013), no. 4, 421–438, DOI 10.1007/s11786-013-0166-5. MR3158080

[4] G. Baumslag, R. Mikhailov, and K. E. Orr, *A new look at finitely generated metabelian groups*, Computational and combinatorial group theory and cryptography, Contemp. Math., vol. 582, Amer. Math. Soc., Providence, RI, 2012, pp. 21–37, DOI 10.1090/conm/582/11551. MR2987381

[5] G. Baumslag, R. Mikhailov, and K. E. Orr, *Localization, metabelian groups, and the isomorphism problem*, Trans. Amer. Math. Soc. **369** (2017), no. 10, 6823–6852, DOI 10.1090/tran/6838. MR3683095

[6] P. Hall, *Some constructions for locally finite groups*, J. London Math. Soc. **34** (1959), 305–319, DOI 10.1112/jlms/s1-34.3.305. MR0162845

[7] P. R. Halmos, *Algebraic logic*, Chelsea Publishing Co., New York, 1962. MR0131961

[8] K. Hulek, *Elementary algebraic geometry*, Student Mathematical Library, vol. 20, American Mathematical Society, Providence, RI, 2003. Translated from the 2000 German original by Helena Verrill. MR1955795

[9] O. Kharlampovich and A. Myasnikov, *Elementary theory of free non-abelian groups*, J. Algebra **302** (2006), no. 2, 451–552, DOI 10.1016/j.jalgebra.2006.03.033. MR2293770

[10] S. Lang, *Algebra*, Graduate Texts in Math. **211**, Springer-Verlag, New York, 2002.

[11] S. Mac Lane, *Categories for the working mathematician*, Springer-Verlag, New York-Berlin, 1971. Graduate Texts in Mathematics, Vol. 5. MR0354798

[12] A. I. Mal′cev, *On free soluble groups* (Russian), Soviet Math. Dokl. **1** (1960), 65–68. MR0117274

[13] D. Marker, *Model theory: An introduction*, Graduate Texts in Mathematics, vol. 217, Springer-Verlag, New York, 2002. MR1924282

[14] A. Myasnikov and N. Romanovskiy, *Krull dimension of solvable groups*, J. Algebra **324** (2010), no. 10, 2814–2831, DOI 10.1016/j.jalgebra.2010.07.013. MR2725203

[15] A. G. Myasnikov and N. S. Romanovskiĭ, *Universal theories for rigid solvable groups* (Russian, with English and Russian summaries), Algebra Logika **50** (2011), no. 6, 802–821, 853–854, 857, DOI 10.1007/s10469-012-9164-y; English transl., Algebra Logic **50** (2012), no. 6, 539–552. MR2953279

[16] A. G. Myasnikov and N. S. Romanovskiĭ, *Logical aspects of the theory of divisible rigid groups* (Russian), Dokl. Akad. Nauk **459** (2014), no. 2, 154–155; English transl., Dokl. Math. **90** (2014), no. 3, 697–698. MR3409993

[17] A. Myasnikov and N. Romanovskii, *Characterization of finitely generated groups by types*, Internat. J. Algebra and Comput., 2018, to appear.

[18] A. Ould Houcine, *Homogeneity and prime models in torsion-free hyperbolic groups*, Confluentes Math. **3** (2011), no. 1, 121–155, DOI 10.1142/S179374421100028X. MR2794551

[19] C. Perin and R. Sklinos, *Homogeneity in the free group*, Duke Math. J. **161** (2012), no. 13, 2635–2668, DOI 10.1215/00127094-1813068. MR2988905

[20] B. Plotkin, *Universal algebra, algebraic logic, and databases*, Mathematics and its Applications, vol. 272, Kluwer Academic Publishers Group, Dordrecht, 1994. Translated from the 1991 Russian original by J. Cīrulis, A. Nenashev and V. Pototsky and revised by the author. MR1273136

[21] B. Plotkin, *Varieties of algebras and algebraic varieties.* part B, Israel J. Math. **96** (1996), no. part B, 511–522, DOI 10.1007/BF02937320. MR1433704

[22] B. Plotkin, *Algebraic geometry in first-order logic* (Russian, with Russian summary), Sovrem. Mat. Prilozh. **22, Algebra i Geom.** (2004), 16–62, DOI 10.1007/s10958-006-0288-2; English transl., J. Math. Sci. (N.Y.) **137** (2006), no. 5, 5049–5097. MR2462070

[23] B. I. Plotkin, *Problems in algebra inspired by universal algebraic geometry* (Russian, with English and Russian summaries), Fundam. Prikl. Mat. **10** (2004), no. 3, 181–197, DOI 10.1007/s10958-006-0390-5; English transl., J. Math. Sci. (N.Y.) **139** (2006), no. 4, 6780–6791. MR2123349

[24] B. Plotkin, *Some results and problems related to universal algebraic geometry*, Internat. J. Algebra Comput. **17** (2007), no. 5-6, 1133–1164, DOI 10.1142/S0218196707003986. MR2355690

[25] B. Plotkin, E. Aladova, and E. Plotkin, *Algebraic logic and logically-geometric types in varieties of algebras*, J. Algebra Appl. **12** (2013), no. 2, 1250146, 23, DOI 10.1142/S0219498812501460. MR3005597

[26] B. Plotkin and E. Plotkin, *Multi-sorted logic and logical geometry: some problems*, Demonstr. Math. **48** (2015), no. 4, 578–619, DOI 10.1515/dema-2015-0042. MR3430891

[27] B. Plotkin, E. Plotkin, and G. Zhitomirski, *Type of a point in universal geometry and in model theory*, Algebra Discrete Math. **19** (2015), no. 1, 87–100. MR3376343

[28] T. Plotkin and B. Plotkin, *Geometrical aspect of databases and knowledge bases*, Algebra Universalis **46** (2001), no. 1-2, 131–161, DOI 10.1007/PL00000333. The Viktor Aleksandrovich Gorbunov memorial issue. MR1835790

[29] B. Plotkin and T. Plotkin, *An algebraic approach to knowledge base models informational equivalence*, Acta Appl. Math. **89** (2005), no. 1-3, 109–134 (2006), DOI 10.1007/s10440-005-9008-z. MR2220200

[30] B. Plotkin and T. Plotkin, *Categories of elementary sets over algebras and categories of elementary algebraic knowledge*, Pillars of computer science, Lecture Notes in Comput.

Sci., vol. 4800, Springer, Berlin, 2008, pp. 555–570, DOI 10.1007/978-3-540-78127-1_30. MR2478796

[31] B. Plotkin and G. Zhitomirski, *Some logical invariants of algebras and logical relations between algebras*, Algebra i Analiz **19** (2007), no. 5, 214–245, DOI 10.1090/S1061-0022-08-01023-6; English transl., St. Petersburg Math. J. **19** (2008), no. 5, 829–852. MR2381947

[32] P. Rogers, H. Smith, and D. Solitar, *Tarski's problem for solvable groups*, Proc. Amer. Math. Soc. **96** (1986), no. 4, 668–672, DOI 10.2307/2046323. MR826500

[33] Z. Sela, *Diophantine geometry over groups. VI. The elementary theory of a free group*, Geom. Funct. Anal. **16** (2006), no. 3, 707–730, DOI 10.1007/s00039-006-0565-8. MR2238945

[34] I. R. Shafarevich, *Basic algebraic geometry*, Springer-Verlag, New York-Heidelberg, 1974. Translated from the Russian by K. A. Hirsch; Die Grundlehren der mathematischen Wissenschaften, Band 213. MR0366917

[35] G. Zhitomirski, *On logically-geometric types of algebras*, preprint, (2012), arXiv: 1202.5417v1.

[36] G. Zhitomirski, *On types of points and algebras*, Internat. J. Algebra and Comput., 2018, to appear.

Department of Mathematics, Bar-Ilan University, 5290002, Ramat Gan, Israel

Current address: Federal University of Rio Grande do Norte, University Campus Lagoa Nova, CEP 59078-970, Natal/RN, Brazil

Email address: aladovael@mail.ru

Contemporary Mathematics
Volume **726**, 2019
https://doi.org/10.1090/conm/726/14607

Syntax versus semantics in knowledge bases II

E. Aladova and T. Plotkin

Abstract. This paper is the second in a series of papers. Here special attention is given to the interaction between syntax and semantics in knowledge bases theory. The first paper includes preparatory material which allows to introduce in detail the given approach to a knowledge base model.

The considered model of a knowledge base involves various ideas of universal algebra, algebraic logic and algebraic geometry that was introduced by B.I. Plotkin.The main novelty of this approach is that a knowledge base is considered as an algebraic structure.

The main goal of this paper is to show that syntax and semantics are tightly connected in the constructed knowledge base model.

1. Introduction

This paper is devoted to the algebraic model of a knowledge base which was introduced and developed by B.I. Plotkin in [**16**], [**17**], [**18**]; see also references therein. The main peculiarity of this approach is that a knowledge base is considered as a certain algebraic structure. The considered model of a knowledge base involves various ideas of universal algebra, algebraic logic and algebraic geometry.

Knowledge bases theory provides an important example of the field where applications of universal algebra and algebraic logic look very natural and their interaction with practical problems arising in computer science might be very productive. There are many specialized knowledge bases and it is desirable to determine their characteristic properties without referring to their complicated structure and without studying their detailed architecture.

The mathematical model of a knowledge base represented as an algebraic structure helps us to understand the nature of a concrete real knowledge base and it enables to solve various problems in knowledge bases theory. For example, one of the main problem in database theory is *the equivalence problem.* Various equivalences and results in this area are obtained in [**2**], [**18**], [**23**], [**24**].

This paper is the second in a series of papers. Here special attention is given to the interaction between syntax and semantics in knowledge bases theory. We put syntax and semantics into appropriate algebraic structures, that is, we use techniques and structures of algebraic logic (see [**3**], [**9**], and [**16**] for details). We should

2010 *Mathematics Subject Classification.* 08A70, 68T30.

Research of E. Aladova was partially supported by the Israel Science Foundation grants No. 533/14, No. 1207/12, by the Emmy Noether Institute and the Gelbart Institute for Mathematical Sciences of Bar-Ilan University.

note that here we present a slightly different approach to the main components of a knowledge base than in the previous papers (cf. [**2**], [**23**], [**25**]) and this gives us some advantages, namely, the result of Theorem 4.1.

The paper is organized as follows. In Section 2, we give a brief review of basic notions and notations from universal algebraic geometry. In particular, we define Halmos categories and construct the Galois correspondence which is very important in our considerations. The material of this section can be found in the papers of B. Plotkin (see [**17**], [**18**], [**20**]); see also [**3**] for some detailed proofs. In Section 3, we introduce a knowledge base model. To do this, we recall the design of the category of knowledge content $D_\Theta(\mathcal{H})$, of the category of knowledge description $F_\Theta(\mathcal{H})$ and the knowledge functor $Ct_\mathcal{H}$ from [**3**]. Section 4 deals with the interaction between syntax and semantics in knowledge bases. We show that syntax and semantics are tightly connected in the constructed knowledge base model, namely, categories of knowledge content $D_\Theta(\mathcal{H})$ and of knowledge description $F_\Theta(\mathcal{H})$ are dually isomorphic.

2. Preliminaries

2.1. Basic notions and notations. Let $X^0 = \{x_1, \dots, x_n, \dots\}$ be an infinite set of variables. Denote by Γ the collection of all finite subsets X of X^0.

Let Θ be a *variety of algebras*, that is, a class of algebras satisfying a set of identities (see, for instance, [**12**], [**16**]). We denote by $\mathrm{Var}(H)$ the variety generated by the algebra H.

Denote by $W(X)$ the *free algebra* in the variety Θ with free generating set X, $X \in \Gamma$. All free algebras $W(X) \in \Theta$ form a *category of free algebras* Θ^0 with homomorphisms $s : W(X) \to W(Y)$ as morphisms, $X, Y \in \Gamma$.

By a *model* $\mathcal{H}$ we mean a triple (H, Ψ, f), where H is an algebra from Θ, Ψ is a set of relation symbols φ, f is an interpretation of all φ in H (see, for instance, [**5**], [**13**], [**16**]).

Take an algebra H in Θ. A *point* $(a_1, \dots, a_n)$ from the n-th Cartesian power of H can be represented as a map $\mu : X \to H$ such that $a_i = \mu(x_i)$. This map can be extended up to homomorphism of algebras $\mu : W(X) \to H$. Thus, the point $(a_1, \dots, a_n)$ can be also viewed as a homomorphism $\mu : W(X) \to H$.

Denote by $\mathrm{Hom}(W(X), H)$ the set of all homomorphisms from $W(X)$ to H. We will regard $\mathrm{Hom}(W(X), H)$ as an *affine space*.

All affine spaces $\mathrm{Hom}(W(X), H)$, $X \in \Gamma$, constitute the *category* $\Theta^0(H)$ *of affine spaces* with morphisms

$$\widetilde{s} : \mathrm{Hom}(W(X), H) \to \mathrm{Hom}(W(Y), H),$$

for each homomorphism of free algebras $s : W(Y) \to W(X)$. The map $\widetilde{s}$ is defined as $\widetilde{s}(\mu) = \mu s$, where $\mu : W(X) \to H$, $\widetilde{s}(\mu) : W(Y) \to H$.

The categories Θ^0 and $\Theta^0(H)$ are very important for further considerations. Moreover, the following theorem holds.

THEOREM 2.1 ([**14**]). *The categories* Θ^0 *and* $\Theta^0(H)$ *are dual if and only if* $\mathrm{Var}(H) = \Theta$.

2.2. Halmos categories. Halmos categories were introduced in papers of B.I. Plotkin ([**17**], [**18**]). Halmos categories are related to the first-order logic in a way analogous to the relationship between boolean algebras and propositional logic. The immediate advantage of this phenomenon is that we can view queries to a knowledge base and replies to these queries as objects of the same nature, i.e., objects of Halmos categories. Then the transition query-reply can be treated as a functor (for details see Section 3).

We start from the notion of an existential quantifier on a boolean algebra. Let B be a boolean algebra. The existential quantifier on B is a unary operation $\exists : B \to B$ such that the following conditions hold:

(1) $\exists\, 0 = 0$,
(2) $a \leq \exists a$,
(3) $\exists(a \wedge \exists b) = \exists a \wedge \exists b$.

The universal quantifier $\forall : B \to B$ is dual to $\exists : B \to B$, and they are related by $\forall a = \neg(\exists(\neg a))$.

DEFINITION 2.2. Let a set of variables $X = \{x_1, \ldots, x_n\}$ and a set of relations Ψ be given. A boolean algebra B is called an extended boolean algebra over $W(X)$ relative to Ψ if

1. the existential quantifier $\exists x$ is defined on B for all $x \in X$, and $\exists x \exists y = \exists y \exists x$ for all $x, y \in X$;
2. to every relation symbol $\varphi \in \Psi$ of arity n_φ and a collection of elements $w_1, \ldots, w_{n_\varphi}$ from $W(X)$ there corresponds a nullary operation (a constant) of the form $\varphi(w_1, \ldots, w_{n_\varphi})$ in B.

Thus, the signature L_X of an extended boolean algebra consists of the boolean connectives, existential quantifiers $\exists x$ and constants $\varphi(w_1, \ldots, w_{n_\varphi})$:

$$L_X = \{\vee, \wedge, \neg, \exists x, M_X\},$$

where M_X is a set of all $\varphi(w_1, \ldots, w_{n_\varphi})$.

We give two important examples of extended boolean algebras.

EXAMPLE 2.3. Let a model $\mathcal{H} = (H, \Psi, f)$ be given. Let $\mathrm{Bool}(W(X), H)$ be the boolean algebra of all subsets in $\mathrm{Hom}(W(X), H)$. One can equip the boolean algebra $\mathrm{Bool}(W(X), H)$ with the structure of an extended boolean algebra ([**2**], [**3**], [**16**]). We denote this extended boolean algebra by $Hal_\Theta^X(\mathcal{H})$.

EXAMPLE 2.4. Another important example of an extended boolean algebra is presented by the algebra of formulas $\Phi_0(X) = \mathfrak{L}_X/\tau_X$, where $\mathfrak{L}_X$ is the absolutely free algebra in the signature L_X over the set M_X, τ_X is a congruence relation on $\mathfrak{L}_X$ defined by the rule: $u\tau_X v$ if and only if $\vdash (u \to v) \wedge (v \to u)$, $u, v \in \mathfrak{L}_X$. Boolean operations and quantifiers on $\Phi_0(X)$ are naturally inherited from $\mathfrak{L}_X$. For more details see [**2**], [**3**], [**16**], [**18**], [**21**].

Now we define *a Halmos category* which plays a very important role in further considerations.

DEFINITION 2.5. A category $\mathbb{H}$ is a Halmos category if:

(1) every one of its objects has the form $\mathbb{H}(X)$, where $\mathbb{H}(X)$ is an extended boolean algebra over $W(X)$.
(2) the morphisms in $\mathbb{H}$ correspond to morphisms in the category Θ^0. To every morphism $s : W(X) \to W(Y)$ in Θ^0 corresponds a morphism $s_* : \mathbb{H}(X) \to \mathbb{H}(Y)$ in $\mathbb{H}$ in such a way that
    (a) the transitions $W(X) \to \mathbb{H}(X)$ and $s \to s_*$ determine a covariant functor from Θ^0 to $\mathbb{H}$.

(b) $s_* : \mathbb{H}(X) \to \mathbb{H}(Y)$ is a homomorphism of corresponding boolean algebras.

(3) there are special identities controlling the interaction of morphisms with quantifiers and constant (for details see [**18**], [**21**], [**22**]).

The next two examples of Halmos categories are based on examples of extended boolean algebras above.

EXAMPLE 2.6. **Category** $Hal_\Theta(\mathcal{H})$. Objects of this category are extended boolean algebras $Hal_\Theta(\mathcal{H})$ from Example 2.3 for various $X \in \Gamma$. Morphisms

$$s_* : Hal_\Theta^X(\mathcal{H}) \to Hal_\Theta^Y(\mathcal{H})$$

are defined as follows:

$$\mu \in s_* A \Leftrightarrow \ \mu s \in A,$$

where $\mu : W(Y) \to H$, $A \in \mathrm{Bool}(W(X), H)$, $s : W(X) \to W(Y)$.

REMARK 2.7. A homomorphism $s : W(X) \to W(Y)$ gives rise to a map

$$\widetilde{s} : \mathrm{Hom}(W(Y), H) \to \mathrm{Hom}(W(X), H)$$

by the rule $\widetilde{s}(\mu) = \mu s$, which is a morphism in the category of affine spaces $\Theta^0(H)$ (see Section 2.1). Let a subset A from $\mathrm{Hom}(W(X), H)$ be given. Then $s_* A$ is the full pre-image of A under $\widetilde{s}$.

EXAMPLE 2.8. **Category** $\widetilde{\Phi}$. Objects of the category $\widetilde{\Phi}$ are constructed using extended boolean algebras $\Phi_0(X)$, $X \in \Gamma$. Denote by $[\varphi(w_1, \dots, w_{n_\varphi})]_{\tau X}$ the image of the element $\varphi(w_1, \dots, w_{n_\varphi}) \in M_X$ under the homomorphism $\mathfrak{L}_X \to \Phi_0(X) = \mathfrak{L}_X / \tau_X$. Let $[M_X]_{\tau X}$ be a set of all $[\varphi(w_1, \dots, w_{n_\varphi})]_{\tau X}$, $\varphi \in \Psi$.

A homomorphism of free algebras $s : W(X) \to W(Y)$ induces the map $s_* : [M_X]_{\tau X} \to [M_Y]_{\tau_Y}$, by the rule

$$s_*\big([\varphi(w_1, \dots, w_{n_\varphi})]_{\tau X}\big) = [\varphi(sw_1, \dots, sw_{n_\varphi})]_{\tau Y}.$$

This map can be extended up to homomorphism of boolean algebras $s_* : \Phi_0(X) \to \Phi_0(Y)$.

Note that a morphism of a Halmos category s_* correlates with quantifiers under the certain rules (see [**18**], [**21**]) and they are not homomorphisms of extended boolean algebras. Thus, the extended boolean algebras $\Phi_0(X)$ cannot be an object of the category $\widetilde{\Phi}$. We should to add to each $\Phi_0(X)$ all formulas of the form $s_* u$, $u \in \Phi_0(X)$. Denote objects of the category $\widetilde{\Phi}$ by $\Phi(X)$. So, the category $\widetilde{\Phi}$ is a category with objects of the form $\Phi(X)$ and morphisms $s_* : \Phi(X) \to \Phi(Y)$, $X, Y \in \Gamma$.

The next remark is connected with Remark 2.7.

REMARK 2.9. Let a homomorphism $s : W(X) \to W(Y)$ be given. In parallel to the map $\widetilde{s} : \mathrm{Hom}(W(Y), H) \to \mathrm{Hom}(W(X), H)$, we define a map from the set of subsets in $\Phi(Y)$ to the set of subsets in $\Phi(X)$. We denote it by the same symbol $\widetilde{s}$ and define

$$\widetilde{s}T = \{u \in \Phi(X) \mid s_* u \in T\},$$

where $T \subset \Phi(Y)$. Then $\widetilde{s}T$ is the full pre-image of T under s_*. Moreover, one can see that

$$s_* \widetilde{s} T \subset T.$$

Halmos categories $\widetilde{\Phi}$ and $Hal_\Theta(\mathcal{H})$ are tightly connected via homomorphism of extended boolean algebras

$$Val_{\mathcal{H}}^X : \Phi(X) \to \mathrm{Bool}(W(X), H).$$

Intuitively, the image of a formula $u \in \Phi(X)$ under the homomorphism $Val_{\mathcal{H}}^X$ is a value of u in the algebra H, i.e., $Val_{\mathcal{H}}^X u$ is a set of point in $\mathrm{Hom}(W(X), H)$ satisfied u. For details see [**17**], [**18**], [**21**].

REMARK 2.10. The homomorphism $Val_{\mathcal{H}}^X$ gives us the bridge between syntax and semantics, namely, between the algebra $\Phi(X)$ and the algebra $\mathrm{Hal}_\Theta^X(H)$.

Syntax	$\widetilde{\Phi}$	$\Phi(X)$
		$\downarrow \mathrm{Val}_{\mathcal{H}}^X$
Semantics	$\mathrm{Hal}_\Theta(H)$	$\mathrm{Hal}_\Theta^X(H)$

FIGURE 1. Syntax and semantics: categories $\widetilde{\Phi}$ and $\mathrm{Hal}_\Theta(H)$.

Let μ be a point from the affine space $\mathrm{Hom}(W(X), H)$.

DEFINITION 2.11. The logical kernel $\mathrm{LKer}(\mu)$ of a point μ is the set of all formulas $u \in \Phi(X)$ which hold true on the point μ, that is,

$$\mathrm{LKer}(\mu) = \{u \in \Phi(X) \mid \mu \in \mathrm{Val}_{\mathcal{H}}^X(u)\}.$$

Note that the logical kernel $\mathrm{LKer}(\mu)$ of a point μ is a boolean ultrafilter (maximal filter) in the algebra $\Phi(X)$ (see [**19**]).

2.3. Galois correspondence. Now we define a correspondence between sets of formulas in the algebra $\Phi(X)$ and subsets of points from the affine space $\mathrm{Hom}(W(X), H)$.

Let T be a set of formulas from $\Phi(X)$. We define a set of points $T_{\mathcal{H}}^L$ in $\mathrm{Hom}(W(X), H)$ as

$$T_{\mathcal{H}}^L = \{\mu : W(X) \to H \mid T \subset \mathrm{LKer}(\mu)\}.$$

That is, $T_{\mathcal{H}}^L$ is a set of all points $\mu \in \mathrm{Hom}(W(X), H)$ satisfying all formulas from $T \subset \Phi(X)$. The set $T_{\mathcal{H}}^L$ can be written as follows:

$$T_{\mathcal{H}}^L = \bigcap_{u \in T} \mathrm{Val}_{\mathcal{H}}^X(u).$$

Take a set of points $A \subset \mathrm{Hom}(W(X, H))$ and define a set of formulas $A_{\mathcal{H}}^L$ in $\Phi(X)$:

$$A_{\mathcal{H}}^L = \{u \in \Phi(X) \mid A \subset \mathrm{Val}_{\mathcal{H}}^X(u)\}.$$

The set $A_{\mathcal{H}}^L$ is the set of all formulas $u \in \Phi(X)$ that hold true at all points from A. One can present the set $A_{\mathcal{H}}^L$ as follows:

$$A_{\mathcal{H}}^L = \bigcap_{\mu \in A} \mathrm{LKer}(\mu).$$

The defined above correspondence between sets of formulas and sets of points is the Galois correspondence (see [**11**]). In the case of the Galois correspondence, one can speak about Galois closures. In particular, subsets $T^L_{\mathcal{H}} \subset \mathrm{Hom}(W(X), H)$ and $A^L_{\mathcal{H}} \subset \Phi(X)$ are Galois-closed.

We call the subset $T^L_{\mathcal{H}} \subset \mathrm{Hom}(W(X), H)$ a *definable set* presented by the set of formulas T. The set $A^L_{\mathcal{H}}$ is a boolean filter in the algebra $\Phi(X)$, as an intersection of boolean filters $\mathrm{LKer}\,\mu$. It is called a *$\mathcal{H}$-closed filter.*

The Galois correspondence constructed above gives us a bijection between definable sets in $\mathrm{Hom}(W(X), H)$ and $\mathcal{H}$-closed filters in the extended boolean algebra $\Phi(X)$.

The following proposition is known.

PROPOSITION 2.12 ([**19**]). *The intersection of $\mathcal{H}$-closed filters is an $\mathcal{H}$-closed filter.*

The next proposition describes one more property of $\mathcal{H}$-closed filters ([**3**], [**18**]).

PROPOSITION 2.13. *Let a homomorphism of free algebras $s : W(X) \to W(Y)$ be given. If T is an $\mathcal{H}$-closed filter in $\Phi(Y)$, then $\widetilde{s}T$ is an $\mathcal{H}$-closed filter in $\Phi(X)$.*

COROLLARY 2.14. *Let A be a set of points in* $\mathrm{Hom}(W(Y), H)$ *and $s : W(X) \to W(Y)$ be a homomorphism of free algebras. Then*

$$\widetilde{s}(A^L_{\mathcal{H}}) = \left(\widetilde{s}A\right)^L_{\mathcal{H}}.$$

The next proposition describes the relation between the Galois correspondence and morphisms in the categories $\mathrm{Hal}_\Theta(\mathcal{H})$ and $\widetilde{\Phi}$ (see [**3**], [**18**]).

PROPOSITION 2.15. *Let T be a set of formulas from $\Phi(X)$, A be a set of points in* $\mathrm{Hom}(W(X), H)$ *and $s : W(X) \to W(Y)$ be a homomorphism of free algebras. Then*

(1) $(s_*T)^L_{\mathcal{H}} = s_*T^L_{\mathcal{H}}$,
(2) $s_*A^L_{\mathcal{H}} \subseteq (s_*A)^L_{\mathcal{H}}$.

COROLLARY 2.16. *If A is a definable set in* $\mathrm{Hom}(W(X), H)$, *then s_*A is also a definable set.*

A similar relation takes place between definable sets, $\mathcal{H}$-closed filters and maps $\widetilde{s}$ ([**3**], [**17**], [**18**]).

PROPOSITION 2.17. *Let T be a set of formulas from $\Phi(Y)$, A be a set of points in* $\mathrm{Hom}(W(Y), H)$. *Let a homomorphism of free algebras $s : W(X) \to W(Y)$ be given. Then*

(1) $\widetilde{s}(A^L_{\mathcal{H}}) = \left(\widetilde{s}A\right)^L_{\mathcal{H}}$,
(2) $\widetilde{s}(T^L_{\mathcal{H}}) \subseteq \left(\widetilde{s}T\right)^L_{\mathcal{H}}$.

3. Knowledge base model

In this section, we will introduce in detail our approach to a *knowledge base model* and we start with a discussion what is *knowledge*.

3.1. What is knowledge? Knowledge is a quite general and abstract notion. Although it is one of the most familiar concepts, the fundamental question remains:

"What is knowledge?"

Rarely has this question been answered directly. Numerous papers introduce a definition of knowledge, depending on the needs of a particular research and field of interest (see, for example, [**4**], [**6**], [**7**], [**8**], [**10**], [**15**], etc.)

In our discussion here, we say that a representation of knowledge has the following three components:

(1) *Subject area of knowledge*
(2) *Description of knowledge*
(3) *Content of knowledge*

Let us describe these components in more detail.

Subject area of knowledge is presented by a model $\mathcal{H} = (H, \Psi, f)$, where

- H is an algebra in a fixed variety of algebras Θ.
- Ψ is a set of relation symbols φ.
- f is an interpretation of each symbol φ in H.

Description of knowledge presents a syntactical component of knowledge. From the algebraic viewpoint, description of knowledge is a set of formulas T, more precisely, it is a boolean $\mathcal{H}$-closed filter in the algebra of formulas $\Phi(X)$, $X = \{x_1, \dots, x_n\}$.

Content of knowledge is a subset in H^n, where H^n is the Cartesian power of H. Each content of a knowledge A corresponds to the description of a knowledge $T \subset \Phi(X)$, $|X| = n$. If we regard H^n as an affine space, then this correspondence can be treated geometrically via Galois correspondence.

In order to describe the dynamic nature of a knowledge base, two categories and a functor are introduced:

- *the category of knowledge description* $F_\Theta(\mathcal{H})$,
- *the category of knowledge content* $D_\Theta(\mathcal{H})$,
- *the knowledge functor* $Ct_{\mathcal{H}}$.

These categories are defined using the machinery of logical geometry (cf. [**2**], [**25**]).

3.2. Knowledge description and knowledge content. A detailed description of these categories and the knowledge functor is given in [**3**]. Let us recall these constructions.

3.2.1. *Category of knowledge description* $F_\Theta(\mathcal{H})$. An object $F_\Theta^X(\mathcal{H})$ of the category of knowledge description $F_\Theta(\mathcal{H})$ is the lattice of all $\mathcal{H}$-closed filters in the algebra $\Phi(X)$, $X \in \Gamma$.

REMARK 3.1. We cannot say that the usual set-theoretical union of $\mathcal{H}$-closed filters is an $\mathcal{H}$-closed filter. To constitute a lattice of $\mathcal{H}$-closed filters in $\Phi(X)$, the following new operation is introduced:

$$T_1 \overline{\cup} T_2 = (T_1 \cup T_2)_{\mathcal{H}}^{LL}.$$

Then all $\mathcal{H}$-closed filters in $\Phi(X)$ form a lattice with the operation $\overline{\cup}$ and $\cap$ (for details see [**3**], [**18**], [**19**]).

Let a homomorphism $s : W(X) \to W(Y)$ and $\mathcal{H}$-closed filters $T_1 \in \Phi(X)$ and $T_2 \in \Phi(Y)$ be given. We say that a map $[s_*] : T_1 \to T_2$ is *admissible* if $s_* T_1 \subseteq T_2$.

A morphism between objects $F_\Theta^X(\mathcal{H})$ and $F_\Theta^Y(\mathcal{H})$

$$[s_*] : F_\Theta^X(\mathcal{H}) \to F_\Theta^Y(\mathcal{H})$$

is determined if $[s_*] : T_1 \to T_2$ is admissible for every $T_1 \in F_\Theta^X(\mathcal{H})$. A composition of morphisms $[s_*^1] : F_\Theta^X(\mathcal{H}) \to F_\Theta^Y(\mathcal{H})$ and $[s_*^2] : F_\Theta^Y(\mathcal{H}) \to F_\Theta^Z(\mathcal{H})$ is defined by the rule: $[s_*^2] \circ [s_*^1] = [s_*^2 s_*^1]$.

3.2.2. *Category of knowledge content* $D_\Theta(\mathcal{H})$. Define now the category $D_\Theta(\mathcal{H})$ of knowledge content. An object $D_\Theta^X(\mathcal{H})$ of $D_\Theta(\mathcal{H})$ is the lattice of all definable sets in the affine space $\mathrm{Hom}(W(X), H)$, $X \in \Gamma$.

Let a homomorphism $s : W(X) \to W(Y)$ and definable sets $A_2 \in \mathrm{Hom}(W(Y), H)$ and $A_1 \in \mathrm{Hom}(W(X), H)$ be given. We say that a map $[\widetilde{s}] : A_2 \to A_1$ is *admissible* if $\widetilde{s}A_2 \subseteq A_1$.

A morphism between objects $D_\Theta^Y(\mathcal{H})$ and $D_\Theta^X(\mathcal{H})$

$$[\widetilde{s}] : D_\Theta^Y(\mathcal{H}) \to D_\Theta^X(\mathcal{H})$$

is determined if $[\widetilde{s}] : A_2 \to A_1$ is admissible for every $A_2 \in D_\Theta^Y(\mathcal{H})$. Composition of morphisms $[\widetilde{s^1}] : F_\Theta^Z(\mathcal{H}) \to D_\Theta^Y(\mathcal{H})$ and $[\widetilde{s^2}] : D_\Theta^Y(\mathcal{H}) \to D_\Theta^X(\mathcal{H})$ is defined as follows: $[\widetilde{s^2}] \circ [\widetilde{s^1}] = [\widetilde{s^2}\widetilde{s^1}]$.

3.3. The knowledge functor $Ct_\mathcal{H}$. Categories of knowledge description $F_\Theta(\mathcal{H})$ and knowledge content $D_\Theta(\mathcal{H})$ are connected by the knowledge functor:

$$Ct_\mathcal{H} : F_\Theta(\mathcal{H}) \to D_\Theta(\mathcal{H}),$$

which is defined on objects by

$$Ct_\mathcal{H}(F_\Theta^X(\mathcal{H})) = D_\Theta^X(\mathcal{H}),$$

and on morphisms by

$$Ct_\mathcal{H}([s_*]) = [\widetilde{s}],$$

where $[s_*] : F_\Theta^X(\mathcal{H}) \to F_\Theta^Y(\mathcal{H})$ and $[\widetilde{s}] : D_\Theta^Y(\mathcal{H}) \to D_\Theta^X(\mathcal{H})$, for a given homomorphism of free algebras $s : W(X) \to W(Y)$. More precisely, if $[s_*] : F_\Theta^X(\mathcal{H}) \to F_\Theta^Y(\mathcal{H})$ is a morphism, such that $[s_*] : T_1 \to T_2$, then $[\widetilde{s}] : D_\Theta^Y(\mathcal{H}) \to D_\Theta^X(\mathcal{H})$ is a morphism, such that $[\widetilde{s}] : (T_2)_\mathcal{H}^L \to (T_1)_\mathcal{H}^L$.

The following theorem was proved in [**3**].

THEOREM 3.2. *The correspondence $Ct_\mathcal{H}$ is a contravariant functor.*

3.4. Definition of a knowledge base. Now we are at the point to give a definition of knowledge base model. Let a model $\mathcal{H} = (H, \Psi, f)$ be given.

DEFINITION 3.3. A knowledge base $KB = KB(H, \Psi, f)$ is a triple $\big(F_\Theta(\mathcal{H}), D_\Theta(\mathcal{H}), Ct_\mathcal{H}\big)$, where $F_\Theta(\mathcal{H})$ is the category of knowledge description, $D_\Theta(\mathcal{H})$ is the category of knowledge content, and

$$Ct_\mathcal{H} : F_\Theta(\mathcal{H}) \to D_\Theta(\mathcal{H})$$

is the contravariant functor.

REMARK 3.4. We will use the term “a knowledge base” instead of the more precise “a knowledge base model”.

One can say that a defined knowledge base model is a sort of automaton (see [**16**]), where queries are objects of the category of knowledge descriptions $F_\Theta(\mathcal{H})$, replies are objects of the category of knowledge content $D_\Theta(\mathcal{H})$. In order for such an automaton to be a knowledge base, connection with a particular data (information) is assumed. This information is held in the subject area presented by a model $\mathcal{H} = (H, \Psi, f)$.

The knowledge functor $Ct_{\mathcal{H}}$ gives a dynamical passage between queries and replies, namely, between categories $F_\Theta(\mathcal{H})$ and $D_\Theta(\mathcal{H})$.

Moreover, as we will see in the next section this passage is a one-to-one correspondence (Theorem 4.1).

4. Interaction between syntax and semantics in knowledge bases

Categories of knowledge descriptions $F_\Theta(\mathcal{H})$ and of knowledge content $D_\Theta(\mathcal{H})$ are defined using the machinery of logical geometry (see [**2**], [**25**]).

We presuppose that logic has two important components:

syntax and semantics.

Applying techniques and structures of algebraic logic (see [**9**], [**16**], Section 2.2), we put syntax and semantics into appropriate algebraic structures, namely, the category of knowledge description $F_\Theta(\mathcal{H})$ and the category of knowledge content $D_\Theta(\mathcal{H})$, respectively.

Syntax	$F_\Theta(\mathcal{H})$
	$\downarrow Ct_{\mathcal{H}}$
Semantics	$D_\Theta(\mathcal{H})$

FIGURE 2. Syntax and semantics: categories $F_\Theta(\mathcal{H})$ and $D_\Theta(\mathcal{H})$.

The next theorem shows that the knowledge functor $Ct_{\mathcal{H}}$ defines a dual isomorphism between categories of knowledge descriptions $F_\Theta(\mathcal{H})$ and of knowledge content $D_\Theta(\mathcal{H})$.

THEOREM 4.1. *The category of knowledge description $F_\Theta(\mathcal{H})$ is dual to the category of knowledge content $D_\Theta(\mathcal{H})$.*

PROOF. We have already defined the knowledge functor

$$Ct_{\mathcal{H}} : F_\Theta(\mathcal{H}) \to D_\Theta(\mathcal{H}).$$

Our goal is to construct the inverse functor

$$Ct_{\mathcal{H}}^{-1} : D_\Theta(\mathcal{H}) \to F_\Theta(\mathcal{H}),$$

and to show that it is indeed a functor. Let us define the correspondence $Ct_{\mathcal{H}}^{-1}$ on objects by

$$Ct_{\mathcal{H}}^{-1}(D_\Theta^Y(\mathcal{H})) = F_\Theta^Y(\mathcal{H}),$$

and on morphisms by

$$Ct_{\mathcal{H}}^{-1}([\widetilde{s}]) = [s_*],$$

where $s : W(X) \to W(Y)$ is a given homomorphism of free algebras, $[\widetilde{s}] : D^Y_\Theta(\mathcal{H}) \to D^X_\Theta(\mathcal{H})$ is a morphism in $D_\Theta(\mathcal{H})$, such that $[s_*] : A_2 \to A_1$, and $[s_*] : F^X_\Theta(\mathcal{H}) \to F^Y_\Theta(\mathcal{H})$ is a morphism in $F_\Theta(\mathcal{H})$, such that $[s_*] : (A_1)^L_\mathcal{H} \to (A_2)^L_\mathcal{H}$. Note that the map $[s_*] : (A_1)^L_\mathcal{H} \to (A_2)^L_\mathcal{H}$ is admissible for $(A_1)^L_\mathcal{H}$ and $(A_2)^L_\mathcal{H}$. Indeed, since $\widetilde{s}A_2 \subseteq A_1$, then

$$(A_1)^L_\mathcal{H} \subseteq (\widetilde{s}A_2)^L_\mathcal{H}.$$

By Proposition 2.17, we have $(\widetilde{s}A_2)^L_\mathcal{H} = \widetilde{s}(A_2)^L_\mathcal{H}$. Hence

$$s_*(A_1)^L_\mathcal{H} \subseteq s_*\widetilde{s}(A_2)^L_\mathcal{H} \subseteq (A_2)^L_\mathcal{H}.$$

For the last inclusion see Remark 2.9.

By construction of $Ct^{-1}_\mathcal{H}$, the property

$$Ct^{-1}_\mathcal{H}(id_{D^Y_\Theta(\mathcal{H})}) = id_{Ct^{-1}_\mathcal{H}(D^Y_\Theta(\mathcal{H}))}$$

holds, where $id_{D^Y_\Theta(\mathcal{H})}$ is the identity morphism of $D^Y_\Theta(\mathcal{H})$ and $id_{Ct^{-1}_\mathcal{H}(D^Y_\Theta(\mathcal{H}))}$ is the identity morphism of $Ct^{-1}_\mathcal{H}(D^Y_\Theta(\mathcal{H})) = F^Y_\Theta(\mathcal{H})$.

It remains to check that

$$Ct^{-1}_\mathcal{H}([\widetilde{s^2}] \circ [\widetilde{s^1}]) = Ct^{-1}_\mathcal{H}([\widetilde{s^1}]) \circ Ct^{-1}_\mathcal{H}([\widetilde{s^2}]),$$

where $[\widetilde{s^1}] : D^Z_\Theta(\mathcal{H}) \to D^Y_\Theta(\mathcal{H})$, $[\widetilde{s^2}] : D^Y_\Theta(\mathcal{H}) \to D^X_\Theta(\mathcal{H})$, and $s_2 : W(X) \to W(Y)$, $s_1 : W(Y) \to W(Z)$.

Let A_3 be a definable set in $D^Z_\Theta(\mathcal{H})$. Then, $[\widetilde{s^1}] : A_3 \to A_2$, $[\widetilde{s^2}] : A_2 \to A_1$, and, by the definition of morphisms in $D_\Theta(\mathcal{H})$, we have $\widetilde{s^1}A_3 \subseteq A_2$, $\widetilde{s^2}A_2 \subseteq A_1$. Thus, we have

$$[\widetilde{s^2}] \circ [\widetilde{s^1}] : A_3 \to A_1$$

and

$$Ct^{-1}_\mathcal{H}([\widetilde{s^2}] \circ [\widetilde{s^1}]) : (A_1)^L_\mathcal{H} \to (A_3)^L_\mathcal{H}. \tag{1}$$

On the other hand,

$$Ct^{-1}_\mathcal{H}([\widetilde{s^1}]) : (A_2)^L_\mathcal{H} \to (A_3)^L_\mathcal{H},$$
$$Ct^{-1}_\mathcal{H}([\widetilde{s^2}]) : (A_1)^L_\mathcal{H} \to (A_2)^L_\mathcal{H}.$$

Hence

$$Ct^{-1}_\mathcal{H}([\widetilde{s_1}]) \circ Ct^{-1}_\mathcal{H}([\widetilde{s_2}]) : (A_1)^L_\mathcal{H} \to (A_3)^L_\mathcal{H}. \tag{2}$$

Therefore, by equations (1) and (2), we have

$$Ct^{-1}_\mathcal{H}([\widetilde{s^2}] \circ [\widetilde{s^1}]) = Ct^{-1}_\mathcal{H}([\widetilde{s^1}]) \circ Ct^{-1}_\mathcal{H}([\widetilde{s^2}]),$$

and $Ct^{-1}_\mathcal{H}$ is a contravariant functor.

One can check that $Ct_\mathcal{H} \circ Ct^{-1}_\mathcal{H} = id_{D_\Theta(\mathcal{H})}$ and $Ct^{-1}_\mathcal{H} \circ Ct_\mathcal{H} = id_{F_\Theta(\mathcal{H})}$. Thus, $Ct^{-1}_\mathcal{H}$ is a functor inverse to $Ct_\mathcal{H}$. The theorem is proved. □

5. Conclusion

As we noted in the Introduction, one of the main problems in database and knowledge bases theory is the equivalence problem. In view of Theorem 4.1, our next goal is to revise our previous results in this area (see [**2**], [**18**], [**23**], [**24**]). One can expect that using the considered approach to a knowledge base model, we can get new helpful results on knowledge bases equivalence.

References

[1] E. Aladova, A. Gvaramiya, B. Plotkin, and T. Plotkin, *Multisorted logic, models and logical geometry* (Russian, with English and Russian summaries), Fundam. Prikl. Mat. **19** (2014), no. 3, 5–22; English transl., J. Math. Sci. (N.Y.) **214** (2016), no. 6, 742–754. MR3431877

[2] E. Aladova, E. Plotkin, and T. Plotkin, *Isotypeness of models and knowledge bases equivalence*, Math. Comput. Sci. **7** (2013), no. 4, 421–438, DOI 10.1007/s11786-013-0166-5. MR3158080

[3] E. Aladova, *Syntax versus semantics in knowledge bases I*, to appear in Int. J. Algebra Comput.

[4] M. Alavi and D. E. Leidner, *Review: Knowledge management and knowledge management systems: Conceptual foundations and research issues*, Management Information Systems Quarterly **25** (1) (2001), 107–136.

[5] C. C. Chang and H. J. Keisler, *Model theory: Studies in Logic and the Foundations of Mathematics*, Vol. 73, North-Holland Publishing Co., Amsterdam-London; American Elsevier Publishing Co., Inc., New York, 1973. MR0409165

[6] T. Davenport and L. Prusak, *Working knowledge: How organizations manage what they know*, Harvard Business School Press, 1998.

[7] R. Davis, H. Shrobe, P. Szolovits, *What is a knowledge representation?*, AI Magazine **14** (1) (1993). 17–33.

[8] W. Frawley, G. Piatetsky-Shapiro, and C. Matheus, *Knowledge discovery in databases: An overview*, AI Magazine **13** (1992), 57–70.

[9] P. R. Halmos, *Algebraic logic*, Chelsea Publishing Co., New York, 1962. MR0131961

[10] M. Jarke, B. Neumann, Y. Vassiliou, W. Wahlster, *KBMS requirements of knowledge-based systems*, Foundations of Knowledge Base Management: Contributions from Logic, Databases, and Artificial Intelligence (J. W. Schmidt and C. Thanos, eds.), Springer, 1989, 391-195.

[11] S. Mac Lane, *Categories for the working mathematician*, Graduate Texts in Mathematics, vol. 5, Springer-Verlag, New York-Berlin, 1971. MR0354798

[12] A. I. Mal′cev, *Algebraic systems*, Springer-Verlag, New York-Heidelberg, 1973. Posthumous edition, edited by D. Smirnov and M. Taĭclin; Translated from the Russian by B. D. Seckler and A. P. Doohovskoy; Die Grundlehren der mathematischen Wissenschaften, Band 192. MR0349384

[13] D. Marker, *Model theory: An introduction*, Graduate Texts in Mathematics, vol. 217, Springer-Verlag, New York, 2002. MR1924282

[14] G. Mashevitzky, B. Plotkin, and E. Plotkin, *Automorphisms of the category of free Lie algebras*, J. Algebra **282** (2004), no. 2, 490–512, DOI 10.1016/j.jalgebra.2003.09.038. MR2101073

[15] C. McInerney, *Knowledge management and the dynamic nature of knowledge*, Journal of the American Society for Information Science and Technology **53** (12) (2002), 1009–1018.

[16] B. Plotkin, *Universal algebra, algebraic logic, and databases*, Mathematics and its Applications, vol. 272, Kluwer Academic Publishers Group, Dordrecht, 1994. Translated from the 1991 Russian original by J. Cīrulis, A. Nenashev and V. Pototsky and revised by the author. MR1273136

[17] B. Plotkin, *Seven lectures on the universal algebraic geometry*, preprint, arXiv:math/0204245, 87 pp, 2002.

[18] B. Plotkin, *Algebraic geometry in first-order logic* (Russian, with Russian summary), Sovrem. Mat. Prilozh. **22, Algebra i Geom.** (2004), 16–62, DOI 10.1007/s10958-006-0288-2; English transl., J. Math. Sci. (N.Y.) **137** (2006), no. 5, 5049–5097. MR2462070

[19] B.I. Plotkin, *Isotyped algebras*, Proc. Steklov Inst. Mathematics **287** (2012), 91–115.

[20] B. Plotkin, *Algebraic logic and logical geometry in arbitrary varieties of algebras*, Group theory, combinatorics, and computing, Contemp. Math., vol. 611, Amer. Math. Soc., Providence, RI, 2014, pp. 151–167, DOI 10.1090/conm/611/12206. MR3204611

[21] B. Plotkin, E. Aladova, and E. Plotkin, *Algebraic logic and logically-geometric types in varieties of algebras*, J. Algebra Appl. **12** (2013), no. 2, 1250146, 23 pp., DOI 10.1142/S0219498812501460. MR3005597

[22] B. Plotkin and E. Plotkin, *Multi-sorted logic and logical geometry: some problems*, Demonstr. Math. **48** (2015), no. 4, 578–619, DOI 10.1515/dema-2015-0042. MR3430891

[23] T. Plotkin and B. Plotkin, *Geometrical aspect of databases and knowledge bases*, Algebra Universalis **46** (2001), no. 1-2, 131–161, DOI 10.1007/PL00000333. MR1835790

[24] B. Plotkin and T. Plotkin, *An algebraic approach to knowledge base models informational equivalence*, Acta Appl. Math. **89** (2005), no. 1-3, 109–134 (2006), DOI 10.1007/s10440-005-9008-z. MR2220200

[25] B. Plotkin and T. Plotkin, *Categories of elementary sets over algebras and categories of elementary algebraic knowledge*, Pillars of computer science, Lecture Notes in Comput. Sci., vol. 4800, Springer, Berlin, 2008, pp. 555–570, DOI 10.1007/978-3-540-78127-1_30. MR2478796

Department of Mathematics, Bar-Ilan University, 5290002, Ramat Gan, Israel

Current address: Federal University of Rio Grande do Norte, University Campus Lagoa Nova, CEP 59078-970, Natal/RN, Brazil

Email address: aladovael@mail.ru

Department of Computer Science, Bar Ilan University, 5290002, Ramat Gan, Israel

Email address: plot@macs.biu.ac.il

Contemporary Mathematics
Volume **726**, 2019
https://doi.org/10.1090/conm/726/14608

Polyadic algebras with terms: A signature-free approach

Jānis Cīrulis

Abstract. Two familiar versions of polyadic algebras with terms in the literature are specialized Halmos algebras and superpolyadic algebras. They are polyadic algebras with additional operations determined by some relatively free algebra W of any signature generated by the underlying set of variables: instead of the usual transformations of the set of variables, either endomorphisms of W or (equivalently) substitutions of elements of W for variables act on a polyadic algebra as Boolean endomorphisms (and interact also with quantifiers). Elements of W are then considered as terms of the expanded polyadic algebra. In fact, only the monoid of endomorphisms, respectively, of substitutions of W, plays any role rather than the signature and algebraic properties of W. Recently, a notion of a freeoid as an abstraction from couples $(W, \mathrm{End}(W))$ has been introduced. In the paper, superpolyadic algebras are defined replacing the relatively free algebra in the above scheme by a freeoid. Basic properties of superpolyadic and superpolyadic equality algebras are discussed. It is shown that, if the set of variables is infinite, every superpolyadic algebra is isomorphic to a functional algebra and embeddable into an equality superpolyadic algebra, and that a locally finite-dimensional superpolyadic equality algebra is representable in the sense that it admits a homomorphism into a superpolyadic set algebra with the so-called functional equality.

1. Introduction

Algebraization of (full) first-order logic began with the introduction of cylindric algebras by A. Tarski and his collaborators at the beginning of the 1950's [**30**]; these algebras are the algebraic counterpart of quantifier logic with equality. Later, P. Halmos proposed polyadic algebras, another form of algebraized first order logic (without equality), in [**11, 12**]; it seems that also the term 'algebraic logic' was coined by him. See also [**18, 19**].

Every polyadic algebra is related to a set of elements I, informally thought of as variables of some first-order logical language L, and is defined, roughly, as a quantifier algebra (a Boolean algebra together with a family of operations called quantifiers) A on which the transformation semigroup of I acts by Boolean endomorphisms; these endomorphisms also interact with the quantifiers in a certain way.

2010 *Mathematics Subject Classification.* Primary 03G12.

Key words and phrases. freeoid, locally finite, polyadic algebra, representation, superpolyadic algebra.

This work was supported by the Latvian Council of Science, Grant No. 271/2012. The author is grateful to the anonymous referee for his/her remarks and sugestions.

The underlying quantifier algebra is, in a sense, an algebra of "squeezed" (considered up to logical equivalence) formulas of L; each transformation of I represents a replacement of variables, and then the mentioned Boolean endomorphisms correspond to substitutions of variables for variables into formulas. Terms other than variables (even if they are present in L) do not appear in this scheme. However, there is a certain involved algebraic procedure for extracting operations and terms from a polyadic algebra, which is the algebraic counterpart of the way how function symbols are introduced in a logical language by definitions; see [**16**] and [**8**] for details.

However, it seems to be natural that, together with an algebra of squeezed formulas, also an algebra of squeezed terms should appear explicitly in an algebraic counterpart of a first-order theory. The idea of such a two-sorted algebra of first-order logic was proposed by H. Andréka and I. Sain already in [**2**]. In the context of cylindric algebras, it was realized by N. Feldman in [**10**] and the present author in [**4, 5**]. In both versions, the natural operations on terms correspond to substitutions of a term for a variable.

In a series of papers written since 1983 (see, e.g., [**23, 26, 27**]), B. Plotkin developed an algebraic model of a relational database as an automaton. The algebras of queries and replies in this model initially were polyadic algebras, later renamed Halmos algebras by him. Involving data algebras in the model forced also enriching Halmos algebras by a certain algebra of squeezed terms. As a result, Plotkin introduced a concept of Halmos algebra related to a variety of algebras; these algebras were called specialized Halmos algebras. A specialized Halmos algebra presupposes, besides the set of variables I, also a given variety of algebras Θ. The algebra W of this variety freely generated by I plays the role of the algebra of terms. Now not only variables, but arbitrary terms, i.e., elements of W, can be substituted for variables; correspondingly, arbitrary endomorphisms of W act, by Boolean endomorphisms, on the underlying quantifier algebra. See Chapter 9 of [**28**] for details.

With the same motivation, the present author independently found a similar but simpler notion of expanded polyadic algebras, called superpolyadic in [**3, 31**]. A more recent, preliminary attempt to define polyadic algebras with terms is [**1**]. We do not mention here approaches leading to categorical logic.

For two reasons, both coming out of the spirit of algebraic logic, neither specialized Halmos algebras nor superpolyadic algebras can be considered as a final version of polyadic algebras with terms. First, they are not abstract to a sufficient degree: a free algebra of a variety presupposes one or another set of operation symbols of the language, while this is not the case with the quantifier algebra and the set of predicate (or relation) symbols of L. Secondly, structures of algebraic logic do not exclude infinitary relations and, hence, should not exclude also infinitary operations. The first consideration suggests that the free algebra W should be replaced by an abstract clone; the other one requires the clone to be infinitary. One more reason: both specialized Halmos algebras and superpolyadic algebras are, in fact, the usual polyadic algebras expanded by additional operations connected with terms – rather than full-scale two-sorted algebras, consisting of two separate domains and thus admitting also operations of mixed types. Cf. [**10**], where cylindric algebras with terms are presented as such two-sorted algebras; in [**4**], both forms of supercylindric algebras are discussed.

In this paper, we shall not concern ourselves with this third problem: in distinction to the cylindric paradigm, several technical problems arise when realizing the idea in the framework of polyadic algebras. As to the two former ones (signature-freeness and infinitary operations), a part of the job has already been done in [**6**, **31**, **32**].

Proceeding via the approach initiated in [**32**], the author has proposed in [**6**] a version of infinitary clones, so-called freeoids, appropriate for the aforementioned purposes (see Section 1 of that paper for a critical discussion, from this point of view, of several other kinds of clones already known in the literature). Also an algebraic theory of freeoids and their models has been developed there. Roughly, a freeoid is an abstraction of a relatively free infinitary algebra, as far as we are interested only in the endomorphism monoid of the latter (not in its operations); its elements may be thought of as sqeezed terms, and it provides a monoid of substitutions of terms for variables, which may act, by Boolean endomorphisms, on a quantifier algebra. Further, the class (in fact a category) of the so-called models of a freeoid is the counterpart of the variety generated by a relatively free algebra. In the present paper, we make the next step in developing the notion of a signature-free polyadic algebra with terms, forming from quantifier algebras and freeoids united structures, (general) superpolyadic algebras. We also transfer to these new algebras several results obtained for superpolyadic algebras in the restricted sense of [**31**, **32**].

The next section contains the background information about polyadic algebras and their reducts known as quantifier algebras and transformation algebras. Likewise, the necessary information on freeoids, their substitution monoids an models is presented in Section 3. Section 4 contains the bulk of the new material; we introduce here notions of substitution algebra and substitution quantifier algebra (also, with an equality) and define superpolyadic algebras as particular substitution quantifier algebras. Examples and some preliminary results also are presented here. Two representation theorems for superpoliadic algebras are proved in the final section, Section 5.

Before ending this introduction, we note that a polyadic algebra can be turned into a superpolyadic one in various ways, i.e., supplementing it with non-isomorphic freeoids. Therefore, relations between the terms provided by an adjoined freeoid and terms in the sense of Halmos [**16**] extracted from the given polyadic algebra itself requires a further investigation.

2. Polyadic algebras

In what follows, X is a set which we assume to be infinite, though, in several places, it could be nonempty or containing at least two elements. Further, $\mathcal{P}X$ is its powerset and $Tr := (X^X, \cdot, \varepsilon)$ is the monoid of transformations of X; here, ε is the identity transformation, and $\cdot$, the composition operation. In expressions, the composition of two mappings will usually be indicated by juxtaposition of their notations.

Elements of X will be called *variables*. Thus, we partly change the notation appearing in the Introduction and use the letter 'X' instead of 'I', the standard symbol for the set of variables in the early papers on polyadic algebras.

2.1. Quantifiers and quantifier algebras [13–15].

DEFINITION 2.1. An (existential) *quantifier* on a Boolean algebra A is a unary operation $\exists$ satisfying the axioms (0 is the zero element of A)

$$\exists 0 = 0, \quad a \leq \exists a, \quad \exists(a \wedge \exists b) = \exists a \wedge \exists b. \tag{2.1}$$

An equivalent set of axioms is

$$a \leq \exists a, \quad \text{if } a \leq b, \text{ then } \exists a \leq \exists b, \quad \exists(-\exists a) = -\exists a, \tag{2.2}$$

where $-$ is the complementation in A. Every quantifier is actually a completely additive closure operator on A. Its range, $\exists(A)$, is a Boolean subalgebra of A.

Quantifiers are known also as *cylindrifications* [**19**]. For example, the identity map id_A of A is always a quantifier. It is easily seen (cf. [**24**, Lemma 3.4] and [**20**, Proposition 2]) that the composition of two quantifiers is a quantifier if and only if they commute. Thus, quantifiers form a partial commutative monoid $(M, \cdot, \mathsf{id}_A)$ (w.r.t. the composition operation).

DEFINITION 2.2. A *quantifier algebra* is a pair $\mathsf{A} := (A, \boldsymbol{\exists})$, where A is a Boolean algebra and $\boldsymbol{\exists}$ is a homomorphism from the semilattice $(\mathcal{P}X, \cup, \varnothing)$ to M (we write $\exists_J$ for $\boldsymbol{\exists}(J)$):

$$\exists_\varnothing = \mathsf{id}_A, \quad \exists_{J \cup K} = \exists_J \exists_K \text{ for all } J, K \in \mathcal{P}X. \tag{2.3}$$

A quantifier algebra may be presented also as an algebra $(A, \exists_J)_{J \in \mathcal{P}X}$ of a more traditional form.

A subset $J \subseteq X$ is said to be a *support* of an element a (in symbols, $J \operatorname{spp} a$) of a quantifier algebra A if $\exists_{\overline{J}} a = a$, where $\overline{J}$ stands for the complement of J. The set of all supports of an element a is a filter on X. An element a is said to be *finite-dimensional* if it has a finite support. If every element of A has a finite support, the algebra is said to be *locally finite*, more precisely, *locally finite-dimensional.*

An element a is said to be *independent* of a variable x, in symbols, $a \operatorname{ind} x$, if x does not belong to any support of a, i.e., $\overline{x} \operatorname{spp} a$ (we write $\overline{x}$ for $\overline{\{x\}}$).

In every quantifier algebra, $\overline{J} \operatorname{spp} \exists_J a$. More generally,

$$\text{if } K \operatorname{spp} a, \text{ then } (K \smallsetminus J) \operatorname{spp} \exists_J a. \tag{2.4}$$

THEOREM 2.3. *Let A be a quantifier algebra, and let $a \in A$. If J is a support of a, then, for any subsets K, L of X,*

$$\text{if } J \cap K = J \cap L, \text{ then } \exists_K a = \exists_L a. \tag{2.5}$$

PROOF. Assume that $J \operatorname{spp} a$ (hence, $\overline{K \smallsetminus J} \operatorname{spp} a$) and suppose that $J \cap K = J \cap L$. Then $\exists_K a = \exists_{(K \cap J) \cup (K \smallsetminus J)} a = \exists_{K \cap J} \exists_{K \smallsetminus J} a = \exists_{K \cap J} a$. Likewise $\exists_L a = \exists_{L \cap J} a$, and then $\exists_K a = \exists_L a$. □

EXAMPLE 2.4 (*functional quantifier algebras*). Let Q be a non-empty set, and let B be a Boolean algebra. The set $F := F_Q(B)$ of all functions $f \colon Q^X \to B$ also inherits the structure of a Boolean algebra. For $J \subseteq X$, define a function $\exists_J f \in F_Q(B)$ by setting

$$\exists_J f(\varphi) := \bigvee (f(\psi) \colon \psi | \overline{J} = \varphi | \overline{J}), \tag{*}$$

provided the indicated supremum exists for every φ in Q^X. If A is a Boolean subalgebra of F such that the function $\exists_J f$ is defined and belongs to A for all

$f \in A$, then we come to an operation $\exists_J$ on A; this operation is a quantifier and is called a *functional quantifier*. If such a quantifier can be defined for every J, then the algebra $\mathsf{A} := (A, \exists_J)_{J \in \mathcal{P}X}$ is a quantifier algebra, called a *functional quantifier algebra*. The Boolean algebra B is its *algebra of values*, and the set Q is its *domain*. If B is complete, then all functional quantifiers on F exist.

Notice that a support of a function $f \in A$ may be characterized as follows:

$$(2.6) \qquad J \operatorname{spp} f \quad \text{iff} \quad f(\varphi) = f(\psi) \text{ for all } \varphi, \psi \in Q^X \text{ with } \varphi|J = \psi|J.$$

Necessity of the condition: if $\exists_{\overline{J}} f = f$, then, by virtue of (*), $f(\psi) \leq f(\varphi)$ whenever $\psi|J = \varphi|J$, and likewise $f(\varphi) \leq f(\psi)$. Its sufficiency follows from another simple observation: if the right-hand part of (2.6), with $\overline{J}$ instead of J, holds, then all terms of the join in the likewise modified equality (*) equal to $f(\varphi)$.

Functional algebras with a two-element value algebra $B := \{0, 1\}$ are of special interest. Elements of such an algebra may be considered as characteristic functions of subsets of the domain; owing to this, two-valued functional algebras can be presented as a set algebras (this comment concerns also similar situations below).

EXAMPLE 2.5 (*quantifier set algebras*). Given a nonempty set Q, consider the Boolean algebra of functions $C := \mathcal{P}(Q^X)$. For every $\Phi \in C$ and $J \subseteq X$, let

$$(2.7) \qquad \exists_J \Phi := \{\varphi \in Q^X : \varphi|\overline{J} = \psi|\overline{J} \text{ for some } \psi \in \Phi\}.$$

Then each $\exists_J$ is a quantifier on C, also called *functional*, and $(C, \exists_J)_{J \in \mathcal{P}(X)}$ is a quantifier algebra, called the *full quantifier set algebra* with domain Q. Any subalgebra of such algebra is said to be a *quantifier set algebra*. In such an algebra,

$$(2.8) \qquad J \operatorname{spp} \Phi \quad \text{iff} \quad \text{for all } \varphi, \psi \in Q^X \text{ with } \varphi|J = \psi|J, \ (\varphi \in \Phi \text{ iff } \psi \in \Phi).$$

Definition (2.7) can be rewritten as

$$\exists_J \Phi := \{\psi\alpha \colon \psi \in \Phi \text{ and } \alpha|\overline{J} = \varepsilon|\overline{J}\}.$$

In the case when J is a singleton, this reduces to

$$(2.9) \qquad \exists_x \Phi = \{\psi^x_u \colon \psi \in \Phi \text{ and } u \in Q\}.$$

2.2. Transformation algebras [7, 14, 21].

DEFINITION 2.6. A *transformation algebra* is a pair (A, S), where

- A is a Boolean algebra,
- S is a homomorphism $Tr \to \operatorname{End}(A)$ (we write S_α for $\mathsf{S}(\alpha)$):

$$(2.10) \qquad S_\varnothing = \mathsf{id}_A, \quad S_{\alpha\beta} = S_\alpha S_\beta \text{ for all } \alpha, \beta \in Tr.$$

It can be presented also in a more traditional form as an algebra $(A, S_\alpha)_{\alpha \in T}$.

A subset J of X is said to be a *support* of an element a if $S_\alpha a = a$ for every α which agrees with ε on J. A transformation algebra is said to be *locally finite-dimensional*, if each of its element has a finite support, and also the notations $X \operatorname{spp} a$ and $a \operatorname{ind} x$ are defined as in the preceding subsection. If J is a support of an element $a \in A$, then

$$(2.11) \qquad \alpha(J) \operatorname{spp} S_\alpha a,$$

$$(2.12) \qquad \text{if } \alpha|J = \beta|J, \text{ then } S_\alpha a = S_\beta a$$

(see Lemma 3.5 in [9], resp., Theorem 8.10 in [7]).

EXAMPLE 2.7 (*functional transformation algebras*). Consider again the Boolean algebra $F_Q(B)$ of all functions $f\colon Q^X \to B$ from Example 2.4. For every $\alpha \in Tr$ and all $f \in F_Q(B)$, $\varphi \in Q^X$, set

$$(S_\alpha f)(\varphi) := f(\varphi\alpha).$$

Then $(F_Q(B), S_\alpha)_{\alpha\in Tr}$ is a transformation algebra, called the *full functional transformation algebra* with a value algebra B and a domain Q. A *functional transformation algebra* is any subalgebra of a full functional algebra.

In a functional transformation algebra,

$$J \operatorname{spp} f \quad \text{iff} \quad f(\varphi) = f(\psi) \text{ whenever } \varphi|J = \psi|J. \tag{2.13}$$

Indeed, if the condition on the right is fulfilled and β is a transformation such that $\beta|J = \varepsilon|J$, then $f(\varphi\beta) = f(\varphi)$ for all φ and, further, $S_\beta f = f$. If, conversely, the latter equality is fulfilled whenever $\beta|J = \varepsilon|J$, and if $\varphi|J = \psi|J$, consider first the case $J \neq \varnothing$. Then, for any transformation β such that $\beta(X) \subseteq J$ and $\beta|J = \varepsilon|J$, $\varphi\beta = \psi\beta$ and, further,

$$f(\varphi) = S_\beta f(\varphi) = f(\varphi\beta) = f(\psi\beta) = S_\beta f(\psi) = f(\psi).$$

If $J = \varnothing$, then $S_\beta f = f$ for all transformations β. We can choose transformations β_1, β_2 such that $\beta_1 X \cap \beta_2 X = \varnothing$ and $\chi \in Q^X$ which agrees with φ on $\beta_1 X$ and with ψ on $\beta_2 X$. Now,

$$f(\varphi) = (S_{\beta_1} f)(\varphi) = f(\varphi\beta_1) = f(\chi\beta_1) = (S_{\beta_1} f)(\chi) = f(\chi)$$

and likewise, $f(\psi) = f(\chi)$.

Every transformation algebra is isomorphic to a functional transformation algebra; see [**21**, Theorem 1.1] or [**9**, Theorem 3.1].

EXAMPLE 2.8 (*transformation set algebras*). Given a nonempty set Q, consider the Boolean algebra of sets of valuations $C := \mathcal{P}(Q^X)$ from Example 2.5. For every $\Phi \in A$ and $\alpha \in Tr$, set

$$S_\alpha \Phi := \{\varphi \in Q^X\colon \varphi\alpha \in \Phi\}. \tag{2.14}$$

Then $(A, S_\alpha)_{\alpha\in Tr}$ is a transformation algebra, called a *full transformation set algebra with domain Q*. A *transformation set algebra* is any subalgebra of such a full set algebra.

It follows from (2.13) that, in a transformation set algebra, $J \operatorname{spp} \Phi$ if and only if ($\varphi \in \Phi$ iff $\psi \in \Phi$) whenever $\varphi|J = \psi|J$.

2.3. Polyadic algebras [7, 14, 15]. For more recent developments in the theory of polyadic algebras, see [**29**].

DEFINITION 2.9. A *polyadic algebra* is a triple $(A, \exists, \mathsf{S})$, where

- $(A, \exists)$ is quantifier algebra,
- (A, S) is a transformation algebra,
- the homomorphisms $\exists$ and S interact in this way:

for every $J \in \mathcal{P}X$ and all $\alpha, \beta \in Tr$,

$$S_\alpha \exists_J = S_\beta \exists_J \text{ whenever } \alpha|\overline{J} = \beta|\overline{J}, \tag{2.15}$$

$$\exists_J S_\alpha = S_\alpha \exists_{\alpha^{-1}(J)} \text{ whenever } \alpha \text{ is injective on } \alpha^{-1}(J) \tag{2.16}$$

(here, $\alpha^{-1}(J) = \{x \in X\colon \alpha x \in J\}$). It can be presented also as an algebra $(A, \exists_J, S_\alpha)_{J\in\mathcal{P}X, \alpha\in Tr}$.

Let us consider standard examples.

EXAMPLE 2.10 (*functional polyadic algebras*). Let $(A, S_\alpha)_{\alpha\in Tr}$ be a functional transformation algebra with domain Q (Example 2.7). If all possible functional quantifiers $\exists_J$ on A are defined and give rise to a quantifier algebra $(A, \exists_J)_{J\in\mathcal{P}X}$ (Example 2.4), then $(A, S_\alpha, \exists_J)_{\alpha\in Tr, J\in\mathcal{P}X}$ is a polyadic algebra, known as a *functional polyadic algebra*, also with domain Q.

EXAMPLE 2.11 (*polyadic set algebras*). Let C be the Boolean algebra $\mathcal{P}(Q^X)$ for some set Q; see Examples 2.5 and 2.8. If $\exists_J$ and S_α are defined as in these examples, then the obtained algebra $(A, \exists, \mathsf{S})$ is a polyadic algebra, called a *full polyadic set algebra* (with domain Q). A *polyadic set algebra* is any subalgebra of such a set algebra.

As (2.6) and (2.13) show, the two possible support relations in functional polyadic algebras agree. This is the case also in general: by [**7**, Theorem 8.10] or [**9**, Lemma 4.1], if $|X| \geq 1$, then, for every element a of an arbitrary polyadic algebra, its supports in the sense of subsection 2.1 and of subsection 2.2 coincide. Thus we can freely use the relevant notation, definitions and properties of the relations spp and ind also in the context of polyadic algebras. In particular, a polyadic algebra is *locally finite-dimensional* if each of its elements has finite supports.

By [**21**, Corollary 1.2], any transformation algebra can be embedded into a polyadic algebra.

It follows from the results of Section 4 in [**24**] that the pair of axioms (2.15) and (2.16) of polyadic algebras can be replaced by a triple of simpler axioms. For a transformation α, put

$$\operatorname{edm}\alpha := \{x\colon \alpha x \neq x\}, \quad \operatorname{ern}\alpha := \{\alpha(x)\colon x \in J\}.$$

A transformation α of X is said to be a *projection* if it is idempotent. Notice that this is the case if and only if $\operatorname{edm}\alpha \cap \operatorname{ern}\alpha = \varnothing$.

PROPOSITION 2.12. *Suppose that $(A, \exists)$ is a quantifier algebra and (A, S) is a transformation algebra. Then $(A, \exists, \mathsf{S})$ is a polyadic algebra if and only if the following conditions are fulfilled for every projection α:*

$$S_\alpha\exists_J = \exists_J, \ \exists_J S_\alpha = S_\alpha, \quad \textit{where } J = \operatorname{edm}\alpha, \tag{2.17}$$

$$\textit{if } J \cap (\operatorname{edm}\alpha \cup \operatorname{ern}\alpha) = \varnothing, \ \textit{then } \exists_J S_\alpha = S_\alpha \exists_J, \tag{2.18}$$

It is worth noting that the supposition in (2.18) is equivalent to the requirement that, for every $x \in J$ and all $y \in X$, $x = y$ whenever $\alpha(x) = \alpha(y)$ ([**24**, Lemma 3.3]).

A simple variant of this theorem for locally finite-dimensional polyadic algebras can be derived from results of Section 3 of [**25**]. (In that paper, the term 'quantifier algebra' has a meaning quite different from that assumed in subsection 2.1 above.)

PROPOSITION 2.13. *Suppose that $(A, \exists)$ is a locally finite-dimensional quantifier algebra and (A, S) is a like transformation algebra. Then $(A, \exists, \mathsf{S})$ is a (locally finite-dimensional) polyadic algebra if and only if the following conditions are fulfilled for all x, y, z:*

$$S_y^x\exists_x = \exists_x, \quad \exists_x S_y^x = S_y^x \ \textit{if } x \neq y,$$
$$S_y^x\exists_z = \exists_z S_y^x \ \textit{if } z \neq x, y.$$

2.4. Equality for transformation and polyadic algebras [7,17,21]. Following [**16**, Section 2], we define a *predicate* on a transformation algebra A as any function $p\colon X^n \to A$ invariant under transformations in the sense that, for every $\alpha \in Tr$,

$$S_\alpha p(x_1, x_2, \ldots, x_n) = p(\alpha x_1, \alpha x_2, \ldots, \alpha x_n),$$

where all variables are supposed to be mutually distinct. For a predicate p,

$$\{x_1, x_2, \ldots, x_n\} \operatorname{spp} p(x_1, x_2, \ldots, x_n). \tag{2.19}$$

DEFINITION 2.14. An *equality* for a transformation algebra A is a predicate $e\colon X^2 \to A$ such that

$$\text{(i) } e(x,x) = 1, \quad \text{(ii) } a \wedge e(x,y) \le S^x_y a, \tag{2.20}$$

where S^x_y is the transformation that sends x onto y and agrees with ε elsewhere. A *transformation equality algebra* is a transformation algebra with an equality.

An equality for a polyadic algebra, and *polyadic equality algebras* are defined similarly.

Any equality predicate may be replaced by a family e_{xy} of elements of the algebra, where e_{xy} stands for $e(x,y)$. Then a transformation equality algebra can be presented as a usual algebra $(A, S_\alpha, e_{xy})_{\alpha\in Tr,\, x,y\in X}$ (the short form: $(A, \exists, \mathsf{S}, \mathsf{e})$), which is characterized by equational axioms.

Equality is *symmetric* and *transitive*:

$$e(x,y) = e(y,x), \quad e(x,y) \wedge e(y,z) \le e(x,z).$$

EXAMPLE 2.15 (*functional equality*). In a full functional transformation algebra $\mathsf{F} := (F_Q(B), \mathsf{S})$, the function $e\colon X^2 \to F_Q(B)$ defined by

$$e(x,y)(\varphi) := \begin{cases} 1 & \text{if } \varphi x = \varphi y, \\ 0 & \text{if } \varphi x \ne \varphi y \end{cases}$$

is an equality, called the *functional equality for* F. Notice that it does not depend on the choice of B. If a subalgebra A of F contains all elements $e(x,y)$, then e is an equality also for the subalgebra. However, A may have an equality which is not functional ([**17**, Section 6]). We shall not deal with such equalities, and retain the standard notation 'e' for the functional equality. In transformation set algebras, the functional equality is characterized by

$$e(x,y) = \{\varphi \in Q^X : \varphi(x) = \varphi(y)\}.$$

Similarly for polyadic algebras.

Not every polyadic algebra has an equality, but if it does have, then it has only one (Lemma (3.1) of [**17**]). However, it can be embedded into an algebra with equality.

PROPOSITION 2.16 ([**17**, Theorem 7.14]). *Every (locally finite-dimensional) polyadic algebra is a polyadic subalgebra of a (locally finite-dimensional) polyadic equality algebra.*

3. Substitutions and assignments

We still consider X as a set of variables. Let W be any superset of X. Elements of W will be referred to as (abstract) *terms*, and those of W^X, as *substitutions* (of terms for variables: a substitution α substitutes αx for x). In particular, every transformation (Section 2) is a substitution. (Possibly, it would be better in this context to change the traditional name 'transformation' to '*replacement*' (of variables).) Let, as before, ε stand for the identity transformation, and let id_W be the identity map on W.

Where α is a substitution, let $\alpha^{x_1, x_2, \ldots, x_m}_{w_1, w_2, \ldots, w_m}$ be the substitution which agrees with α outside $\{x_1, x_2, \ldots, x_m\}$ and substitutes w_i for a variable x_i, $i = 1, 2, \ldots, m$. The *effective domain* $\operatorname{edm} \alpha$ of α is the set of all variables x with $\alpha x \neq x$. A substitution is said to be *finite* if its effective domain is finite. If $\{x_1, x_2, \ldots, x_m\}$ is the effective domain of α, and if $w_i := \alpha x_i$ for all i, then α will usually be coded as $[w_1/x_1, w_2/x_2, \ldots, w_m/x_m]$; however, sometimes it will be more convenient to write such α as $[{}^{x_1, x_2, \ldots, x_m}_{w_1, w_2, \ldots, w_m}]$. In any case, $[\,]$ stands for ε. A substitution of type $[w/x]$ is said to be *one-dimensional*.

An *assignment* (of values to variables) into a nonempty set Q is defined to be a mapping $X \to Q$: an assignment φ assigns φx to x. For instance, every substitution can be regarded as an assignment into W. For any sequence of elements $q_1, q_2, \ldots, q_m$ of Q, $\varphi^{x_1, x_2, \ldots, x_m}_{q_1, q_2, \ldots, q_m}$ is the assignment which agrees with φ outside $\{x_1, x_2, \ldots, x_m\}$ and assigns the value q_i to a variable x_i, $i = 1, 2, \ldots, m$.

Both substitutions and assignments should act also on arbitrary terms. Extension sets, freeoids and their models introduced in [**6**] provide appropriate means to extend assignments, including substitutions, to the whole set W.

An *extension set* for Q is a subset of Q^W containing just one extension of every assignment into Q. A *freeoid* is defined to be a pair $\mathsf{W} := (W, E)$, where E is some extension set for W which contains id_W and is closed under composition; functions in E are called *extended substitutions*. A *model* for W is a pair $\mathsf{Q} := (Q, H)$, where H is an E-closed extension set for Q (i.e., $HE \subset H$); functions in H are called *extended assignments*. In particular, a freeoid is its own model. A typical example of an freeoid comes from universal algebra: if $\mathbf{W} := (W, \ldots)$ is a relatively free algebra on the set of generators X, then the pair $F(\mathbf{W}) := (W, \operatorname{End}(\mathbf{W}))$ is a freeoid [**6**, Example 1]. Freeoids of this kind were said there to be *algebraic*. For other examples of freeoids, see Section 3 of [**6**]. Further, let $\mathcal{K}$ be the variety generated by $\mathbf{W}$, and let $\mathbf{Q} = (Q, \ldots)$ be an algebra from $\mathcal{K}$. Then the set $\operatorname{Hom}(\mathbf{W}, \mathbf{Q})$ is an $\operatorname{End}(\mathbf{W})$-closed extension set for Q^X, so that $(Q, \operatorname{Hom}(\mathbf{W}, \mathbf{Q}))$ is a model for $F(\mathbf{W})$. Conversely, if $\mathcal{K}$ is a class of algebras $\mathbf{Q}$ (of a given signature), for which $(Q, \operatorname{Hom}(\mathbf{W}, \mathbf{Q}))$ is a model of $F(\mathbf{W})$, then $\mathcal{K}$ is a variety and $\mathbf{W}$ is a free algebra in $\mathcal{K}$ and generates this variety (see [**6**, Example 5]; recall that X is infinite).

Let $\mathsf{W} := (W, E)$ be a freeoid. Given a substitution α, let $\widetilde{\alpha}$ be the single extension of α in E. The extension map $\widetilde{\ }$ satisfies identities

$$\text{(3.1)} \qquad \text{(i) } \widetilde{\alpha}\varepsilon = \alpha, \quad \text{(ii) } \widetilde{\varepsilon} = \mathsf{id}_W, \quad \text{(iii) } \widetilde{\alpha}\widetilde{\beta} = \widetilde{\widetilde{\alpha}\beta}.$$

Moreover, the map is injective (this follows also from (3.1)(i)), and its range is E. Likewise, if $\mathsf{Q} := (Q, H)$ is a model of the freeoid, then let $\widetilde{\varphi}$ be the single extension of the assignment φ in H. This extension map $\widetilde{\ }$ satisfies similar identities

$$\text{(3.2)} \qquad \text{(i) } \widetilde{\varphi}\varepsilon = \varphi, \quad \text{(ii) } \widetilde{\varphi}\widetilde{\beta} = \widetilde{\widetilde{\varphi}\beta},$$

also is injective, and H is its range.

Now the set of all substitutions becomes a monoid $Sb := (W^X, \cdot, \varepsilon)$, where the operation $\cdot$ is defined by

$$\alpha \cdot \beta := \widetilde{\alpha}\beta. \tag{3.3}$$

Then, due to (3.1),

$$\widetilde{\alpha \cdot \beta} = \widetilde{\alpha}\widetilde{\beta}, \tag{3.4}$$

i.e., the extension map $\widetilde{}$ is an isomorphism from E onto Sb. We call Sb the *substitution system* (also *substitution monoid*) of W. In its turn, the set H may be considered as a right E-act under composition. Then the set of all assignments Q^X becomes a right Sb-act $As_Q := (Q^X, \circ)$, where the operation $\circ$ is defined by

$$\varphi \circ \beta := \widetilde{\varphi}\beta; \tag{3.5}$$

thus,

$$\varphi \circ \varepsilon = \varphi, \quad \varphi \circ (\alpha \cdot \beta) = (\varphi \circ \alpha) \circ \beta.$$

Due to (3.2),

$$\widetilde{\varphi \circ \beta} = \widetilde{\varphi}\widetilde{\beta}, \tag{3.6}$$

i.e., the extension map $\widetilde{}$ is a right-act isomorphism from H onto As_Q. We call As_Q the *assignment system* of Q.

If β is a finite substitution $[w_1/x_1, w_2/x_2, \ldots, w_m/x_m]$, the term $\widetilde{\beta}w$ will be written also as $[w_1/x_1, w_2/x_2, \ldots, w_m/x_m]w$ (thus omitting $\widetilde{}$). Observe that, for this β and any $\alpha \in Sb$, $\varphi \in As_Q$,

$$\alpha \cdot \beta = \alpha^{x_1,\ x_2,\ \ldots, x_m}_{\widetilde{\alpha}w_1, \widetilde{\alpha}w_2, \ldots, \widetilde{\alpha}w_m}, \qquad \varphi \circ \beta = \varphi^{x_1,\ x_2,\ \ldots, x_m}_{\widetilde{\varphi}w_1, \widetilde{\varphi}w_2, \ldots, \widetilde{\varphi}w_m}.$$

Following [**6**, Definition 4.1], a subset $K \subseteq X$ is called a *support* of a term w if $\widetilde{\alpha}w = w$ for every transformation α of X that agrees with ε on K. We again write $J \operatorname{spp} w$ to mean that J is a support of w, and $w \operatorname{ind} x$, that w is *independent* of x, i.e., $\overline{x} \operatorname{spp} w$ or, equivalently, $[v/x]w = w$ for any v. By Proposition 4.3 of [**6**], for any model Q of W, all assignments $\varphi, \psi \in Q$ and $w \in W$,

$$\text{if } J \operatorname{spp} w \text{ and } \varphi|J = \psi|J, \text{ then } \widetilde{\varphi}w = \widetilde{\psi}w; \tag{3.7}$$

of course, this concerns also assignments into W, i.e., substitutions. It is easily seen that

$$\text{if } J \operatorname{spp} w \text{ and } K \operatorname{spp} \alpha x \text{ for every } x \in J, \text{ then } K \operatorname{spp} \widetilde{\alpha}w. \tag{3.8}$$

Indeed, let β be any transformation identical on K. Then, in particular, $\widetilde{\beta}(\alpha x) = \alpha x$ for every $x \in J$, whence $\widetilde{\beta}\alpha|J = \alpha|J$. Then $\widetilde{\beta}(\widetilde{\alpha}w) = \widetilde{\alpha}w$ in view of (3.1)(iii).

Notice that a substitution α is idempotent if and only $\overline{\operatorname{edm} \alpha} \operatorname{spp} \alpha x$ for every $x \in \operatorname{edm} \alpha$.

We say that the freeoid W is *locally finite*, or *locally finite-dimensional*, if each of its elements has a finite support. Significance of such freeoids is shown by Theorem 4.6 of [**6**], which states that a freeoid is algebraic if and only if it is locally finite. Thus, we may think of a locally finite freeoids as an abstraction of relatively free algebras, as far as we are interested only in the endomorphism monoids of the latter.

PROPOSITION 3.1. *Suppose that* W *is locally finite. Then*

(a) *for every* $w \in W$ *and* $\alpha \in W^X$, *there is a finite substitution* β *such that* $\widetilde{\alpha}w = \widetilde{\beta}w$,

(b) *every finite substitution is a product (in* Sb*) of idempotent one-dimensional substitutions.*

PROOF. Let α be any substitution.

(a) If J is a finite support of w, let β be a substitution which agrees with α on J and with ε outside J. Then β is finite, and $\widetilde{\alpha}w = \widetilde{\beta}w$ by (3.7).

(b) Suppose that $K = \{x_1, x_2, \ldots, x_m\}$, $m \geq 0$, with all x_i mutually distinct, is the effective domain of α. Put $w_i := \alpha x_i$ for all i; then there is a finite subset L of $\overline{K}$ such that $L \operatorname{spp} \alpha x_i$ for all i. Choose mutually distinct variables $y_1, y_2, \ldots, y_m$ outside $K \cup L$; then each transformation $[y_i/x_i]$ and substitution $[w_i/y_i]$ is idempotent. The identity

$$\alpha = [w_m/y_m][w_{m-1}/y_{m-1}]\cdots[w_1/y_1][y_m/x_m][y_{m-1}/x_{m-1}]\cdots[y_1/x_1]$$

($\cdot$ omitted) is now easily verified. □

4. Superpolyadic algebras

In the rest of the paper, let W be a fixed freeoid, with the corresponding extension operation $\widetilde{\ }$, substitution monoid Sb and extension system E. Recall that the set of variables X is assumed to be infinite.

4.1. Substitution algebras.

DEFINITION 4.1. A *substitution algebra* is a pair $\mathsf{A} := (A, \mathsf{S})$, where

- A is a Boolean algebra,
- S is a homomorphism $Sb \to \operatorname{End}(A)$:

$$S_\varepsilon = \mathsf{id}_A, \quad S_{\alpha\cdot\beta} = S_\alpha S_\beta \text{ for all } \alpha, \beta \in S \tag{4.1}$$

(we still write S_α for $\mathsf{S}(\alpha)$). It can be presented also in more traditional form as an algebra $(A, S_\alpha)_{\alpha\in Sb}$ satisfying the identities (4.1).

Therefore, every substitution algebra is an expansion of a transformation algebra. In particular, the definitions of support and independence, and of local finiteness (subsection 2.2), as well as the relevant notation, can be used in any substitution algebra. The subsequent theorems show that the support relation spp behaves properly with respect to arbitrary substitutions; cf. the properties (2.11) and (2.12), respectively.

THEOREM 4.2. *In a substitution algebra,*

$$\textit{if } J \operatorname{spp} a \textit{ and } K \operatorname{spp} \alpha x \textit{ for all } x \in J, \textit{ then } K \operatorname{spp} S_\alpha a. \tag{4.2}$$

PROOF. Suppose that $J \operatorname{spp} a$ and β is a transformation such that $\beta|K = \varepsilon|K$. For any $x \in J$, K is a support of αx (by definition), so, $\widetilde{\beta}(\alpha x) = \alpha x$ (by (3.7)), whence $(\beta \cdot \alpha)|J = \alpha|J$. Then $S_\beta(S_\alpha a) = S_{\beta\cdot\alpha} a = S_\alpha a$. □

Notice that, unless W is locally finite, it may happen that no K in (4.2) is finite even if J is. Also, a substitution algebra is locally finite only if such is W.

THEOREM 4.3. *Let* A *be a substitution algebra and let* $a \in A$. *For any two substitutions* α, β,

$$\text{(4.3)} \qquad \text{if } J \operatorname{spp} a \text{ and } \alpha|J = \beta|J, \text{ then } S_\alpha a = S_\beta a.$$

PROOF. Assume that $J \operatorname{spp} a$ and suppose that $\alpha|J = \beta|J$. If $J \neq \varnothing$ and γ is a transformation such that $\gamma|J = \varepsilon|J$ and $\gamma(I) \subseteq J$, then $S_\gamma a = a$ (by the assumption) and $\alpha \cdot \gamma = \beta \cdot \gamma$. So $S_\alpha a = S_\alpha S_\gamma a = S_{\alpha\cdot\gamma} a = S_{\beta\cdot\gamma} a = S_\beta S_\gamma a = S_\beta a$. If $J = \varnothing$, then $S_\gamma a = a$ for every γ. There are transformations γ_1, γ_2 such that $\gamma_1(X) \cap \gamma_1(X) = \varnothing$, and a substitution δ such that $\delta|\gamma_1(X) = \alpha|\gamma_2(X)$ and $\delta|\gamma_2(X) = \beta|\gamma_2(X)$. So, $\delta\gamma_1 = \alpha\gamma_1$, $\delta\gamma_2 = \beta\gamma_2$ and, consequently,

$$S_\alpha a = S_\alpha S_{\gamma_1} a = S_{\alpha\gamma_1} a = S_{\delta\gamma_1} a = S_\delta S\gamma_1 = S_\delta a.$$

Likewise $S_\beta a = S_\delta a$. □

The assumption $J \operatorname{spp} a$ was used in the proof only once, to conclude in its first part that $S_\gamma a = a$. Notice that actually γ was there a projection. This suggests that the definition of a support in transformation algebras (subsection 2.2) could be weakened as in the corollary below, while (the proof of) the above theorem shows that this weakening has no aftereffect.

COROLLARY 4.4. *In any transformation algebra,* $J \operatorname{spp} a$ *if and only if* $S_\beta a = a$ *for every idempotent transformation* β *with* $\beta|J = \varepsilon|J$.

EXAMPLE 4.5 (*functional substitution algebras*). Let Q be a model of W. Given a Boolean algebra B, consider again the Boolean algebra $F_Q(B)$ of all functions $f\colon Q^X \to B$ (see Example 2.7) and define the action of a substitution α on $F_Q(B)$ by

$$\text{(4.4)} \qquad (S_\alpha f)(\varphi) := f(\varphi \circ \alpha) \ [= f(\widetilde{\varphi}\widetilde{\alpha})].$$

Then $(F_Q(B), S_\alpha)_{\alpha \in Sb}$ is a substitution algebra, called the *full functional* (B-valued) *substitution algebra* (with domain Q). A *functional substitution algebra* is any subalgebra of such a full algebra. Of course, any functional substitution algebra is an expansion of a functional transformation algebra. Conversely, each full functional transformation algebra can be expanded, using (4.4), to a (full) functional substitution algebra when its domain is converted in any way into a model.

It was noted in subsection 2.2 that every transformation algebra is isomorphic to a functional transformation algebra. Owing to this result and the latter observation in the above example, we come to the following conclusion.

PROPOSITION 4.6. *Every transformation algebra is a reduct of a substitution algebra.*

Like transformation algebras in subsection 2.2, a functional B-valued substitution algebra with B a two-element Boolean algebra can be reorganized into a substitution algebra of sets.

EXAMPLE 4.7 (*substitution set algebras*). Given a model Q, consider the Boolean algebra of sets of functions $C := \mathcal{P}(Q^X)$ (Example 2.8). For every $\Phi \in C$ and $\alpha \in S$, set

$$\text{(4.5)} \qquad S_\alpha \Phi := \{\varphi \in Q^X : \varphi \circ \alpha \in \Phi\}.$$

Then $(C, S_\alpha)_{\alpha\in Sb}$ is a substitution algebra, the *full substitution set algebra* with domain Q. A *substitution set algebra* is any subalgebra of such a full functional algebra. Of course, each substitution set algebra is an expansion of the transformation set algebra with the same domain and value algebra.

4.2. Superpredicates on a substitution algebra. Predicates on a transformation algebra were mentioned in subsection 2.4. Clearly, predicates can be considered also on substitution algebras. In this section we, following Section 4.2 of [**32**], specialize this notion by taking into account arbitrary substitutions from Sb.

Let A be a substitution algebra. We define a *superpredicate* on it to be a function $p\colon W^n \to A$ such that, for every $\alpha \in Sb$,

$$S_\alpha p(w_1, w_2, \ldots, w_n) = p(\widetilde{\alpha} w_1, \widetilde{\alpha} w_2, \ldots, \widetilde{\alpha} w_n).$$

Observe that

$$\text{if } K \operatorname{spp} w_i \text{ for every } i, \text{ then } K \operatorname{spp} p(w_1, w_2, \ldots, w_n):$$

if β is a transformation identical on K, then

$$S_\beta p(w_1, w_2, \ldots, w_n) = p(\widetilde{\beta} w_1, \widetilde{\beta} w_2, \ldots, \widetilde{\beta} w_n) = p(w_1, w_2, \ldots, w_n);$$

see Theorem 4.3.

Thus, a superpredicate is always an extension of a predicate. On the other hand, any predicate q on a substitution algebra can be extended to a superpredicate $\widetilde{q}$ by setting

$$(*)\qquad \widetilde{q}(w_1, w_2, \ldots, w_n) := S_\alpha q(x_1, x_2, \ldots, x_n),$$

where the substitution α and elements $w_1, w_2, \ldots, w_n \in W$ are such that $w_i = \alpha x_i$ for all i. The extension does not depend on the choice of α: if, for all i, $\beta x_i = w_i$, then $\beta|\{x_1, x_2, \ldots, x_n\} = \alpha|\{x_1, x_2, \ldots, x_n\}$ and $S_\beta q(x_1, x_2, \ldots, x_n) = S_\alpha q(x_1, x_2, \ldots, x_n)$, as $\{x_1, x_2, \ldots, x_n\}$ supports $q(x_1, x_2, \ldots, x_n)$; see (2.19). It does not depend also on the choice of variables x_i: if $w_i = \alpha y_i$ for some $y_1, y_2, \ldots, y_n$ and β is a transformation that takes every x_i into y_i, then

$$\begin{aligned} S_\alpha q(y_1, y_2, \ldots, y_n) &= S_\alpha q(\beta x_1, \beta x_2, \ldots, \beta x_n) \\ &= S_\alpha S_\beta q(x_1, x_2, \ldots, x_n) = S_\alpha q(x_1, x_2, \ldots, x_n), \end{aligned}$$

for $\{x_1, x_2, \ldots, x_n\}$ is a support of $q(x_1, x_2, \ldots, x_n)$ and the substitutions $\alpha \cdot \beta$ and α agree on it. Thus, the definition (*) is correct. It can be equivalently rewritten as

$$\widetilde{q}(\alpha x_1, \alpha x_2, \ldots, \alpha x_n) := S_\alpha q(x_1, x_2, \ldots, x_n),$$

where x_i and α are now arbitrary.

We thus have come to the following result.

LEMMA 4.8. *For every predicate q, its extension $\widetilde{q}$ is a superpredicate. The resulting correspondence between predicates and superpredicates is bijective.*

PROOF. The function $\widetilde{q}$ is indeed a superpredicate: with an appropriate substitution β,

$$\begin{aligned} S_\alpha\widetilde{q}(w_1, w_2, \ldots, w_n) &= S_\alpha\widetilde{q}(\beta x_1, \beta x_2, \ldots, \beta x_n) \\ &= S_\alpha S_\beta q(x_1, x_2, \ldots, x_n) \\ &= S_{(\alpha\cdot\beta)} q(x_1, x_2, \ldots, x_n) \\ &= \widetilde{q}((\alpha\cdot\beta)x_1, (\alpha\cdot\beta)x_2, \ldots, (\alpha\cdot\beta)x_n) \\ &= \widetilde{q}(\widetilde{\alpha}\beta x_1, \widetilde{\alpha}\beta x_2, \ldots, \widetilde{\alpha}\beta x_n) \\ &= \widetilde{q}(\widetilde{\alpha} w_1, \widetilde{\alpha} w_2, \ldots, \widetilde{\alpha} w_n). \end{aligned}$$

The transfer from q to $\widetilde{q}$ is evidently injective. On the other hand, every superpredicate p is the $^\sim$-extension of its restriction to X^n. □

In view of this lemma, we may simplify our notation and omit the extension sign over a predicate, thus denoting it and its extension by the same symbol.

4.3. Equality in substitution algebras. We now apply this construction to the equality predicate. By an equality for a substitution algebra, we mean an equality for its transformational part.

DEFINITION 4.9. A *superequality* for a substitution algebra is the $^\sim$-extension of an equality (when the latter exists) for it, i.e., a superpredicate e satisfying (2.20). A *substitution (super)equality algebra* is a substitution algebra with a (super)equality.

Therefore, a substitution algebra may have at most one superequality. Notice the particular case of an observation made in the preceding subsection for superpredicates:

(4.6) $$\text{if } K \operatorname{spp} v, w, \text{ then } K \operatorname{spp} e(v, w).$$

EXAMPLE 4.10 (*functional superequality*). Let F be a full functional substitution algebra with domain Q. The functional equality e for the transformational reduct of F defined as in Example 2.15 is said to be the functional equality also for F. Evidently, the function e from W^2 to Q^X defined as follows

$$e(v, w)(\varphi) := 1 \text{ if } \widetilde{\varphi}(v) = \widetilde{\varphi}(w);\ e(v, w)(\varphi) := 0 \text{ if } \widetilde{\varphi}(v) \neq \widetilde{\varphi}(w)$$

is a superequality for F; it is said to be the *functional superequality*. If a subalgebra of F contains all elements e_{xy}, resp., e_{vw}, then e is the equality, resp., superequality also for this subalgebra.

In a full substitution set algebra, the superequality is characterized by

$$e(v, w) = \{\varphi \in Q^W : \widetilde{\varphi}(v) = \widetilde{\varphi} w\}.$$

Lemma 4.8 allows us to admit a slight ambiguity in our language: where this cannot cause any misunderstanding, we shall use the more common term 'equality' also for a superequality.

We now present the most basic properties of equalities. Remarkably, the properties (c)–(f) essentially are consequences of (a) and (b).

THEOREM 4.11. *The superequality e on a substitution algebra A has the following properties:*

(a) $S^x_v a \wedge e(v, w) \leq S^x_w a$ *(substitutivity),*
(b) $e(w, w) = 1$ *(reflexivity),*

(c) $e(v, w) = e(w, v)$ *(symmetry)*,
(d) $e(u, v) \wedge e(v, w) \leq e(u, w)$ *(transitivity)*,
(e) $S^x_v a \wedge e(v, w) = S^x_w a \wedge e(v, w)$ *(full substitutivity)*,
(f) $e(v, w) \leq e([v/x]u, [w/x]u)$ *(extensionality)*.

PROOF. Recall that e is a superpredicate by definition.

(a) Let us consider first the case when $a \operatorname{ind} y$ for some y distinct from x. Then $S^x_v a \wedge e(v, w) = S^{xy}_{vw}(a \wedge e(x, y)) \leq S^{xy}_{vw} S^x_y a = S^x_w a$ as needed. If there is not such y, i.e., X is the single support of a, we need beforehand to rename some variables.

Let γ be an injective transformation such that $\gamma x = x$ and $y \notin \gamma(\overline{x})$ for some y distinct from x. By (2.11), $S_\gamma a \operatorname{ind} y$, for $\gamma X \subseteq \overline{y}$. So, in view of the case of (a) just considered,

$$(*) \qquad S^x_{\widetilde{\gamma} v}(S_\gamma a) \wedge e(\widetilde{\gamma} v, \widetilde{\gamma} w) \leq S^x_{\widetilde{\gamma} w}(S_\gamma a).$$

Now let δ be any substitution inverse to γ on γX; then $\delta \cdot \gamma = \varepsilon$ and $\widetilde{\delta}\widetilde{\gamma} = \widetilde{\delta \cdot \gamma} = \widetilde{\varepsilon}$, see (3.4). It is easily seen that $\delta \cdot [\widetilde{\gamma} v/x] \cdot \gamma = [v/x]$: if $z = x$, then

$$(\delta \cdot [\widetilde{\gamma} v/x] \cdot \gamma)z = (\delta \cdot [\widetilde{\gamma} v/x])x = \widetilde{\delta}(\widetilde{\gamma} v) = \widetilde{\varepsilon} v = v,$$

and if $z \neq x$, then $\gamma z \neq x$ and $(\delta \cdot [\widetilde{\gamma} v/x] \cdot \gamma)z = (\delta \cdot \gamma)z = z$. Likewise $\delta \cdot [\widetilde{\gamma} w/x] \cdot \gamma = [w/x]$, and $S_\delta e(\widetilde{\gamma} v, \widetilde{\gamma} w) = \widetilde{e}(v, w)$, for $\widetilde{\delta}\widetilde{\gamma} = \widetilde{\varepsilon}$. Now an application of S_δ to both sides of (*) yields the needed inequality.

(b) Evidently, $e(w, w) = e([w/x]x, [w/x]x) = S^x_w e(x, x) = S^x_w 1 = 1$.

(c) Apply (a) and (b) to $a := e(x, v)$.

(d) Apply (a) to $a := e(u, x)$.

(e) By symmetry, (a) gives a variant $S^x_w a \wedge e(v, w) \leq S^x_v a$. Together, the two inequalities lead us to (e).

(f) If $v \operatorname{ind} x$, then $e([v/x]u, [w/x]u) = S^x_w e([v/x]u, u)$ and $S^x_v e([v/x]u, u) = e([v/x]u, [v/x]u) = 1$ by (b), and it remains to apply (a) with $a := e([v/x]u, u)$. Otherwise, choose an injective transformation α such that $x \notin \alpha X$. Then $\alpha v \operatorname{ind} x$, so that likewise $e(\widetilde{\alpha} v, \widetilde{\alpha} w) \leq e([\widetilde{\alpha} v/x]u, [\widetilde{\alpha} w/x]u)$. Now let β be a transformation inverse to α on αX; apply S_β to both sides of the latter inequality. □

4.4. Substitution quantifier algebras.

DEFINITION 4.12. A *substitution quantifier algebra* is a triple $(A, \exists, \mathsf{S})$, where

- $(A, \exists)$ is a quantifier algebra,
- (A, S) is a substitution algebra,
- every element $a \in A$ has the same supports in both algebras:

$$\exists_{\overline{J}} a = a \text{ iff } S_\alpha a = a \text{ for all } \alpha \in Tr \text{ with } \alpha|J = \varepsilon|J.$$

EXAMPLE 4.13 (*functional substitution quantifier algebras*). Given a model Q and a Boolean algebra B, let A be any algebra on $F_{\mathsf{Q}}(B)$ which is a combination of a functional quantifier algebra and a functional substitution algebra. As seen from (2.6), (2.13), A is a *functional substitution quantifier algebra* (with domain Q).

EXAMPLE 4.14 (*substitution quantifier set algebras*). Given a model Q, consider any algebra on $\mathcal{P}X(Q^X)$ which is a combination of a quantifier and a substitution set algebra. Then it is a *substitution quantifier set algebra* (with domain Q).

Substitution quantifier algebras are, in fact, equationally definable.

THEOREM 4.15. *Let* $(A, \exists)$ *be a quantifier algebra, and let* (A, S) *be a substitution algebra. Then* $(A, \exists, \mathsf{S})$ *is a substitution quantifier algebra if and only if the equations* (2.17) *hold for every projection* α.

PROOF. The proof of necessity of the condition is straightforward. Let α be a projection with $\operatorname{edm} \alpha = J$. As $\overline{J} \operatorname{spp} \exists_J a$ in $(A, \exists)$ (by (2.4)) and $\alpha|\overline{J} = \varepsilon|\overline{J}$, we get the first of the two equations by the definition of spp in (A, S). On the other hand, $\alpha(X) \subseteq \overline{J}$. Then $\overline{J} \operatorname{spp} \alpha x$ in (A, S) for every $x \in X$. By Theorem 4.2, now $\overline{J} \operatorname{spp} S_\alpha a$, whence, by the definition of spp (in $(A, \exists)$), the second equation of (2.17) follows.

To prove its sufficiency, assume that the triple $(A, \exists, \mathsf{S})$ consists of a quantifier and a substitution algebra, with spp_q and spp_s the respective support relations, and that (2.17) is fulfilled. First suppose that $K \operatorname{spp}_s a$; we should show that $\exists_{\overline{K}} a = a$. The case $K = X$ is trivial in view of the quantifier algebra axiom $\exists_\varnothing a = a$. If $K \neq X$, choose a projection α with $\operatorname{edm} \alpha = K$; then $\alpha|\overline{K} = \varepsilon|\overline{K}$ and, by the supposition and (4.3), $S_\alpha a = a$. In view of the second axiom in (2.17), now $\exists_{\overline{K}} a = \exists_{\overline{K}} S_\alpha a = S_\alpha a = a$.

Now suppose that $K \operatorname{spp}_q a$ and $\alpha|K = \varepsilon|K$ for some transformation α; we then should show that $S_\alpha a = a$. By Corollary 4.4, we may assume that α is projection. Let $J := \operatorname{edm} \alpha$; then $J \subseteq \overline{K}$ and, in view of the supposition, $a = \exists_{\overline{K}} a = \exists_J \exists_{\overline{K}} a$; see (2.3). In view of the first axiom in (2.17), $S_\alpha \exists_J = \exists_J$, then $S_\alpha a = S_\alpha \exists_J \exists_{\overline{K}} a = \exists_J \exists_{\overline{K}} a = a$. □

Next we state some properties of quantifiers, substitutions and equality which can be proved without help of (2.18).

LEMMA 4.16. *In every substitution quantifier algebra, if* $L \operatorname{spp} a$, *then*

(a) $S_\alpha \exists_J a = S_\beta \exists_J a$, *if* $\alpha|(L \setminus J) = \beta|(L \setminus J)$,

(b) $\exists_J S_\alpha a = \exists_K S_\alpha a$, *if* $J \cap M = K \cap M$ *for some* M *such that* $M \operatorname{spp} \alpha x$ *for every* $x \in L$.

(c) $S_\alpha a \leq \exists_J a$ *if* $\alpha|L \setminus J = \varepsilon|L \setminus J$,

(d) $a \leq \exists_{\alpha J} S_\alpha a$ *if* α *embeds* J *into* X, $\alpha|L \setminus J = \varepsilon|L \setminus J$ *and* $\alpha J \cap (L \setminus J) = \varnothing$,

(e) $\exists_{\alpha J} S_\alpha a = \exists_J a$ *for* α *and* J *as in* (d).

PROOF. Let $L \operatorname{spp} a$ for some $a \in A$.

(a) By (4.3), as $\exists_J \exists_J a = \exists_J a$ and, consequently, $\overline{J} \operatorname{spp} \exists_J a$; see (2.4).

(b) From Theorems 4.2 and 2.3.

(c) By (2.1) and (a), $S_\alpha a \leq S_\alpha \exists_J a = \exists_J a$.

(d) Suppose that α satisfies the condition. Then, in particular, $(\alpha J \cup (L \setminus J)) \operatorname{spp} S_\alpha a$; see (4.2). Choose a transformation β inverse to α on αJ and such that $\beta|(L \setminus J) = \varepsilon|(L \setminus J)$; then $(\beta \cdot \alpha)|L = \varepsilon|L$. As $S_\alpha a \leq \exists_{\alpha J} S_\alpha a$ by (2.1), and then $(L \setminus J) \operatorname{spp} \exists_{\alpha J} S_\alpha a$ by (2.4) and (4.3), further

$$a = S_{\beta \cdot \alpha} a = S_\beta (S_\alpha a) \leq S_\beta (\exists_{\alpha J} S_\alpha a) = \exists_{\alpha J} S_\alpha a.$$

(e) By (c), (2.2), (2.4)

$$\exists_{\alpha J} S_\alpha a \leq \exists_{\alpha J} (\exists_J a) = \exists_J a,$$

and then, by (d), (2.2) and the last identity,

$$\exists_J a \leq \exists_J (\exists_{\alpha J} S_\alpha a) = \exists_{\alpha J} S_\alpha a$$

(recall that quantifiers in a quantifier algebra commute). □

LEMMA 4.17. *In a substitution quantifier algebra with equality,*

(a) $\exists_x e(x,w) = 1$ *if* $w \operatorname{ind} x$,
(b) $S^x_w a = \exists_x(a \wedge e(x,w))$ *if* $w \operatorname{ind} x$,
(c) $e(v,w) = \exists_x(e(v,x) \wedge e(x,w))$ *if* $v, w \operatorname{ind} x$,
(d) $S^x_w \exists_J = \exists_J S^x_w$ *if* $w \operatorname{ind} x, \overline{J} \operatorname{spp} w$ *and* $x \notin J$.

PROOF. (a) By Lemma 4.16(c) and superpredicativity and reflexivity of e, $\exists_x e(x,w) \geq S^x_w e(x,w) = e(w,w) = 1$ when $w \operatorname{ind} x$. Thus, $\exists_x e(x,w) = 1$.

(b) If $w \operatorname{ind} x$, then $S^x_w a \operatorname{ind} x$, and then $\exists x S^x_w a = S^x_w a$. In virtue of full substitutivity of e (Theorem 4.11 with $v := x$), (2.1) and (a), now

$$\begin{aligned}\exists_x(a \wedge e(x,w)) &= \exists_x(S^x_w a \wedge e(x,w)) = \exists_x(\exists_x S^x_w a \wedge e(x,w)) \\ &= \exists_x S^x_w a \wedge \exists_x e(x,w) = S^x_w a \wedge 1 = S^x_w a,\end{aligned}$$

provided that $w \operatorname{ind} x$.

(c) By (b), $e(v,w) = S^x_w e(v,x) = \exists_x(e(v,x) \wedge e(x,w))$.

(d) By (b), (2.1) and the particular case $\overline{J} \operatorname{spp} e(x,w)$ of (4.6),

$$\begin{aligned}S^x_w \exists_J a = \exists_x(\exists_J a \wedge e(x,w)) &= \exists_x(\exists_J a \wedge \exists_J e(x,w)) \\ &= \exists_x \exists_J(a \wedge \exists_J e(x,w)) = \exists_J \exists_x(a \wedge e(x,w)) = \exists_J S^x_w a\end{aligned}$$

for any a. □

Let us consider the following particular case of (d) (and of (2.18)):

$$S^x_y \exists_J = \exists_J S^x_y \text{ if } x \neq y \text{ and } x, y \notin J. \tag{4.7}$$

Note that it can be obtained even in an transformation quantifier algebra (i.e., substitution quantifier algebra with $W = X$) with equality. Furthermore, it covers the last axiom in Theorem 2.13. Therefore, in view of Proposition 2.13 and Theorem 4.15, we come to the following conclusion.

LEMMA 4.18. *A locally finite-dimensional transformation quantifier algebra with equality is a polyadic (equality) algebra.*

The next proposition follows from the results of [**17**, Section 7]. They are stated for polyadic algebras; however, the full strength of polyadic algebras has not been exploited in the proofs.

PROPOSITION 4.19. *Every transformation quantifier algebra* A *satisfying* (4.7) *is a subalgebra of a transformation quantifier algebra* A^+ *with equality, and every Boolean endomorphism (quantifier) of A extends, in a standard way, to a Boolean endomorphism (quantifier) of A^+. This extension preserves compositions. The locally finite part of* A *is a subalgebra of the locally finite part of* A^+.

PROOF. Assume that A is a transformation quantifier algebra. We first note that if A is a subreduct of some transformation quantifier algebra A^+, then Theorem 4.15 guarantees that the support relation of A agrees with that of A^+.

If A satisfies (4.7), then we can almost literally repeat the arguments of Section 7 in [**17**] for proving Theorem 7.14 there, cited above as Proposition 2.16. The identity (4.7) is the single property of quantifiers used in the proof, and Lemma 7.11 from [**17**] allows us to extend to the embedding algebra of A *all* Boolean endomorphisms and quantifiers living on the latter. If A is locally finite-dimensional, then it is included in its subalgebra consisting of the finite-dimensional elements. □

These observations lead us to an embedding theorem that generalizes Proposition 2.16. However, observe that, in view of Proposition 2.13, the identity (2.18) turns any locally finite-dimensional transformation quantifier algebra into a polyadic algebra.

THEOREM 4.20. *If a (locally finite-dimensional) substitution quantifier algebra satisfies* (4.7), *then it is a subalgebra of a (locally finite-dimensional) substitution quantifier algebra with equality.*

4.5. Superpolyadic algebras. The subsequent definition is an abstract version of a definition of a superpolyadic algebra *with respect to a relatively free algebra* (of any signature). These algebras were introduced in [**3**] and studied in [**31**].

DEFINITION 4.21. An algebra *superpolyadic with respect to* W or just a W-*polyadic algebra* is a triple $(A, \exists, \mathsf{S})$, where

- (A, S) is a substitution algebra,
- the reduct $(A, \exists, \mathsf{S}')$, where S' is a restriction of S to Tr, is a polyadic algebra.

A W-*polyadic equality algebra* is a W-polyadic algebra with equality.

Therefore, superpolyadic algebras are particular substitution quantifier algebras. In fact, a superpolyadic algebra is a substitution quantifier algebra in which the axiom (2.18) is fulfilled. The following theorem follows immediately from Lemma 4.18.

THEOREM 4.22. *A locally finite-dimensional substitution quantifier algebra with equality is a superpolyadic (equality) algebra.*

Chapter 9 in B. Plotkin's monograph [**28**] is devoted to *specialized Halmos algebras* (over a variety of algebras) – generalized polyadic algebras with stronger axioms instead of (2.15), (2.16) (see also the description of an early version of specialized polyadic algebras in [**31**]). In terms of the present paper, these are superpolyadic algebras, where also endomorphisms S_α with α an arbitrary substitution are involved in these axioms, as in (4.8), (4.9) below.

DEFINITION 4.23. A *strong superpolyadic algebra* is a triple $(A, \exists, \mathsf{S})$, where

- $(A, \exists)$ is a quantifier algebra,
- (A, S) is a substitution algebra,
- the homomorphisms $\exists$ and S interact in the following way:

for every $J \in \mathcal{P}X$ and all $\alpha, \beta \in Sb$,

$$S_\alpha \exists_J = S_\beta \exists_J \text{ whenever } \alpha|\overline{J} = \beta|\overline{J}; \tag{4.8}$$

$$\exists_J S_\alpha = S_\alpha \exists_{\alpha^{-1}(J)} \text{ whenever } \alpha \in S \text{ injective on } \alpha^{-1}(J), \tag{4.9}$$

$$\text{and for every } x \notin \alpha^{-1}(J),\ \overline{J} \text{ is a support of } \alpha(x).$$

EXAMPLE 4.24 (*functional superpolyadic algebras*). Simple calculations (similar to those necessary in Example 2.10) show that functional substitution quantifier algebras of Example 4.13 are strong superpolyadic algebras.

EXAMPLE 4.25 (*superpolyadic set algebras*). It is now a consequence of the conclusion in the preceding example that every substitution quantifier set algebra is also an example of a strong superpolyadic algebra.

Lemma 4.16(a) shows that the identity (4.9) actually holds in every substitution quantifier algebra. Even more: every superpolyadic algebra is strong. For locally finite-dimensional algebras with W algebraic, this result was announced in [**3**], and asserted also in Theorem 2.3 of [**31**], proved by direct calculations. However, there is a defect in the proof of an important lemma (see Lemma 2.4 in [**31**]: the assumptions ($\tau 3$) and ($\tau 4$), as stated, are not compatible in any algebra). We shall obtain the mentioned result in the next section for an arbitrary superpolyadic algebra as a consequence of a functional representation theorem for these algebras.

5. Representation theorems

In [**15**, Theorem 10.9], Halmos proved that every locally finite-dimensional polyadic algebra of infinite degree (i.e., with an infinite set of variables) is isomorphic to a functional polyadic algebra; cf. Theorem 9.5 in [**7**]. The restriction that an algebra has to be locally finite-dimensional was removed by Daigneault and Monk in [**9**, Theorem 6.3]. We now extend the representation theorem of this kind to superpolyadic algebras. Recall that we also concern ourselves only with an infinite set of variables X.

THEOREM 5.1. *A* W*-polyadic algebra* A *with* X *infinite is isomorphic to an* A*-valued functional algebra with domain* W.

PROOF. Assume that A is a superpolyadic algebra. We first show that the substitutional part A_s of the latter is isomorphic to a functional algebra.

Let F be the full functional substitution algebra with domain W and value algebra A itself. Notice that its underlying Boolean algebra F is the set of functions $W^X \to A$ and recall that the freeoid W is its own model with the semigroup $(Sb, \cdot)$ as the corresponding assignment system. Define a mapping $\mu\colon A \to F$ by

$$\text{for all } \alpha \in Sb,\ (\mu a)(\alpha) := S_\alpha a.$$

Then μ is a Boolean homomorphism $\mathsf{A} \to \mathsf{F}$ which is injective (if $\mu a = \mu b$, then $a = S_\varepsilon a = S_\varepsilon b = b$) and commutes with every S_β: in view of (4.4),

$$(\mu(S_\beta a))(\alpha) = S_\alpha(S_\beta a) = S_{\alpha\cdot\beta} a = (\mu a)(\alpha \cdot \beta) = (S_\beta(\mu a))(\alpha)$$

for every $\alpha \in Sb$ and $a \in A$. Thus, A_s is isomorphic to the subalgebra of F determined by the range of μ.

Of course, μ simultaneously provides a representation also for the transformational part of A. Moreover, μ is exactly the representative mapping used in proofs of the above-mentioned theorems on functional representation of polyadic algebras. This means, in particular, that all functional quantifiers $\exists_J$ on $\mu(A)$ exist, and the mapping preserves all quantifiers of A; cf. Example 2.10. Therefore, (i) the subset $\mu(A)$ of F is the carrier of a functional polyadic algebra, and (ii) μ is a polyadic monomorphism from A_p, the polyadic part of A, onto this algebra.

It now follows that $\mu(A)$ actually determines a functional superpolyadic algebra and that μ is a superpolyadic monomorphism from A onto the latter. □

We have thus obtained a functional copy A^μ of the initial superpolyadic algebra A. As both axioms (4.8) and (4.9) are fulfilled in any functional substitution quantifier algebra (Example 4.24), A^μ also satisfies these axioms. This observation leads us to the following conclusion.

COROLLARY 5.2. *Every superpolyadic algebra of infinite degree is strong.*

Halmos' representation theorem for locally finite-dimensional polyadic algebras in [**12**] states that every such an algebra admits a homomorphism into a set algebra. Without the restriction of local finiteness, this theorem is proved in [**9**], see Theorem 6.4 therein. According to Theorem 9.3 of [**7**], every locally finite-dimensional polyadic equality algebra of infinite degree admits a homomorphism into a polyadic set algebra with the functional equality. Again, we shall transfer these results to superpolyadic algebras.

The prototype of our last theorem is Theorem 4 announced in [**3**] and proved in [**31**] (Theorem 3.2 therein). Algebras considered in these theorems were superpolyadic with respect to a relatively free algebra. As the freeoid of a relatively free algebra is locally finite, the theorem below could be derived from the theorem of [**31**]. We give it a straightforward proof, leaving out any constructions of relatively free algebras and related ones.

THEOREM 5.3. *Assume that a* W*-polyadic equality algebra* A*, with* X *infinite, is locally finite-dimensional. Then the algebra* A *admits a homomorphism into a superpolyadic set algebra with a functional equality.*

PROOF. Assume that A is as in the theorem and that a polyadic equality algebra A^- is a reduct of A. By the result of [**7**] noted above, there is an equality-preserving homomorphism μ from A^- into a polyadic set algebra C with functional equality. Then, in particular, $(\mu e)|X^2$ is a functional equality for C. Evidently, any homomorphism preserves supports (defined as in quantifier algebras). We shall prove that the domain Q of C can be turned into a model Q of W, and the image of A in C, into a W-polyadic set algebra with a W-equality so that μ becomes a superequality-preserving superpolyadic algebra homomorphism.

We first prove that, with $w \operatorname{ind} x$, the rule

$$\widetilde{\varphi}w = q :\equiv \varphi^x_q \in \mu e(x, w) \tag{5.1}$$

defines an extension $\widetilde{\varphi}$ of φ on W. First of all, $\widetilde{\varphi}w$ is always defined, i.e., for all x and w,

$$\text{there is } q \text{ such that } \varphi^x_q \in \mu e(x, w). \tag{A}$$

To see why, notice that in C

$$\exists_x \mu e(x, w) = \mu\exists_x e(x, w) = \mu 1 = Q^X.$$

By the definition of a quantifier on a set algebra (2.7), this means that, to any $\varphi \in Q^X$, there is $\psi \in \mu e(x, w)$ such that $\psi(y) = \varphi(y)$ whenever $y \neq x$, so that we may put $q := \psi(x)$.

Further, the element $\widetilde{\varphi}w$ for any w does not depend on the choice of x: if also $w \operatorname{ind} y$, where $y \neq x$, then by Lemma 4.17(c),

$$\mu e(y, w) = \mu\exists_x(e(y, x) \wedge e(x, w)) = \exists_x(\mu e(y, x) \cap \mu e(x, w)).$$

Consequently,

$$
\begin{aligned}
\varphi^y_q \in \mu e(y,w) \text{ iff there is } &\psi \in \mu e(y,x) \cap \mu e(x,w) \text{ such that} \\
&\quad \psi(z) = \varphi^y_q(z) \text{ whenever } z \neq x, \\
\text{iff there is } &\psi \in Q^X \text{ such that } \psi(y) = \psi(x),\ \psi \in \mu e(x,w) \text{ and} \\
&\quad \psi(y) = q,\ \psi(z) = \varphi(z) \text{ when } z \neq x, y, \\
\text{iff there is } &\psi \in \mu e(x,w) \text{ such that } \psi = \varphi^{xy}_{qq}, \\
\text{iff } &\varphi^x_q \in \mu e(x,w).
\end{aligned}
$$

Since $e(x,w)$ and, hence, also $\mu e(x,w)$ have a support K not containing y (see (4.6)), the last step here is justified by (3.7) and the observation that φ^{xy}_{qq} agrees on K with φ^x_q. It thus follows that

(B) $$\varphi^y_q \in \mu e(y,w) \text{ iff } \varphi^x_q \in \mu e(x,w),$$

as needed.

Now let us ascertain that q in (5.1) is unique:

(C) $$\text{if } \varphi^x_q \in \mu e(x,w) \text{ and } \varphi^x_r \in \mu e(x,w), \text{ then } q = r.$$

If also $w \operatorname{ind} y$ for some $y \neq x$, then, by virtue of (B) and (2.8),

$$
\begin{aligned}
\varphi^x_q \in \mu e(x,w) &\text{ and } \varphi^x_r \in \mu e(x,w) \\
&\text{iff } \varphi^x_q \in \mu e(x,w) \text{ and } \varphi^y_r \in \mu e(y,w) \\
&\text{iff } \varphi^{xy}_{qr} \in \mu e(x,w) \text{ and } \varphi^{xy}_{qr} \in \mu e(y,w) \\
&\text{iff } \varphi^{xy}_{qr} \in \mu e(x,w) \cap \mu e(y,w) \\
&\text{iff } \varphi^{xy}_{qr} \in \mu(e(x,w) \wedge e(y,w)).
\end{aligned}
$$

The last relationship implies, by symmetry and transitivity of the equality μe in A^-, that $\varphi^{xy}_{qr} \in \mu e(x,y)$, whence $q = r$, as μe is the functional equality in C (by the choice of μ).

Thus, (A), (B) and (C) demonstrate that the rule (5.1) correctly defines a function $\widetilde{\varphi}\colon W \to Q$. It is indeed an extension of φ: with $y \neq x$, $\widetilde{\varphi}y = q$ iff $\varphi^x_q \in \mu e(x,y)$ iff $q = \varphi^x_q(y) = \varphi y$. To be sure that the set of all obtained extended assignments $\widetilde{\varphi}$ turn Q into a model of W, we have also to check that any composition $\widetilde{\varphi}\widetilde{\beta}$ is the extension of an appropriate assignment. As W, together with A, is locally finite, it suffices to consider only finite, in fact, only one-dimensional idempotent substitutions β (Proposition 3.1), for a composition of substitutions can be managed as follows:

$$\widetilde{\varphi}\widetilde{(\alpha \cdot \beta)} = \widetilde{\varphi}\widetilde{\alpha}\widetilde{\beta} = \widetilde{(\widetilde{\varphi}\alpha)}\widetilde{\beta} = \widetilde{\widetilde{(\widetilde{\varphi}\alpha)}\beta}.$$

We thus shall show that

(D) $$\widetilde{\varphi}\widetilde{\left[{x \atop v}\right]} = \widetilde{\widetilde{\varphi}\left[{x \atop v}\right]},$$

where $v \operatorname{ind} x$. First observe that the following equivalence follows immediately from (5.1):

(E) $$\varphi \in \mu e(x,w) \text{ iff } \varphi x = \widetilde{\varphi} w.$$

We shall need also to know that, for any y such that $w, v \operatorname{ind} y$,

(F) $$\varphi^{y\,x}_{q\,\widetilde{\varphi}v} = \left(\widetilde{\varphi}\left[{x \atop v}\right]\right)^y_q.$$

Indeed,

$$(\varphi^{y\,x}_{q\widetilde{\varphi}v})z = \begin{cases} \widetilde{\varphi}v & \text{if } z = x, \\ q & \text{if } z = y, \\ \varphi z & \text{otherwise,} \end{cases} \qquad (\widetilde{\varphi}[{}^x_v])^y_q)z = \begin{cases} \widetilde{\varphi^y_q}v & \text{if } z = x, \\ q & \text{if } z = y, \\ \varphi z & \text{otherwise.} \end{cases}$$

But, using (E),

$$\widetilde{\varphi^y_q}v = \widetilde{\varphi}v \text{ iff } (\varphi^y_q)^y_{\widetilde{\varphi}v} \in \mu e(y,v) \text{ iff } \varphi^y_{\widetilde{\varphi}v} \in \mu e(y,w) \text{ iff } \widetilde{\varphi}v = \widetilde{\varphi}v.$$

Now we can obtain (D): using Lemma 4.17(c), for $y \neq x$, all w and any $q \in Q$,

$$\begin{aligned} \widetilde{\varphi}(\widetilde{[{}^x_v]w}) = q \text{ iff } & \varphi^y_q \in \mu e(y, \widetilde{[{}^x_v]w}) \\ \text{iff } & \varphi^y_q \in \mu S^x_v e(y,w) \\ \text{iff } & \varphi^y_q \in \mu \exists_x (e(y,w) \wedge e(x,v)) \\ \text{iff } & \varphi^y_q \in \exists_x (\mu(y,w) \cap \mu e(x,v)) \\ \text{iff } & \text{for some } p, \varphi^{yx}_{qp} \in \mu e(y,w) \text{ and } \varphi^{yx}_{qp} \in \mu e(x,v) \\ \text{iff } & \text{for some } p, \varphi^{yx}_{qp} \in \mu e(y,w) \text{ and } \widetilde{\varphi}^y_q v = p \quad \text{(see (E))} \\ \text{iff } & \varphi^{y\,x}_{q\widetilde{\varphi}v} \in \mu e(y,w) \qquad (\text{as } \widetilde{\varphi}^y_q v = \widetilde{\varphi}v), \\ \text{iff } & (\widetilde{\varphi}[{}^x_v])^y_q \in \mu e(y,w) \qquad \text{(see (F))} \\ \text{iff } & \widetilde{\widetilde{\varphi}[{}^x_v]}w = q. \end{aligned}$$

As q and w are arbitrary, (D) immediately follows. Therefore, Q together with the set of all extended assignments $\widetilde{\varphi}$ form a model.

The equivalence (E) can be extended – in fact, μe is the functional superequality on C: due to Lemma 4.17(c) and (D), if $v, w \operatorname{ind} x$, then

$$\begin{aligned} \varphi \in \mu e(v,w) \text{ iff } & \varphi \in \mu \exists_x (e(x,v) \wedge e(x,w)) \\ \text{iff } & \varphi \in \exists_x (\mu e(x,v) \cap \mu e(x,w)) \\ \text{iff } & \text{for some } q \in Q, \varphi^x_q \in \mu e(x,v) \text{ and } \varphi^x_q \in \mu e(x,w) \\ \text{iff } & \text{for some } q \in Q, \widetilde{\varphi}v = q \text{ and } \widetilde{\varphi}w = q \\ \text{iff } & \widetilde{\varphi}v = \widetilde{\varphi}w. \end{aligned}$$

Thus, the homomorphism preserves the superequality e:

$$\mu e(v,w) = \{\varphi \in Q^X : \widetilde{\varphi}v = \widetilde{\varphi}w\}.$$

Finally, μ is a superpolyadic algebra homomorphism. It preserves quantifiers by definition, and preserves also the action of substitutions:

$$\text{(E)} \qquad \mu S_\alpha = S_\alpha \mu \text{ for every } \alpha \in Sb.$$

As A is locally finite-dimensional, it suffices to prove this only for finite, hence (by Proposition 3.1), for one-dimensional idempotent substitutions α. With $w \operatorname{ind} x$, and using Theorem 4.17(c),

$$\begin{aligned}\varphi \in \mu S^x_w a \text{ iff } &\varphi \in \mu\exists_x(a \wedge e(x,w))\\ \text{iff } &\varphi \in \exists_x(\mu a \cap \mu e(x,w))\\ \text{iff } &\text{for some } q \in Q, \varphi^x_q \in \mu a \text{ and } \varphi^x_q \in \mu e(x,w)\\ \text{iff } &\text{for some } q \in Q, \varphi^x_q \in \mu a \text{ and } \widetilde{\varphi} w = q\\ \text{iff } &\varphi^x_{\widetilde{\varphi} w} \in \mu a\\ \text{iff } &\varphi \circ [w/x] \in \mu a\\ \text{iff } &\varphi \in S^x_w \mu a.\end{aligned}$$

The proof is completed. □

COROLLARY 5.4. *Every locally finite-dimensional superpolyadic algebra of infinite degree is representable.*

PROOF. Theorem 4.20 allows us to embed any locally finite-dimensional superpolyadic algebra into an equality algebra. Then the preceding theorem applies. □

Of course, both the theorem and its corollary hold true for strong superpolyadic algebras. Our method of proof of Theorem 5.3 clearly does not permit removing the condition of local finiteness. We conjecture, that, like polyadic algebras, any superpolyadic algebra is nevertheless representable in this sense.

References

[1] K. H. Alzubaidu and N. B. Bennour. *Locally finite functional polyadic algebra with terms*, Bull. Math. Sci. Appl. **1** (2012), 70–77.

[2] H. Andreka and I. Shain, *Connections between algebraic logic and initial algebra semantics of CF languages*, Mathematical Logic in Computer Science. Colloq. Math. Soc.J. Bolyai **28**, Budapest, North Holland (1981), 25–83.

[3] J. Cīrulis, *Generalizing the notion of polyadic algebra*, Polish Acad. Sci. Inst. Philos. Sociol. Bull. Sect. Logic **15** (1986), no. 1, 2–9. MR878298

[4] J. Cīrulis, *An algebraization of first order logic with terms*, Algebraic logic (Budapest, 1988), Colloq. Math. Soc. János Bolyai, vol. 54, North-Holland, Amsterdam, 1991, pp. 125–146. MR1153421

[5] J. Cīrulis, *Corrections to my paper: "An algebraization of first order logic with terms"* in [4] (English, with English, Russian and Latvian summaries), Mathematics, Latv. Univ. Zināt. Raksti, vol. 595, Latv. Univ., Riga, 1994, pp. 49–52. MR1380052

[6] J. Cīrulis, *Freeoids: a semi-abstract view on endomorphism monoids of relatively free algebras*, Algebra Universalis **65** (2011), no. 1, 41–60, DOI 10.1007/s00012-011-0116-5. MR2784236

[7] A. Daigneault, *Théorie des modèles en logique mathématique* (French), Deuxième édition. Séminaire de Mathématiques Supérieures, No. 6, Les Presses de l'Université de Montréal, Montreal, Que., 1967. MR0246761

[8] A. Daigneault, *Operations in polyadic algebras*, Trans. Amer. Math. Soc. **158** (1971), 219–229, DOI 10.2307/1995783. MR0285368

[9] A. Daigneault and D. Monk, *Representation theory for polyadic algebras*, Fund. Math. **52** (1963), 151–176, DOI 10.4064/fm-52-2-151-176. MR0151373

[10] N. Feldman, *Cylindric algebras with terms*, J. Symbolic Logic **55** (1990), no. 2, 854–866, DOI 10.2307/2274669. MR1056393

[11] P. Halmos, *Polyadic Boolean algebras*, Proc. Int. Congr. Math. Amsterdam, vol. 2 (1954), 402–403; reprinted in [**18**], 261.

[12] P. Halmos, *Polyadic Boolean algebras*, Proc. Nat. Acad. Sci. **40** (1954), 296–301; reprinted in [**18**], 243–248.

[13] P. R. Halmos, *Algebraic logic. I. Monadic Boolean algebras*, Compositio Math. **12** (1956), 217–249. reprinted in [**18**], 37–72. MR0078304

[14] P. R. Halmos, *The basic concepts of algebraic logic*, Amer. Math. Monthly **63** (1956), 363–387, DOI 10.2307/2309396. reprinted in [**18**], 9–33. MR0086028

[15] P. R. Halmos, *Algebraic logic. II. Homogeneous locally finite polyadic Boolean algebras of infinite degree*, Fund. Math. **43** (1956), 255–325. reprinted in [**18**], 97–166. MR0086029

[16] P. R. Halmos, *Algebraic logic. III. Predicates, terms, and operations in polyadic algebras*, Trans. Amer. Math. Soc. **83** (1956), 430–470, DOI 10.2307/1992883. reprinted in [**18**], 169–209. MR0086030

[17] P. R. Halmos, *Algebraic logic. IV. Equality in polyadic algebras*, Trans. Amer. Math. Soc. **86** (1957), 1–27, DOI 10.2307/1992863. Reprinted in [**18**], 213–248. MR0090564

[18] P. R. Halmos, *Algebraic logic*, Chelsea Publishing Co., New York, 1962. MR0131961

[19] L. Henkin, J. D. Monk, and A. Tarski, *Cylindric algebras. Part I*, Studies in Logic and the Foundations of Mathematics, vol. 64, North-Holland Publishing Co., Amsterdam, 1985. With an introductory chapter: General theory of algebras; Reprint of the 1971 original. MR781929

[20] P.-F. Jurie, *Notion de quasi-somme amalgamée: Premières applications à l'algèbre booléienne polyadique* (French), C. R. Acad. Sci. Paris Sér. A-B **264** (1967), A1033–A1036. MR0228394

[21] L. LeBlanc, *Transformation algebras*, Canad. J. Math. **13** (1961), 602–613, DOI 10.4153/CJM-1961-049-1. MR0132011

[22] L. LeBlanc, *Représentabilité et définissabilité dans les algèbres transformationnelles et dans les algèbres polyadiques* (French), Séminaire de Mathématiques Supérieures, No. 24 (Été, 1966), Les Presses de l'Université de Montréal, Montreal, Que., 1966. MR0250858

[23] E. S. Maftsir and B. I. Plotkin, *The automorphism group of a database* (Russian), Ukrain. Mat. Zh. **40** (1988), no. 3, 335–345, 406, DOI 10.1007/BF01061307; English transl., Ukrainian Math. J. **40** (1988), no. 3, 286–294 (1989). MR952122

[24] C. Pinter, *A simpler set of axioms for polyadic algebras*, Fund. Math. **79** (1973), no. 3, 223–232, DOI 10.4064/fm-79-3-223-232. MR0321735

[25] C. C. Pinter, *A simple algebra of first order logic*, Notre Dame J. Formal Logic **14** (1973), 361–366. MR0321736

[26] B. I. Plotkin, *An algebraic model for database automata* (Russian), Latv. Mat. Ezhegodnik **27** (1983), 216–232. MR720034

[27] B. I. Plotkin, *Halmos (polyadic) algebras in database theory*, Algebraic logic (Budapest, 1988), Colloq. Math. Soc. János Bolyai, vol. 54, North-Holland, Amsterdam, 1991, pp. 503–518. MR1153438

[28] B. I. Plotkin, *Universal algebra, algebraic logic, and databases* (Russian), Nauka, Moscow, 1991; English translation: Kluwer Acad. Publ., 1994.

[29] G. Sági, *Polyadic algebras*, Cylindric-like algebras and algebraic logic, Bolyai Soc. Math. Stud., vol. 22, János Bolyai Math. Soc., Budapest, 2012, pp. 367–389, DOI 10.1007/978-3-642-35025-2_18. MR3156409

[30] A. Tarski and F.B. Thompson, *Some general properties of cylindric algebras*, Bull. Amer. Math. Soc. **58** (1952), 65.

[31] Ya. P. Tsirulis, *Two generalizations of the notion of a polyadic algebra* (Russian), Izv. Vyssh. Uchebn. Zaved. Mat. **12** (1988), 39–49; English transl., Soviet Math. (Iz. VUZ) **32** (1988), no. 12, 61–78. MR1002860

[32] Ya. P. Tsirulis, *Algebraic Structures Arising in Relational Database Theory* (Russian), Cand. Sci. thesis, Leningrad State University, 1988.

Institute of Mathematics and Computer Science, University of Latvia, Riga, Latvia
Email address: janis.cirulis@lu.lv

Contemporary Mathematics
Volume **726**, 2019
https://doi.org/10.1090/conm/726/14609

Geometric equivalence of π-torsion-free nilpotent groups

R. Lipyanski

Abstract. In this paper, we study the notion of geometric equivalence of groups introduced by B. Plotkin [**P1, P2**]. Sufficient and necessary conditions are presented for a π-torsion-free nilpotent group to be geometrically equivalent to its π-completion.

We prove that a relatively free nilpotent π-torsion-free group and its π-completion define the same quasi-variety. Examples of π-torsion-free nilpotent groups that are geometrically equivalent to their π-completions are given.

1. Introduction

I attended Professor Plotkin's lectures about 45 years ago as a student at the University of Latvia. Over the years, I also participated in Plotkin's seminars on group theory in Latvia and Israel. I am grateful to Boris Isaakovich Plotkin for revealing to me the world of group theory in all its diversity. The problems posed by Plotkin at seminars and in his numerous papers have always been interesting to me. I would like to wish my teacher further creative success in Algebra in the next 30 years.

The foundation of universal algebraic geometry for different algebraic systems were laid by B. Plotkin [**P1, P2, P3**]. Within the framework of universal algebraic geometry (UAG), the important notion of geometric equivalence of universal algebras was introduced in [**P1, P2**]. Geometric equivalence of two algebras means, in some sense, equal possibilities for solving systems of equations in these algebras.

Algebraic geometry over groups was considered together with B. Plotkin by G. Baumslag, A. Miasnikov, V. Remeslennikov [**B1, B2**] and others.

In [**P3**], B. Plotkin posed a question: Under what conditions is a nilpotent torsion-free group geometrically equivalent to its Mal'tsev completion? Sufficient conditions for this equivalence were presented in [**T**]. A criterion for a nilpotent torsion-free group to be geometrically equivalent to its Mal'tsev completion was given by V. Bludov and B. Guskov [**BG**]. Based on this criterion, they also provided an example of a torsion-free three-step nilpotent group that is not geometrically equivalent to its Mal'tsev completion.

There exists a natural generalization of the concept of a torsion-free group (see [**L1**]). Let π be a (non empty) set of primes. An integer is said to be a π-number if it is a product of powers of primes from π. A group G is π-divisible if any

2010 *Mathematics Subject Classification.* Primary 20F18, 20D15; Secondary 14A22, 14A20.
Key words and phrases. Geometric equivalence, nilpotent groups, $\mathbb{Q}_\pi$-powered groups.

equation of the form $x^n = a$, where $a \in G$ and n is a π-number, has a solution in G. A group G is π-torsion-free if any equation of the form $x^n = 1$, where n is a π-number, has only the trivial solution in G. By definition, a π-divisible π-torsion-free group is called π-complete. It is known that every π-torsion-free nilpotent group G can be embedded as a subgroup into a π-complete nilpotent group (see [**Kh2**]).The minimal π-complete nilpotent group containing the group G is called the π-completion of G. If π consists of all primes, the π-completion of group G coincides with its Mal'tsev completion.

Here we study the problem related to geometric equivalence of a nilpotent π-torsion-free group to its π-completion. We give necessary and sufficient conditions for this geometric equivalence. The proof of this criterion uses the theory of π-isolators of groups, as well as the approach proposed by V. Bludov and B. Guskov, to study the problem of geometric equivalence of a torsion-free nilpotent group to its Mal'tsev completion. Based on this criterion, we prove that a relatively free nilpotent π-torsion-free group is geometrically equivalent to its π-completion. If π is the set of all primes, this result has already been proven by A. Tsurkov in [**T**].

In [**BG**], it was proven that every torsion-free two-step nilpotent group is geometrically equivalent to its Mal'tsev completion. This result corresponds to the case when π is the set of all prime numbers. If π is a set of primes, the situation becomes more complicated and requires additional consideration. We prove that if G is a finitely generated π-torsion-free two-step nilpotent group such that the group $G/I_\pi(G')$ is torsion free, then G is geometrically equivalent to its π-completion. This assertion implies the above result on the geometric equivalence of an arbitrary two-step nilpotent torsion-free group to its Mal'tsev completion.

The question of the geometric equivalence of an arbitrary π-torsion-free two-step nilpotent group to its π-completion remains open.

2. Preliminaries

2.1. Geometric equivalence of groups. Let us recall the main definitions from [**P1**].

Let $F = F(X)$ be a free group generated by a finite set X and G be a group. The set $\mathrm{Hom}(F, G)$ of all homomorphisms from F to G can be treated as an affine space. It is possible to define a Galois correspondence $'$ between subsets in $\mathrm{Hom}(F, G)$ and subsets in F. For a subset $T \subseteq F$ define T'_G, G-closure of T

$$T'_G = \{\varphi \in \mathrm{Hom}(F, G) \mid T \subseteq \mathrm{Ker}\,\varphi\}.$$

On the other hand, for any subset $A \subseteq \mathrm{Hom}(F, G)$ define A'_G, G-closure of A,

$$A'_G = \bigcap_{\mu \in A} \mathrm{Ker}\,\mu.$$

The set $T \subseteq F$ can be treated as a system of equations that we solve in G. Then T'_G is the set of all solutions in G of this system. Similarly, the set A can be treated as an affine space whose points are homomorphisms. Then A'_G is the system of all equations written in F such that A is the set of all solutions of this system.

DEFINITION 2.1 ([**P1**],[**P4**]). A subset $T \subseteq F$ (a subset $A \subseteq \mathrm{Hom}(F, G)$) is called G-closed if $T_G = T''_G$ (respectively, $A_G = A''_G$).

DEFINITION 2.2 ([**P1**],[**P2**]). Two groups G_1 and G_2 are called geometrically equivalent if for any free group $F = F(X)$ of finite rank and for any subset $T \subseteq F$

$$T''_{G_1} = T''_{G_2}.$$

In other words, the groups G_1 and G_2 are geometrically equivalent if and only if every G_1-closed subset of the group F is G_2-closed and vice versa.

PROPOSITION 2.3 ([**P1**]). *Groups G_1 and G_2 are geometrically equivalent if and only if every finitely generated subgroup of G_1 can be approximated by subgroups of G_2 and vice versa.*

DEFINITION 2.4 ([**BG**],[**T**]). Let H and G be groups and $F = F(X)$ be any free group finitely generated by X. If every H-closed subset of F is G-closed, we say that the geometry defined by H is weaker than the geometry defined by G and write $H \preceq G$.

The relation $\preceq$ is a preorder on the class of groups. This preorder defines an order relation on classes of geometrically equivalent groups.

PROPOSITION 2.5 ([**BG**],[**T**]). *$H \preceq G$ if and only if any finitely generated subgroup of the group H can be approximated by subgroups of G.*

PROPOSITION 2.6 ([**BG**],[**T**]). *$H \preceq G$ for any subgroup H of G.*

It is clear that if for groups G_1 and G_2 the conditions $G_1 \preceq G_2$ and $G_2 \preceq G_1$ hold, then they are geometrically equivalent.

2.2. π-completion of a π-torsion-free nilpotent group. Let π be a (non empty) set of primes. An integer is said to be a π-number if it is a product of powers of primes from π. For a subgroup H of a group G, the π-isolator $I_\pi(H)$ of H in G is defined as follows:

$$I_\pi(H) = \{g \in G | g^n \in H \text{ for some } \pi - \text{number } n = n(g)\}.$$

A subgroup A is said to be π-isolated if $I_\pi(A) = A$.

If π is the set of all prime numbers, the π-isolator of H in G is denoted by $I(H)$ and is called the isolator of H in G.

PROPOSITION 2.7 (see [**Kh2**], [**LR**]). *If H is a subgroup of a nilpotent group G and π is a set of prime numbers, then $I_\pi(H)$ is a subgroup.*

A group G is π-divisible if for every π-number n, every element $g \in G$ has an nth root of g in G, i.e., there exists an element $h \in G$ such that $h^n = g$. A group is π-torsion-free if it has no non-identity element whose order is π-number, i.e., $I_\pi(1) = 1$. A π-divisible π-torsion-free group is said to be π-complete.

Let $\mathbb{Q}_\pi$ be the ring of all rational numbers whose denominators are π-numbers and G be a π-divisible π-torsion-free group. A unary operation on G can be defined by taking powers in $\mathbb{Q}_\pi$ satisfying the laws $(x^r)^s = x^{rs}$ for all $r, s \in \mathbb{Q}_\pi$. The group G with this additional unary operation is called the $\mathbb{Q}_\pi$-powered group.

THEOREM 2.8 ([**Kh2**]).

(a) *Every π-torsion-free nilpotent group G of class c can be embedded as a subgroup in the $\mathbb{Q}_\pi$-powered nilpotent group $\widehat{G}^\pi$ of the same class c such that $\widehat{G}^\pi = I_\pi(G)$, i.e., for any element $g \in \widehat{G}^\pi$ there exists a π-number m such that $g^m \in G$.*

(b) *The group* $\widehat{G}^\pi$ *is unique up to isomorphism; moreover every isomorphism* $\varphi : G \to G'$ *extends to an isomorphism of* $\widehat{G}^\pi$ *onto* $\widehat{G'}^\pi$.

The group $\widehat{G}^\pi$ is the minimal π-complete group containing G and it is called the π-completion of G.

From here on, we consider $\mathbb{Q}_\pi$-powered group $\widehat{G}^\pi$ as a π-torsion-free π-divisible abstract group, i.e., as the group with respect to only the usual group operations. In the special case, where π consists of all primes, then $\mathbb{Q}_\pi = \mathbb{Q}$, the property of being π-complete coincides with the property of being complete. In this case, the π-completion $\widehat{G}^\pi$ of the group G coincides with the Mal'tsev completion $\widehat{G}$ of G (see [**M**]).

DEFINITION 2.9. Let G be a group and $H \leq G$ be a subgroup of G. Denote by H^m the subgroup of G generated by mth powers of elements from H.

The subgroup H^m is a verbal subgroup of G defined by the word x^m.

We also need the following lemma, given by A. Mal'tsev [**M**]. The initial proof of this lemma is based on the theory of Lie groups. The proof of this lemma in the framework of group theory (without Lie context) was given by A. Klyachko (Lemma 2.2, in [**K**]). See also the proof of Lemma 1.2 in [**BG**].

LEMMA 2.10 ([**M**]). *Let G be a nilpotent group of nilpotency class k. Fix a natural number n and denote by G^{n^k} a subgroup of G generated by all elements of the form g^{n^k}, $g \in G$. Then the equation $x^n = h$ has a solution in the group G for each $h \in G^{n^k}$.*

3. π-torsion-free nilpotent groups

3.1. Geometric equivalence of π-torsion-free nilpotent groups to their π-completions. Let π be a set of primes. Now we consider some examples of π-torsion-free nilpotent groups for which we can describe their π-completions.

EXAMPLE 3.1. Let H be a finitely generated π-torsion-free abelian group. We wish to describe the structure of the group $\widehat{H}^\pi$.

It is well known (see, e.g., [**Kr**], Ch.6, Sec.20) that the group H is the finite direct product of cyclic π-torsion-free groups, i.e., there exist integers $s, k \geq 0$ and prime numbers $p_1 \leq \cdots \leq p_k$ together with positive integers $\alpha_1, \ldots \alpha_r$ such that

$$H \cong \mathbb{Z}^s \times \mathbb{Z}/p_1^{\alpha_1}\mathbb{Z} \times \cdots \times \mathbb{Z}/p_k^{\alpha_k}\mathbb{Z}, \tag{3.1}$$

where p_i, $i = 1, \ldots, k$, are not π-numbers.

Denote by $\mathbb{Q}_\pi^+$ the additive group of the ring $\mathbb{Q}_\pi$. It is clear that the π-completion $\widehat{\mathbb{Z}}^\pi$ of $\mathbb{Z}$ is equal to $\mathbb{Q}_\pi^+$. If G is a p-group, where $p \notin \pi$, then for every element $g \in G$ and $m \in \pi$ there exists an mth root of g. In fact, the mapping $g \to g^m$ is an automorphism of the cyclic subgroup $\langle g \rangle$. Therefore, every group $\mathbb{Z}/p_i^{\alpha_i}\mathbb{Z}$, $p_i \notin \pi$, is π-complete. It is easy to check that if a group A is π-torsion-free nilpotent such that $A = A_1 \times A_2$, then $\widehat{A}^\pi = \widehat{A_1}^\pi \times \widehat{A_2}^\pi$. As a consequence, we obtain

$$\widehat{H}^\pi \cong (\mathbb{Q}_\pi^+)^s \times \mathbb{Z}/p_1^{\alpha_1}\mathbb{Z} \times \cdots \times \mathbb{Z}/p_k^{\alpha_k}\mathbb{Z}.$$

By Proposition 2.3, H is geometrically equivalent to $\widehat{H}^\pi$.

Denote by T the torsion group of H and by $\widehat{T}^{\pi}$ its π-completion. It is clear that $\widehat{T}^{\pi} = T$, i.e., the group T is π-complete.

Now we wish to select another class of π-complete nilpotent groups.

EXAMPLE 3.2. Let G be a π-torsion-free nilpotent group such that the quotient group G/G' is finite. We will show that the group G is π-complete.

Since the group G is nilpotent and G/G' is finite, G is finite (see [**W**, Sec. 3, p. 9]). Therefore, the group G is the direct product of its Sylow subgroups:

$$G = G(p_1) \times G(p_2) \times \cdots \times G(p_k).$$

Since $G(p_i)$ is a π-torsion-free nilpotent group, $p_i \notin \pi$. As a consequence, the group $G(p_i)$ is π-complete. Hence, the group G, as the direct product of π-complete groups, is π-complete too.

Our goal is to provide a criterion for nilpotent π-torsion-free groups to be geometrically equivalent to their π-completions. For that purpose, we need the result from ([**LR**] Ch. 2, Sec. 3) that was established by A. Mal'tsev ([**M**]) in the case when π is the set of all prime numbers.

LEMMA 3.3. *Let G be a finitely generated nilpotent group, $g_1, \dots, g_n$ be a set of generators of G, and H be a subgroup of G. Suppose that $g_i^{r_i} \in H$ for some positive π-numbers r_i, $i = 1, \dots, n$. Then each element of G has a positive π-power in H and $\mid G : H \mid$ is a finite π-number.*

We shall need the following lemma.

LEMMA 3.4. *Let G be a nilpotent π-torsion-free group and $\widehat{G}^{\pi}$ be its π-completion. Then for any finitely generated subgroup H of $\widehat{G}^{\pi}$ there exists a π-number $r = r(G)$ such that $H^r \leq G$.*

PROOF. Let the group H be generated by $h_1, \dots, h_n$. Then there are π-numbers r_i such that $h_i^{r_i} \in G$. Denote by D the subgroup of H generated by $h_1^{r_1}, \dots, h_n^{r_n}$. The group D is a subnormal subgroup of H, i.e., there is a normal series

$$D = D_1 \leq D_2 \leq \cdots \leq D_{n-1} \leq D_n = H$$

from D to H. By Lemma 3.3, the index $\mid H : D \mid$ is a π-number. Therefore, each factor in this series has an index that is a π-number. If r_{n-1} is the order of the quotient group H/D_{n-1}, then $H^{r_{n-1}} \leq D_{n-1}$. By induction, it is easy to prove that there exists a sequence of π-numbers $r_1, \dots, r_{n-1}$ such that $H^r \leq D$, where $r = r_1 \cdots r_{n-1}$. Since D is a subgroup of G, this completes the proof. □

Now we derive a criterion of geometric equivalence of a π-torsion-free nilpotent group to its π-completion.

THEOREM 3.5. *Let G be a π-torsion-free nilpotent group and $\widehat{G}^{\pi}$ be its π-completion. Then groups G and $\widehat{G}^{\pi}$ are geometrically equivalent if and only if the group G is geometrically equivalent to subgroups G^m for every π-number m.*

PROOF. Our argument goes along the lines of Theorem 3.1 in [**BG**] but we add new assertions related to π-torsion in nilpotent groups.

Let the groups G and $\widehat{G}^{\pi}$ be geometrically equivalent. Let us show that then groups G and G^m are also geometrically equivalent for any π-number m. The group G^m is a subgroup of G. By Proposition 2.6, we have $G^m \preceq G$. Let us prove that

$G \preceq G^m$. Consider a finitely generated subgroup $H = \langle h_1, \dots, h_s \rangle$ of G. Denote by $h_i^{\frac{1}{m}}$ the root of the equation $x^m = h_i$ in $\widehat{G}^\pi$. Denote by $H^{\frac{1}{m}} = \langle h_1^{\frac{1}{m}}, \dots, h_s^{\frac{1}{m}} \rangle$ the subgroup of $\widehat{G}^\pi$ generated by $h_i^{\frac{1}{m}}, i = 1, \dots, s$. It is clear that $H^{\frac{1}{m}} \preceq \widehat{G}^\pi$. Therefore, there exists an embedding $\varphi : H^{\frac{1}{m}} \to \prod_{i \in I} \widehat{G}_i^\pi$, where G_i, $i \in I$, are isomorphic copies of G. Since the groups G and $\widehat{G}^\pi$ are geometrically equivalent, there exists an embedding $\psi : H^{\frac{1}{m}} \to \prod_{i \in I} G_i$. Consequently, H is embedded into a cartesian product of the group G^m. Therefore, $H \preceq G^m$. As a consequence, we obtain $G \preceq G^m$. Therefore, G and G^m are geometrically equivalent.

Conversely, let the groups G and G^m be geometrically equivalent for every π-number m. We will prove that G and $\widehat{G}^\pi$ are geometrically equivalent. Since G is a subgroup of $\widehat{G}^\pi$, we have $G \preceq \widehat{G}^\pi$. Let us prove that $\widehat{G}^\pi \preceq G$. Let H be a subgroup of $\widehat{G}^\pi$ generated by elements $h_1, \dots, h_s$. By Lemma 3.4, there exists a π-number r such that $H^r \leq G$. By Proposition 2.6, $H^r \preceq G$. If k is the nilpotency class of G, then G is geometrically equivalent to G^{r^k}. Therefore, $H^r \preceq G^{r^k}$. Hence, there exists an embedding $\varphi : H^r \to \prod_{i \in I} G_i^{r^k}$, where G_i, $i \in I$, are isomorphic copies of the group G. Denote by φ_i, $i \in I$, the coordinate functions of φ, i.e., $\varphi = (\varphi_i)_{i \in I}$.

It is clear that $\widehat{H^r}^\pi = \widehat{H}^\pi$ and $\widehat{G^{r^k}}^\pi = \widehat{G}^\pi$. By Theorem 3.9, the mapping φ can be extended to the embedding $\bar{\varphi} : \widehat{H}^\pi \to \widehat{G}^\pi$. Then $\bar{\varphi}(h_j^r) = \varphi(h_j^r) = (\varphi_i(h_j^r))_{i \in I}$. By Lemma 2.10, the equation $x^r = \varphi_i({h_j}^r)$ has a solution in G_i. Since G_i is a π-torsion-free group, $\varphi_i(h_j) \in G_i$. Therefore, the mapping $\bar{\varphi}$ is an embedding of H into a cartesian power of the group G. Hence, $\widehat{G}^\pi \preceq G$. As a consequence, the groups $\widehat{G}^\pi$ and G are geometrically equivalent. □

COROLLARY 3.6. *Let G be a π-torsion-free nilpotent group. If every finitely generated subgroup H of G is geometrically equivalent to its π-completion $\widehat{H}^\pi$, then the groups G and $\widehat{G}^\pi$ are geometrically equivalent.*

PROOF. The groups H and $\widehat{H}^\pi$ are geometrically equivalent. Therefore, by Theorem 3.5, H and H^m are geometrically equivalent for every π-number m. As a consequence, $H \preceq H^m$. By Proposition 2.5, there exists an embedding $\mu : H \to \prod_{i \in I} H_i^m$, where H_i, $i \in I$, are isomorphic copies of the group H. The mapping μ can be extended to the embedding $\bar{\mu} : H \to \prod_{i \in I} G_i^m$, where G_i, $i \in I$, are isomorphic copies of the group G. Therefore, $G \preceq G^m$. Since $G^m \preceq G$, the groups G and G^m are geometrically equivalent. By Theorem 3.5, the group G is geometrically equivalent to its π-completion $\widehat{G}^\pi$. □

COROLLARY 3.7. *Let G be a π-torsion-free abelian group. Then G is geometrically equivalent to its π-completion $\widehat{G}^\pi$.*

PROOF. Let H be a finitely generated subgroup of G. According to Example 3.1, the groups $\widehat{H}^\pi$ and H are geometrically equivalent. By Corollary 3.6, the group G is geometrically equivalent to its π-completion. □

Now we give other examples of π-torsion-free nilpotent groups which are geometrically equivalent to their π-completions $\widehat{G}^\pi$.

In what follows, we use the following Theorem of Dick (cf. [**Kr**, Ch. 5, Sec. 18]:

THEOREM 3.8. *Let a group G be represented as follows:*

$$G = \langle x_i \mid r_j(x_{j_1}, \dots x_{j_s}) = 1; \quad i, j_k \in I, j \in J \rangle.$$

Suppose that a group D contains a set $\{d_i, i \in I\}$ such that

$$r_j(d_{j_1}, \dots, d_{j_s}) = 1$$

holds in D for every $j \in J$. Then the map $x_i \to d_i, i \in I$ extends to a homomorphism of G onto D.

We also need the following claim.

PROPOSITION 3.9 (see [**Kh1**]). *Let H be a π-torsion-free nilpotent group. If $x^n \cdot y^m = y^m \cdot x^n$, where $x, y \in H$ and m, n are π-numbers, then $x \cdot y = y \cdot x$.*

Now we are ready to prove

PROPOSITION 3.10. *Let G be a finitely generated π-torsion-free two step nilpotent group such that the group $G/I_\pi(G')$ is torsion-free. Then G is geometrically equivalent to its π-completion $\widehat{G}^\pi$.*

PROOF. First, let us show that $I_\pi(G')$ is an abelian group. If $d_1, d_2 \in I_\pi(G')$, then there are π-numbers k_1, k_2 such that $d_1^{k_1}, d_2^{k_2} \in G'$. Since G is a two step nilpotent group, $d_1^{k_1} \cdot d_2^{k_2} = d_2^{k_2} \cdot d_1^{k_1}$. By Proposition 2.8,$d_1 \cdot d_2 = d_2 \cdot d_1$. Since G is a finitely generated nilpotent group, the subgroup $I_\pi(G')$ is an abelian finitely generated group.

Let us choose a basis $\bar{g}_1, \dots, \bar{g}_s$ in the abelian group $G/I_\pi(G')$. Denote by $g_1, \dots, g_s$ the preimages of these elements in the group G. Now choose a basis $c_1, \dots, c_k$ in the abelian group $I_\pi(G')$. By definition, there are π-numbers p_i, $i = 1, \dots k$, such that $c_i^{p_i} \in G'$.

The group G is generated by elements $g_1, \dots, g_s$, $c_1, \dots, c_k$. Let

$$R(g_1, \dots, g_s, c_1, \dots, c_k) = 1 \tag{3.2}$$

be a relation in the group G. Using the collection process in the group G (see, e.g., [**LR**, Ch. 2, Sec. 2]), the relation (3.2) can be rewritten as follows:

$$g_1^{r_1} \cdot \dots \cdot g_s^{r_s} \cdot c_1^{t_1} \dots \cdot c_k^{t_k} \cdot W_1([g_i, g_j]) \cdot W_2([c_l, g_d]) = 1, \ r_i \geq 0, \ t_i \geq 0, \tag{3.3}$$

where $W_1([h_i, h_j])$ is a word in the commutators $[h_i, h_j]$ and $W_2([c_l, g_d])$ is a word in the commutators $[c_l, g_d]$. In the group $G/I_\pi(G')$ the relation (3.3) has the form

$$\bar{g}_1^{r_1} \cdot \dots \cdot \bar{g}_s^{r_s} = 1.$$

Since the group $G/I_\pi(G')$ is free abelian, $r_1 = \dots = r_s = 0$. Therefore, the relation (3.3) is equivalent to

$$c_1^{t_1} \dots \cdot c_k^{t_k} \cdot W_1([g_i, g_j]) \cdot W_2([c_l, g_d]) = 1. \tag{3.4}$$

Denote by $p = \max_i p_i$. Raising the relation (3.4) to the power p, we have

$$c_1^{pt_1} \dots \cdot c_k^{pt_k} \cdot W_1([g_i, g_j]^p) \cdot W_2([c_l, g_d]^p) = 1. \tag{3.5}$$

Since G is a π-torsion-free group, the relation (3.5) is equivalent to (3.4). Since $[c_l, g_d]^p = [c_l^p, g_d] = 1$, we obtain $W_2([c_l, g_d]^p) = 1$. As a consequence, the relation (3.5) is equivalent to

$$c_1^{pt_1} \dots \cdot c_k^{pt_k} \cdot W_1([g_i, g_j]^p) = 1. \tag{3.6}$$

Note that elements $c_i^{pt_i}$ from (3.6) belong to G'.

For a π-number m, consider the map $\varphi : G \to G^m$ given by the rules:

$$\varphi(g_i) = g_i^m, \ \ \varphi(c_i) = c_i^{m^2}.$$

Since $[g_i^m, g_j^m] = [g_i, g_j]^{m^2}$, the relation (3.6) has the following form

$$c_1^{m^2 pt_1} \cdot \ldots \cdot c_k^{m^2 pt_k} \cdot (W_1([g_i, g_j]^p))^{m^2} = 1. \tag{3.7}$$

Since the group G is π-torsion-free, the relation (3.7) is equivalent to the relation (3.6), and, as a consequence, to the relation (3.2). By Dick's Theorem, the mapping φ can be extended to a homomorphism of G into G^m. Since G is a π-torsion-free group, this homomorphism is a monomorphism G into G^m. Therefore, $G \preceq G^m$. By Theorem 3.5, G is geometrically equivalent to $\widehat{G}^\pi$. □

COROLLARY 3.11 ([**BG**]). *Let G be a torsion-free two-step nilpotent group. Then G is geometrically equivalent to its Mal'tsev completion $\widehat{G}$.*

PROOF. Let H be a finitely generated subgroup of G. The group $H/I(H')$ is torsion-free. By Proposition 3.10, the group G is geometrically equivalent to its Mal'tsev completion $\widehat{G}$. □

EXAMPLE 3.12. Let $\mathbb{H}_3(\mathbb{Z})$ be the Heisenberg group generated by elements x, y, z with the defining relations $zx = xz, yz = zy, [x, y] = z$. Elements of $H_3(\mathbb{Z})$ are all formal expressions of the form $x^l y^s z^n$, where $l, m, n \in \mathbb{Z}$. The group $H_3(\mathbb{Z})$ is torsion-free two-step nilpotent. The commutator $H_3(\mathbb{Z})'$ is generated by the element z. The group $H_3(\mathbb{Z})/H_3(\mathbb{Z})'$ is a free abelian one generated by the elements x and y. By Corollary 3.11, the group $\mathbb{H}_3(\mathbb{Z})$ is geometrically equivalent to its Mal'tsev completion $\widehat{\mathbb{H}_3(\mathbb{Z})}$.

EXAMPLE 3.13. Let G be a group generated by elements a, b, c with the defining relations $ca = ac, cb = bc, [a, b] = c^2, c^4 = 1$. Elements of G are all formal expressions of the form $a^l b^s c^n$, where $l, m, \in \mathbb{Z}$ and $n \in \mathbb{Z}_3$. The group G is two step nilpotent π-torsion-free, where π is the set of all prime numbers without 2. The commutator $G' = \langle c^2 \rangle$ and $I_\pi(G') = \langle c \rangle$. The group $G/I_\pi(G')$ is a free abelian one generated by the elements a and b. By Proposition 3.10, the group G is geometrically equivalent to $\widehat{G}^\pi$.

The following question arises: Is it true that every π-torsion-free two step nilpotent group is geometrically equivalent to its π-completion?

3.2. Geometric equivalence of a relatively free nilpotent π-torsion-free group to its π-completion. It was proved by A. Tsurkov in [**T**] that a torsion-free nilpotent relatively free group is geometrically equivalent to its Mal'tsev completion. His proof is based on the Mal'tsev correspondence technique between nilpotent $\mathbb{Q}$-powered groups and nilpotent Lie algebras over $\mathbb{Q}$ as well as on the Lazard correspondence between groups with a central filtration and Lie rings.

Our goal is to extend Tsurkov's result to the class of relatively free nilpotent π-torsion-free groups. We prove that a relatively free nilpotent π-torsion-free group is geometrically equivalent to its π-completion. Note that there is no good correspondence between nilpotent $\mathbb{Q}_\pi$-powered groups and Lie $\mathbb{Q}_\pi$-algebras (see [**Kh2**]). For this reason, here we use the technique related to the theory of isolators in groups. In the proof of this result, we also apply the criterion for a π-torsion-free nilpotent group to be geometrically equivalent to its π-completion (see Theorem 3.5).

The method of associating a Lie ring to a group was studied in detail by M. Lazard [**L2**] (see also [**B**]).

Let G be a group. Consider the lower central series of G:

$$G \geq \gamma_1(G) \geq \cdots \geq \gamma_i(G) \geq \gamma_{i+1}(G) \geq \cdots$$

Let π be a set of primes. Denote by $G_i = I_\pi(\gamma_i(G))$ the π-isolator the group $\gamma_i(G)$ in G. Note that the group $G_i = I_\pi(\gamma_i(G))$ is a subgroup of G, since it coincides with the preimage of the π-isolator of the identity subgroup of the nilpotent groups $G/\gamma_i(G)$ (see Theorem 2.7).

Therefore, we have a normal series

$$G \geq G_1 \geq \cdots \geq G_i \geq G_{i+1} \geq \cdots \tag{3.8}$$

LEMMA 3.14 (see [**Kh2**]). *For any group G and any natural numbers i, j the following holds*

$$[G_i, G_j] \leq G_{i+j},$$

that is, $\{G_i | i = 1, \dots\}$ is a central filtration of G.

Now consider the additive abelian group

$$L_\pi(G) = \bigoplus_{i=1}^{\infty} G_i/G_{i+1} \quad \text{(direct sum)}.$$

Since $\{G_i | i = 1, \dots\}$ is a central filtration of G, then we can define the structure of the Lie ring on the additive group $L_\pi(G)$: for any $\bar{a}_i \in G_i/G_{i+1}$ and $\bar{a}_j \in G_j/G_{j+1}$, by definition, $[\bar{a}_i, \bar{a}_j] = [a_i, a_j]G_{i+j+1} \in G_{i+j}/G_{i+j+1}$. Then this bracket multiplication is extended to the direct sum $L_\pi(G) = \bigoplus_{i=1}^{\infty} G_i/G_{i+1}$ by the distributive laws. We obtain the Lie ring $L_\pi(G)$ of the group G corresponding to the central filtration (3.8). It is clear that the additive group of the ring $L_\pi(G)$ is π-torsion-free.

Let $F_n = F_n(X)$ be a relatively free n-step nilpotent group generated by X. A group $G = F_n/H$, where H is a π-isolated verbal subgroup of F_n, is relatively free n-step nilpotent π-torsion-free. Note that if T is a verbal subgroup of F_n, then the π-isolator $I_\pi(T)$ is a π-isolated verbal subgroup of F_n.

The main aim of this section is to prove the following

THEOREM 3.15. *Let G be a relatively free nilpotent π-torsion-free group. Then G is geometrically equivalent to its π-completion $\widehat{G}^\pi$.*

PROOF. By Theorem 3.5, it suffices to prove that the group G is geometrically equivalent to the subgroups G^m for every π-number m. Let $(x_i)_{i\in I}, I \subseteq \mathbb{N}$, be free generators of the group G. Consider the map $\varphi : G \to G$ such that $\varphi(x_i) = x_i^m$, $i \in I$, where m is a π-number. Since G is a relatively free group, the map φ can be extended to the endomorphism of G. This endomorphism is also denoted by φ. Let us show that $\operatorname{Ker}\varphi = 0$.

To this end, we consider the Lie ring $L_\pi(G) = \bigoplus_{i=1}^{n-1} G_i/G_{i+1}$, where n is the nilpotency class of G. Since $\varphi(\gamma_i(G)) \leq \gamma_i(G)$, we have $\varphi(G_i) \leq G_i$. Therefore, the endomorphism $\varphi : G \to G$ induces the endomorphism $\bar{\varphi} : L_\pi(G) \to L_\pi(G)$ of the ring $L_\pi(G)$ such that $\bar{\varphi}(g_iG_{i+1}) = \varphi(g_i)G_{i+1}$ for $g_i \in G_i$. Let us write $\bar{\varphi} = (\bar{\varphi}_i)_{i\in J}$, where $\bar{\varphi}_i, i \in J = \{1, \dots, n-1\}$, are the coordinate functions of $\bar{\varphi}$. We first prove that $\operatorname{Ker}\bar{\varphi} = 0$.

Let $\bar{g} = (\bar{g}_i)_{i\in J}$, where $\bar{g}_i = g_iG_{i+1}$, belongs to $\operatorname{Ker}\bar{\varphi}$. Since $g_i \in G_i = I_\pi(\gamma_i(G))$, there exists π-number s_i such that $g_i^{s_i} \in \gamma_i(G)$. In $L_\pi(G)$, we obtain

$$s_ig_i = \sum_k v_{i_k} \mod \gamma_{i+1}(G),$$

where $v_{i_k} = v_{i_k}(x_{t_1}, \dots, x_{t_r})$ are simple commutators in $x_{t_1}, \dots, x_{t_r}$ of length i. Without loss of generality, we assume that $s_i g_i \notin G_{i+1}$. We obtain in $L_\pi(G)$

$$s_i g_i = \sum_k v_{i_k} \mod G_{i+1},$$

i.e., $s_i \bar{g}_i = \sum_k \bar{v}_{i_k}$, where $\bar{v}_{i_k} = v_{i_k} G_{i+1}$. Thus we have

$$\bar{\varphi}_i(s_i \bar{g}_i) = \sum_k \bar{v}_{i_k}(x_{t_1}^m, \dots, x_{t_r}^m) = m^i \sum_k \bar{v}_{i_k}(x_{t_1}, \dots, x_{t_r}) = m^i s_i \bar{g}_i.$$

Therefore, $s_i \bar{\varphi}_j(\bar{g}_i) = s_i m^i \bar{g}_i$. Since s_i and m are π-numbers and the group G_i/G_{i+1} is π-torsion-free, $\bar{g}_i = 0$. Hence, $\operatorname{Ker} \bar{\varphi} = 0$.

Now we prove that $\operatorname{Ker} \varphi = 0$. Let $\varphi(h) = 0$ for some $h \in G_i \setminus G_{i+1}$. Since $\varphi(h) = 0$, we have $\bar{\varphi}(\bar{h}) = 0$, i.e., $h \in G_{i+1}$. This contradiction proves the desired result.

As a consequence, we obtain that φ is a monomorphism G into G^m. Therefore, $G \preceq G^m$ for every π-number m. It is clear that $G^m \preceq G$. Hence, G is geometrically equivalent to G^m for each π-number m. By Theorem 3.5, the group G is geometrically equivalent to its π-completion $\widehat{G}^\pi$. □

In the special case, where π consists of all primes, we obtain

COROLLARY 3.16 ([**T**]). *A relatively free nilpotent torsion-free group is geometrically equivalent to its Mal'tsev completion.*

It is known (see [**P4**],[**MR**]) that two nilpotent groups of the same nilpotency class are geometrically equivalent if and only if they have the same quasiidentities. By Theorem 3.15, we obtain the following

COROLLARY 3.17. *A relatively free nilpotent π-torsion-free group and its π-completion define the same quasi-variety.*

Acknowledgments

I would like to express my appreciation to B. I. Plotkin for many inspiring discussions. I am also very pleased to thank G. I. Mashevitzky, whose suggestions helped make the paper more readable.

References

[B] Yu. A. Bahturin, *Identical relations in Lie algebras*, VNU Science Press, b.v., Utrecht, 1987. Translated from the Russian by Bahturin. MR886063

[B1] G. Baumslag, A. Myasnikov, and V. Remeslennikov, *Algebraic geometry over groups. I. Algebraic sets and ideal theory*, J. Algebra **219** (1999), no. 1, 16–79, DOI 10.1006/jabr.1999.7881. MR1707663

[B2] G. Baumslag, A. Myasnikov, and V. Roman′kov, *Two theorems about equationally Noetherian groups*, J. Algebra **194** (1997), no. 2, 654–664, DOI 10.1006/jabr.1997.7025. MR1467171

[BG] V. V. Bludov and B. V. Gusev, *Geometric equivalence of groups*, Proc. Steklov Inst. Math. **257** (2007), no. suppl. 1, S61–S82, DOI 10.1134/S0081543807050045. MR2338240

[Kh1] E. I. Khukhro, *Nilpotent groups and their automorphisms*, De Gruyter Expositions in Mathematics, vol. 8, Walter de Gruyter & Co., Berlin, 1993. MR1224233

[Kh2] E. I. Khukhro, *p-automorphisms of finite p-groups*, London Mathematical Society Lecture Note Series, vol. 246, Cambridge University Press, Cambridge, 1998. MR1615819

[K] A. A. Klyachko *Group theory*, preprint, http://halgebra.math.msu.su/staff/klyachko/lect4.pdf, 2007.

[Kr] A. G. Kurosh, *The Theory of Groups*, Nauka, Moskow, 1952.

[L1] M. Lazard, *Problèmes d'extension concernant les N-groupes; inversion de la formule de Hausdorff* (French), C. R. Acad. Sci. Paris **237** (1953), 1377–1379. MR0059274

[L2] M. Lazard, *Sur les groupes nilpotents et les anneaux de Lie* (French), Ann. Sci. Ecole Norm. Sup. (3) **71** (1954), 101–190. MR0088496

[LR] J. C. Lennox and D. J. S. Robinson, *The theory of infinite soluble groups*, Oxford Mathematical Monographs, The Clarendon Press, Oxford University Press, Oxford, 2004. MR2093872

[M] A. I. Mal′cev, *Nilpotent torsion-free groups* (Russian), Izvestiya Akad. Nauk. SSSR. Ser. Mat. **13** (1949), 201–212. MR0028843

[MR] A. Myasnikov and V. Remeslennikov, *Algebraic geometry over groups. II. Logical foundations*, J. Algebra **234** (2000), no. 1, 225–276, DOI 10.1006/jabr.2000.8414. MR1799485

[P1] B. Plotkin, *Varieties of algebras and algebraic varieties. Categories of algebraic varieties*, Siberian Adv. Math. **7** (1997), no. 2, 64–97. MR1481222

[P2] B. I. Plotkin, *Algebraic logic, varieties of algebras and algebraic variety,* Proc. Int. Alg. Conf. St. Petersburg, New York, London, 1998.

[P3] B. Plotkin, E. Plotkin, and A. Tsurkov, *Geometrical equivalence of groups*, Comm. Algebra **27** (1999), no. 8, 4015–4025, DOI 10.1080/00927879908826679. MR1700201

[P4] B. I. Plotkin, *Seven lectures on universal algebraic geometry*, preprint, Inst. Math., Hebrew Univ., 2000, Jerusalem.

[T] A. Tsurkov, *Geometric equivalence of nilpotent groups*, J. Math. Sci. **110** (2007), no. 5, 407–409.

[W] R. B. Warfield Jr., *Nilpotent groups*, Lecture Notes in Mathematics, Vol. 513, Springer-Verlag, Berlin-New York, 1976. MR0409661

Department of Mathematics, Ben Gurion University, Beer Sheva 84105, Israel
Email address: lipyansk@math.bgu.ac.il

Contemporary Mathematics
Volume **726**, 2019
https://doi.org/10.1090/conm/726/14610

On the Zariski topology of Ω-groups

R. Lipyanski

Abstract. A number of geometric properties of Ω-groups from a given variety of Ω-groups can be characterized using the notions of domain and equational domain. An Ω-group H of a variety Θ is an equational domain in Θ if the union of algebraic varieties over H is an algebraic variety. We give necessary and sufficient conditions for an Ω-group H in Θ to be an equational domain in this variety.

Let $F = F(X)$ be a free Ω-group finitely generated by X in a variety of Ω-groups Θ and H be an Ω-group in Θ. One of the important questions in the algebraic geometry of varieties of Ω-groups is whether it is possible to equip the space of points $\mathrm{Hom}(F, H)$ with the Zariski topology, whose closed sets are precisely algebraic sets. If this is so, then the Ω-group H is called an equational domain (or *stable* in the terminology of the authors of the papers [**P**], [**BPP**]) in the variety Θ (see [**BMR**]). The same problem arises for a variety $\Theta(G)$ of Ω-groups with given Ω-group of constants G.

An important role in the study of equational domains in varieties of Ω-groups is played by the notion of domain. Necessary and sufficient conditions for linear algebras over an algebra of constants and for groups over a group of constants to be equational domains in terms of domains are given in [**BMR**], [**BPP**], [**DMR**] and [**P**]. Here we continue the study of equational domains in varieties of Ω-groups (without Ω-group of constants). We give necessary and sufficient conditions for an Ω-group H in Θ to be an equational domain in the variety Θ.

The results presented in this paper were partially announced earlier in [**L**].

1. Preliminaries

1.1. Ω-groups. We give the basic facts about Ω-groups (see [**H**]).

Definition 1.1. An Ω-group (multioperator group) G is an additive group (not necessarily commutative) with some additional signature Ω, such that for every $\omega \in \Omega$ of arity $n(\omega) = n > 0$, the condition $\underbrace{00\ldots0}_{n}\omega = 0$ should be fulfilled.

A group is an Ω-group with the empty set of operations Ω; in rings the set Ω consists of a single multiplication; in Lie algebras over a commutative associative

2010 *Mathematics Subject Classification.* Primary 14A22, 16S38; Secondary 07A01, 08B20.
Key words and phrases. Equational domain, variety of Ω-groups, anticommutative Ω-group, the Zariski topology.

ring K with unit, the set Ω consists of the Lie bracket and all elements of K belonging to the set Ω.

It is clear that the class of all Ω-groups forms a variety Υ. Let $F_\Upsilon = F_\Upsilon(X)$ be a free Ω-group finitely generated by X in a variety Υ. We use the functional notation $f(x_1, \dots, x_n), x_i \in X$, for words in $F_\Upsilon(X)$. For brevity, we shall use symbols $\bar{x}, \bar{a}$, to denote finite ordered sets $(x_1, x_2, \dots, x_n), (a_1, a_2, \dots, a_m), x_i \in X, a_i \in G$, and write $f(\bar{x})$ for $f(x_1, x_2, \dots, x_n)$, $f(\bar{a}, \bar{b})$ for $f(a_1, a_2, \dots, a_m, b_1, \dots, b_r)$. If $\bar{x}, \bar{y}$ are ordered sets $(x_1, x_2, \dots, x_n), (y_1, y_2, \dots, y_n)$ with the same number of elements, we shall denote the set $x_1 + y_1, x_2 + y_2, \dots, x_n + y_n$ by $\bar{x} + \bar{y}$. We shall also write $\bar{a} \in G$ when we mean $a_1, \dots, a_m \in G$.

Let $f(\bar{x}, \bar{y})$ be a word in two disjoint sets X and Y of variables $\bar{x}$ and $\bar{y}$, respectively. We shall say that the word $f(\bar{x}, \bar{y})$ is a *commutator word in $\bar{x}$ and $\bar{y}$* if $f(\bar{x}, \bar{0}) = f(\bar{0}, \bar{y}) = 0$. The set of all such words will be denoted by $[X, Y]$.

Let A and B be two subsets of an Ω-group G. The set of all elements $f(\bar{a}, \bar{b})$, where $f(\bar{x}, \bar{y}) \in [X, Y]$, is called the *commutator group of A and B* and is denoted by $[A, B]$. Note that if A and B are Ω-subgroups, then $[A, B]$ is also an Ω-subgroup of G.

LEMMA 1.2 ([**H**]). *If $[A, B] = 0$, then*

$$f(\bar{a}, \bar{b}) = f(\bar{a}, \bar{0}) + f(\bar{0}, \bar{b}) = f(\bar{0}, \bar{b}) + f(\bar{a}, \bar{0})$$

for all words $f(\bar{x}, \bar{y})$ and all $\bar{a} \in A$ and $\bar{b} \in B$.

Let A and B be defined as above, $\bar{a} = (a_1, a_2, \dots, a_n)$, $a_i \in A$, and $\bar{b} = (b_1, b_2, \dots, b_n)$, $b_i \in B$. Then the element

$$[\bar{a}; \bar{b}; \omega] = -\bar{a}\omega - \bar{b}\omega + (\bar{a} + \bar{b})\omega, \omega \in \Omega$$

is called the *ω-commutator of $\bar{a}$ and $\bar{b}$.*

DEFINITION 1.3. Let G be an Ω-group. A subset $U$of G is called an ideal in G if the following conditions are fulfilled:

(1) U is closed with respect to all $\omega \in \Omega$.
(2) U is a normal subgroup in the additive group G.
(3) The ω-commutator $[\bar{a}; \bar{b}; \omega]$ belongs to U if $\bar{a} \in U$, $\omega \in \Omega$ and $\bar{b} \in G$.

Let A and B be two Ω-subgroups of G, and let $\{A, B\}$ be an Ω-subgroup of G generated by A and B. The commutator group $[A, B]$ can be characterized as follows.

PROPOSITION 1.4 ([**H**]). *The commutator group $[A, B]$ is the ideal in $\{A, B\}$ generated by all commutators of the form $[a, b]$, $a \in A$, $b \in B$, and all ω-commutators $[\bar{a}; \bar{b}; \omega]$, where $\bar{a} \in A$ and $\bar{b} \in B, \omega \in \Omega$.*

EXAMPLE 1.5. Let R be an associative ring and U_1, U_2 subrings in R. Then it is easy to show that

$$[U_1, U_2] = U_1U_2 + U_2U_1.$$

In the case of Lie algebras (groups), we have the usual commutator subalgebra (commutator subgroup).

DEFINITION 1.6. An Ω-group G is called abelian if $[G, G] = 0$.

Groups and Lie algebras are abelian in the usual sense, while for associative rings this notion means that the product of any two elements is zero.

Now we turn to the property of anticommutativity for Ω-groups.

DEFINITION 1.7 ([**BPP**]). An Ω-group L is called anticommutative (or anti-abelian in the terminology of the authors in [**BMR**]) if the following conditions are fullfilled:

(1) L has no nontrivial abelian ideal.
(2) Every two nontrivial ideals H_1 and H_2 in L have a nontrivial intersection.

A number of interesting properties of anticommutativity are given for groups in [**BMR**] and for Ω-groups in [**BPP**], [**P**]. It is known that all non-abelian free groups, free associative algebras, and non-abelian free Lie algebras are anticommutative (see [**BMR**], [**BPP**], [**L**]).

1.2. Domains. Now we consider the notion of zero divisors in an Ω-group H. For each $a \in H$, denote by $\mathrm{id}\langle a\rangle$ the ideal in H generated by a and by $\langle a\rangle$ the Ω-subgroup of H generated by a. Let P be an Ω- subgroup of H. We denote by $\mathrm{id_P}\langle a\rangle$ the ideal in P generated by a. In our notation, we have $\mathrm{id_H}\langle a\rangle = \mathrm{id}\langle a\rangle$.

DEFINITION 1.8 ([**P**]). A non-zero element $a \in H$ is called a zero divisor if for some non-zero element $b \in H$ we have

$$[\mathrm{id}\langle a\rangle, \mathrm{id}\langle b\rangle] = 0. \tag{1.1}$$

The Ω-group G is called a *domain* if G has no zero divisors, i.e., for any two elements g_1 and g_2 of G the following holds:

$$[\mathrm{id}\langle g_1\rangle, \mathrm{id}\langle g_2\rangle] = 0 \Rightarrow g_1 = 0 \text{ or } g_2 = 0.$$

EXAMPLE 1.9. Let R be an associative ring. In this case Definition 1.8 looks as follows: a non-trivial element a in R is a zero divisor if there exists a non-trivial element $b \in R$ such that

$$\mathrm{id}\langle a\rangle \cdot \mathrm{id}\langle b\rangle = \mathrm{id}\langle b\rangle \cdot \mathrm{id}\langle a\rangle = 0.$$

Let L be a Lie algebra over an associative commutative ring K with unit. A non-trivial element a in L is a zero-divisor if for some non-zero element $b \in L$ we have

$$[\mathrm{id}\langle a\rangle, \mathrm{id}\langle b\rangle] = 0,$$

where $[,]$ is the Lie bracket.

Let H be a group. A non-trivial element a in H is a zero divisor if for some non-trivial element $b \in H$ we have

$$[g_1^{-1}ag_1, g_2^{-1}bg_2] = 1 \tag{1.2}$$

for all $g_1, g_2 \in H$. Here $[,]$ is the usual commutator bracket in the group H.

REMARK 1.10. In [**BMR**]), it was proved that condition (1.2) is equivalent to the following: a non-trivial element a in H is a zero divisor if for some non-trivial element b in H we have

$$[g^{-1}ag, b] = 1 \tag{1.3}$$

for all $g \in H$.

1.3. Algebraic varieties over groups. Let Θ be a variety of Ω-groups and $F = F(X)$ be a free group in Θ finitely generated by X. Consider an Ω-group H in Θ. Any formula $w \equiv w', w, w' \in F(X)$ can be treated as an equation. Denote it as $w = w'$. Every solution of this equation in H is a homomorphism $\mu : F(X) \to H$ such that $w^\mu = w'^\mu$. One can define a Galois correspondence $'$ between subsets in $\mathrm{Hom}(F(X), H)$ and subsets in $F(X)$. For a subset T in $F(X)$ define T', the *H-closure of T*, by

$$T' = \{\mu \in \mathrm{Hom}(F(X), H) \mid T \subseteq \mathrm{Ker}\,\mu\}.$$

On the other hand, for any subset $A \subseteq \mathrm{Hom}(F(X), H)$ define a set A' in $F(X)$, *H-closure of A*

$$A' = \bigcap_{\mu \in A} \mathrm{Ker}\,\mu$$

The set A' is an ideal in $F(X)$.

DEFINITION 1.11. A subset $A \subseteq \mathrm{Hom}(F(X), H)$ is called an affine algebraic variety over H if there exists a set T in $F(X)$ such that $T' = A$.

If A is an algebraic variety, then A' is called the ideal in $F(X)$ corresponding to A.

The intersection $A \cap B$ of algebraic varieties A and B is also an algebraic variety. The union $A \cup B$ of algebraic varieties is not necessarily an algebraic variety. If $A = T_1'$ and $B = T_2'$, then $A \cup B \subseteq (T_1 \cap T_2)'$.

DEFINITION 1.12 ([**P**], [**DMR**]). An Ω-group H is called an equational domain in Θ if for any free Ω-group $F(X)$ and any two algebraic varieties A and B in the space $\mathrm{Hom}(F(X), H)$ the union $A \cup B$ is also an algebraic variety.

Equational domains play an important role in the theory of algebraic varieties. Following [**BPP**], denote by $\mathrm{Alv}_H(F)$ the set of all algebraic varieties in $\mathrm{Hom}(F, H)$. The set $\mathrm{Alv}_H(F)$ can be considered as a lattice, where the union $A \vee B$ is defined by

$$A \vee B = (A \cup B)'.$$

Denote by $\mathrm{Cl}_H(F)$ the set of all H-closed congruences in F. Lattice operations can be defined in a similar way in the set $\mathrm{Cl}_H(F)$. The lattices $\mathrm{Cl}_H(F)$ and $\mathrm{Alv}_H(F)$ are anti-isomorphic. It is clear that if a Ω-group H is an equational domain, then the lattices $\mathrm{Alv}_H(F)$ and $\mathrm{Cl}_H(F)$ are distributive.

2. Equational domains in a variety of Ω-groups

As before, let Θ be a variety of Ω-groups.

THEOREM 2.1. *An Ω-group H in Θ is a domain if and only if H is an equational domain in Θ.*

PROOF. Necessity has been proved by B. Plotkin (see Theorem 1 in [**P**]). We present this proof for the sake of completeness.

Let $F(X)$ be the free Ω-group in Θ, generated by X. Suppose that the Ω-group H is a domain. Let us take two algebraic varieties A and B in $V = \mathrm{Hom}(F(X), H)$. Let T_1 and T_2 be the ideals in $F(X)$ corresponding to these varieties. We check that $A \cup B = (T_1 \cap T_2)'$. It is obvious that

$$A \cup B \subseteq (T_1 \cap T_2)'.$$

To check the opposite inclusion, it suffices to show that $\mu \notin A \cup B$ implies $\mu \notin (T_1 \cap T_2)'$. Since $\mu \notin A \cup B$, we have $T_1 \notin \operatorname{Ker} \mu$ and $T_2 \notin \operatorname{Ker} \mu$. Hence, there exist $u \in T_1$ and $v \in T_2$ such that $u^\mu = a \neq 0$ and $v^\mu = b \neq 0$. Since H is a domain, $[\mathrm{id}\langle \mathrm{a}\rangle, \mathrm{id}\langle \mathrm{b}\rangle] \neq 0$.

The ideal $[\mathrm{id}\langle \mathrm{a}\rangle, \mathrm{id}\langle \mathrm{b}\rangle]$ is generated by ω-commutators

$$c = [a_1, a_2, \dots, a_n; b_1, b_2, \dots, b_n; \omega]$$

and ordinary commutators $[a', b']$, where $a', a_i \in \mathrm{id}\langle \mathrm{a}\rangle$ and $b', b_i \in \mathrm{id}\langle \mathrm{b}\rangle, \mathrm{i} = 1, \dots, \mathrm{n}$. Hence, there exists a nonzero ω-commutator $c = [a_1, a_2, \dots, a_n; b_1, b_2, \dots, b_n; \omega]$ or a nonzero commutator $[a', b']$, where $a', a_i \in \mathrm{id}\langle \mathrm{a}\rangle$ and $b', b_i \in \mathrm{id}\langle \mathrm{b}\rangle, , \mathrm{i} = 1, \dots, \mathrm{n}$. Suppose that such a non-zero commutator is of the form

$$c = [a_1, a_2, \dots, a_n; b_1, b_2, \dots, b_n; \omega].$$

It is easy to check that $(\mathrm{id}\langle \mathrm{a}\rangle)^\mu = \mathrm{id}\langle \mathrm{a}^\mu\rangle$. Hence, we can take $u_1, u_2, \dots, u_n \in \mathrm{id}_{\mathrm{G}}\langle \mathrm{u}\rangle$ and $v_1, v_2, \dots, v_n \in \mathrm{id}_{\mathrm{G}}\langle \mathrm{v}\rangle$ such that $u_i^\mu = a_i$ and $v_i^\mu = b_i, i = 1, \dots, n$. It is clear that

$$w = [u_1, u_2, \dots, u_n; v_1, v_2, \dots, v_n; \omega] \in [\mathrm{id}\langle \mathrm{u}\rangle, \mathrm{id}\langle \mathrm{v}\rangle]$$

and $w^\mu = c \neq 0$. We have

$$[\mathrm{id}\langle \mathrm{u}\rangle, \mathrm{id}\langle \mathrm{v}\rangle] \subseteq [\mathrm{T}_1, \mathrm{T}_2] \subseteq \mathrm{T}_1 \cap \mathrm{T}_2.$$

Finally, we have that $w \in T_1 \cap T_2$ and $w^\mu \neq 0$. Hence, $\mu \notin (T_1 \cap T_2)'$, as desired.

Now we prove the sufficiency. Suppose that H is an equational domain. Assume to the contrary that H is not a domain. As a consequence, there exist two non-zero elements a, b in H such that

$$[\mathrm{id}\langle \mathrm{a}\rangle, \mathrm{id}\langle \mathrm{b}\rangle] = 0. \tag{2.1}$$

Let $X = \{x, y\}$. Take the free Ω-group $F(X)$. The affine space $V = \mathrm{Hom}(F(X), H)$ over H can be identified with the set $H \times H$. In fact, every point $(h_1, h_2) \in H \times H$ determines the homomorphism $\mu : F(X) \to H$ such that $\mu(x) = h_1$ and $\mu(y) = h_2$ and vice versa.

Consider two subvarieties A and B in V defined by the equations $x = 0$ and $y = 0$, respectively. Since H is an equational domain, $D = A \cup B$ is a subvariety of V. Let $f(x, y)$ be "a polynomial" in $F(X)$ that belongs to D'. Then $f(a, 0) = f(0, b) = 0$. Let $Z = \{X_1, Y_1\}$ be a set in two indeterminates X_1 and Y_1 and $F_\Upsilon(Z)$ be the free algebra in the variety Υ of all Ω-groups generated by Z. Denote by $f(X_1, Y_1)$ the polynomial in $F_\Upsilon(Z)$ corresponding to $f(x, y)$. It is clear that the values of the polynomials $f(x, y)$ and $f(X_1, Y_1)$ in H are equal. Since the formula (2.1) is valid, by Lemma 1.2 we get

$$f(a, b) = f(a, 0) + f(0, b) = 0. \tag{2.2}$$

From (2.2) we obtain that the point $(a, b) \in D''$. Since D is an algebraic variety, $(a, b) \in D$. On the other hand, since a and b are nonzero elements in H, $(a, b) \notin D$. We have arrived at a contradiction. This completes the proof. □

As a consequence, Theorem 2.1 holds for the variety of all linear algebras, for the variety of all groups, and for the variety of all modules (see [**P**], [**BPP**], [**DMR**]). Theorem 2.1 is also true for the so-called CD-variety of Ω-groups (see Theorem 2 in [**P**]).

Definition 2.2 ([**L**]). An Ω-group H is called C-anticommutative (completely anticommutative) if each of its nonzero Ω-subgroups is anticommutative.

It turns out that the concepts of an equational domain and C-anticommutativity are closely related to each other.

In what follows, we use the following Proposition 4 from [**P**].

PROPOSITION 2.3. *An Ω-group H is a domain if and only if H is anticommutative.*

We now give a useful criterion for an Ω-group H in a variety Θ to be an equational domain in this variety.

PROPOSITION 2.4. *A non-trivial Ω-group H in Θ is an equational domain in Θ if and only if H is C-anticommutative.*

PROOF. Suppose that the Ω-group H is C-anticommutative. Let us take two nonzero elements a and b in H. Denote by P the Ω-subgroup of H generated by the elements a and b. Since H is C-anticommutative, P is anticommutative. By Proposition 2.3, P is without zero-divisors, i.e., $[\mathrm{id_P}\langle \mathrm{a}\rangle, \mathrm{id_P}\langle \mathrm{b}\rangle] \neq 0$. From this, it follows that $[\langle a\rangle, \langle b\rangle] \neq 0$. Therefore, $[\mathrm{id}\langle \mathrm{a}\rangle, \mathrm{id}\langle \mathrm{b}\rangle] \neq 0$, i.e., H is a domain. By Theorem 2.1, H is an equational domain in Θ.

Now suppose that the Ω-group H is an equational domain. Let us show that it is C-anticommutative. Assume to the contrary that H is not C-anticommutative. Therefore, there exists an Ω-subgroup H_1 of H which is not anticommutative. Hence, there exist two non-zero elements $a, b \in H_1$ such that

$$[\mathrm{id_{H_1}}\langle \mathrm{a}\rangle, \mathrm{id_{H_1}}\langle \mathrm{b}\rangle] = 0. \tag{2.3}$$

Take the free Ω-group $F(X)$ generated by $X = \{x, y\}$. Denote by

$$V = \mathrm{Hom}(F(X), H)$$

the affine space over H. Consider two subvarieties A and B in V defined by the equations $x = 0$ and $y = 0$, respectively. Since the Ω-group H is an equational domain, $D = A \cup B$ is a subvariety of V. Denote by $V_1 = \mathrm{Hom}(F(X), H_1)$ the subvariety of V in the induced Zariski topology. Let A_1 and B_1 be the subvarieties of V_1 defined by the equations $x = 0$ and $y = 0$, respectively. Then we have

$$A_1 = V_1 \cap A \text{ and } B_1 = V_1 \cap B.$$

Since D is a subvariety of V, $D_1 = A_1 \cup B_1$ is a subvariety of V_1. However, the same arguments given in the proof of the sufficiency of the conditions of Theorem 2.1 show that D_1 is not a variety. We have a contradiction. This completes the proof. □

COROLLARY 2.5. *An Ω-group H in Θ is C-anticommutative if and only if for any non-zero elements a and b in H, $[\langle a\rangle, \langle b\rangle] \neq 0$.*

PROOF. Suppose that the elements a and b in H satisfy the above condition. Therefore, $[\mathrm{id}\langle \mathrm{a}\rangle, \mathrm{id}\langle \mathrm{b}\rangle] \neq 0$. Hence, H has no zero divisors, i.e., H is a domain. By Theorem 2.1, it is an equational domain. According to Proposition 2.4, H is C-anticommutative.

Let H be C-anticommutative. Let P be the Ω-subgroup of H generated by two non-zero elements a and b of H. Since H is C-anticommutative, P is anticommutative. By Proposition 2.3, P is without zero-divisors, i.e., $[\mathrm{id_P}\langle \mathrm{a}\rangle, \mathrm{id_P}\langle \mathrm{b}\rangle] \neq 0$. It follows that $[\langle a\rangle, \langle b\rangle] \neq 0$, as desired. □

EXAMPLE 2.6. Now consider some examples of Ω-groups that are not equational domains in varieties related to them.

(1) Every non-trivial module M over an associative commutative ring K with unit is not an equational domain in the variety of all modules over K, since it is abelian, i.e., $[M, M] = 0$.

(2) Every nontrivial soluble Ω-group G is not an equational domain in the variety Υ of all Ω-groups.

Indeed, G has a nontrivial abelian Ω-subgroup. As a consequence, G is not C-anticommutative.

The following examples were considered earlier in [**DMR**]. However, using Proposition 2.4 and Corollary 2.5, in contrast to the paper [**DMR**], we prove all statements (3)–(5) in a unified way.

(3) Every non-trivial group G is not an equational domain in the variety of all groups.

Indeed, G contains a non-trivial cyclic group which is not anticommutative. Therefore, G is not C-anticommutative. By Proposition 2.4, G is not an equational domain.

(4) Every non-trivial Lie algebra L over an associative commutative ring K is not an equational domain in the variety of all Lie algebras.

In fact, L contains a non-trivial cyclic subalgebra which is also not anticommutative. By Proposition 2.4, L is not an equational domain.

Denote by $L_{ring} = \{+, -, \cdot, 0\}$ the language of associative rings. In the formulation of the following assertion, we use the language L_{ring}.

(5) An associative ring A is an equational domain if and only if A satisfies the formula:

$$\forall x, y((xy = yx = 0) \Rightarrow [(x = 0) \vee (y = 0)]). \tag{2.4}$$

Indeed, suppose that A is an equational domain. By Proposition 2.4, A is C-anticommutative. Suppose that $ab = ba = 0$ for some elements a and b in A. From this it follows that

$$[\langle a\rangle, \langle b\rangle] = 0.$$

By Corollary 2.5, $a = 0$ or $b = 0$.

Conversely, assume that formula (2.4) is true in A. Then for every non-zero elements a and b in A, $[\langle a\rangle, \langle b\rangle] \neq 0$. By Corollary 2.5, A is C-anticommutative. According to Proposition 2.4, A is an equational domain.

3. Acknowledgments

I would like to thank Prof. B. I. Plotkin and Prof. E. B. Plotkin for useful suggestions and comments on this paper.

References

[BMR] G. Baumslag, A. Myasnikov, and V. Remeslennikov, *Algebraic geometry over groups. I. Algebraic sets and ideal theory*, J. Algebra **219** (1999), no. 1, 16–79, DOI 10.1006/jabr.1999.7881. MR1707663

[BPP] A. Berzins, B. Plotkin, and E. Plotkin, *Algebraic geometry in varieties of algebras with the given algebra of constants*, J. Math. Sci. (New York) **102** (2000), no. 3, 4039–4070, DOI 10.1007/BF02673880. MR1797978

[DMR] È. Yu. Daniyarova, A. G. Myasnikov, and V. N. Remeslennikov, *Algebraic geometry over algebraic structures. IV. Equational domains and codomains* (Russian, with English and Russian summaries), Algebra Logika **49** (2010), no. 6, 715–756, 844, 847, DOI 10.1007/s10469-011-9112-2; English transl., Algebra Logic **49** (2011), no. 6, 483–508. MR2828872

[H] P. J. Higgins, *Groups with multiple operators*, Proc. London Math. Soc. (3) **6** (1956), 366–416, DOI 10.1112/plms/s3-6.3.366. MR0082492

[L] R. Lipyanski, *On stable Ω-groups* (English, with English and Latvian summaries), Proc. Latv. Acad. Sci. Sect. B Nat. Exact Appl. Sci. **57** (2003), no. 3-4, 102–105. MR2016270

[P] B. Plotkin, *Zero divisors in group-based algebras. Algebras without zero divisors* (English, with English, Russian and Moldavian summaries), Bul. Acad. Ştiinţe Repub. Mold. Mat. **2** (1999), 67–84, 114, 117. MR1738244

Department of Mathematics, Ben Gurion University, Beer Sheva 84105, Israel
Email address: lipyansk@math.bgu.ac.il

Contemporary Mathematics
Volume **726**, 2019
https://doi.org/10.1090/conm/726/14611

Seven lectures on universal algebraic geometry

Boris Plotkin

2010 *Mathematics Subject Classification.* 08-02, 08C05, 03G15.
Partially supported by the Edmund Landau Center for Research in Mathematical Analysis and Related Areas, sponsored by the Minerva Foundation (Germany).

Lecture 1

What is Universal Algebraic Geometry

Contents

1.1. Generic view. Universal algebraic geometry is a system of algebraic concepts and problems going back to classical algebraic geometry. It is mainly algebra, not quite geometry. However, this algebra inherits some geometric intuition; preserves a certain geometric characteristic which is not necessarily explicit.

The key idea of universal algebraic geometry is that every variety of algebras Θ has its own algebraic geometry. It varies if the variety Θ and an algebra $H \in \Theta$ vary, but the core notions are common. We look at algebras $H \in \Theta$ from the point of view of algebraic geometry in Θ. On the other hand, supposing that an algebra H is given, we consider algebraic varieties and their invariants. This is what is used in the study of classical algebraic geometry.

Universality of geometry means that the variety Θ can be an arbitrary fixed variety of algebras. For every algebra $H \in \Theta$, we consider the category of algebraic sets $K_\Theta(H)$. This is a geometric invariant of the algebra $H \in \Theta$. We are looking for conditions which provide the categories $K_\Theta(H_1)$ and $K_\Theta(H_2)$ to be isomorphic or equivalent for different H_1,H_2 in Θ. This is one of the main problems of the theory. For the whole variety Θ, its geometrical invariant is presented by the special category K_Θ. All $K_\Theta(H)$ are the subcategories in K_Θ. There is a problem of isomorphism or equivalence of the categories K_{Θ_1} and K_{Θ_2} for the different varieties Θ_1 and Θ_2.

In the above setting, the classical algebraic geometry occupies a distinguished place. It is based on the variety of commutative and associative rings with unity and is connected with the corresponding varieties of algebras over fields. Every such variety over a field P is denoted by Var-P and is called a *classical variety over* P.

If we replace the variety of associative and commutative algebras by a variety of associative but not necessarily commutative algebras over different P, then we come to non-commutative algebraic geometry.

On the other hand, in group theory there is a strong interest to investigate equations in groups and, especially, in free groups. The corresponding results give rise to the sketch of the special algebraic geometry. The most interesting results are obtained for groups with a given group of constants G. The corresponding variety is denoted by Grp-G.

Something similar is known for semigroups and other algebraic systems.

Now one can say that universal algebraic geometry stands in the same position with respect to all these special theories as does universal algebra with respect to groups, rings, semigroups, etc.

To every variety of algebras Θ, its elementary logic (i.e., first order logic) is associated. We consider algebraic logic in Θ. The elementary properties of an algebra $H \in \Theta$ are considered from the point of view of logic in Θ. Geometrical and logical properties of algebras in Θ are often well interacted. The same is valid for geometrical and logical relations between different algebras in Θ.

In addition, we will consider algebraic geometry in first order logic, which generalizes usual equational algebraic geometry. This outlet to logic is motivated geometrically and also by various applications, in particular, in computer science.

1.2. Classical algebraic geometry from the point of view of universal algebraic geometry. Let us fix an infinite ground field P. This field generates the variety of algebras which is denoted by Var-P, i.e., the variety of all commutative and associative algebras with unity over the given P. The polynomial algebra $P[X] = P[x_1, \dots, x_n]$ is the free algebra in Var-P over the set X.

A field L is an extension of the field P if there is an injection $P \to L$. Every such field L is also an algebra in Var-P.

To every map $\mu : X \to L$, there is a unique correspondence P-homomorphism $\mu : P[X] \to L$. Namely, if $f = f(x_1, \dots, x_n) \in P[X]$, then $f^\mu = f(x_1^\mu, \dots, x_n^\mu)$ is an element in L. The affine space $L^{(n)} = L^X$ can be viewed as the set of all homomorphisms $\mathrm{Hom}(P[X], L)$. A point $a = (a_1, \dots, a_n) \in L^{(n)}$ is identified with the homomorphism $\mu = \mu_a : P[X] \to L$ by the rule $\mu(x_i) = a_i$. Then $f^\mu = f(a)$ and the point a is a root of the polynomial f, if $f \in \mathrm{Ker}\,\mu$.

We consider affine algebraic sets over the given L. These sets are defined by the systems of equations with coefficients from P and solutions in the field L. Since the algebraic sets are supposed to be subsets in $L^{(n)}$, we can consider them also as subsets in $\mathrm{Hom}(P[X], L)$.

Let A be an arbitrary set of points $\mu : P[X] \to L$. An ideal U in $P[X]$ defined by

$$U = A' = \bigcap_{\mu \in A} \mathrm{Ker}\,\mu$$

corresponds to A. This is the set of all polynomials f, such that every point $\mu \in A$ is a root of f. Every such ideal is called an *L-closed ideal.*

On the other hand, let U be a subset in the algebra $P[X]$. One can consider U as a system of equations. The set of points $A \subset \mathrm{Hom}(P[X], L)$, defined by the rule

$$A = U_L' = U' = \{\mu : P[X] \to L \mid U \subset \mathrm{Ker}\,\mu\}$$

corresponds to U. This means that a point μ belongs to A if μ is a root of every polynomial from U. Such sets A are called *algebraic sets* or *closed sets.*

Define two closures

$$A'' = (A')'; \quad U_L'' = U'' = (U_L')'.$$

DEFINITION 1.1. Two extensions L_1 and L_2 of the ground field P are called *geometrically equivalent*, if for every finite set $X = \{x_1, \dots, x_n\}$ and for every $U \subset P[X]$, the equality

$$U_{L_1}'' = U_{L_2}''$$

takes place.

THEOREM 1.1.

1. *If the field P is algebraically closed, then all its extensions are geometrically equivalent.*
2. *If every two extensions of P are geometrically equivalent, then P is algebraically closed.*

The first part of the theorem follows from Hilbert's Nullstellensatz, the second one is deduced easily from definitions.

We consider the category of algebraic sets over the given L and fixed P. Denote this category by $K_P(L)$. Its objects are algebraic sets A with different X.

To the category $K_P(L)$ corresponds the dual category $C_P(L)$, whose objects are P-algebras of the form $P[X]/A'$. We later define these categories in an arbitrary variety Θ.

If the fields L_1 and L_2 are geometrically equivalent, then the categories $C_P(L_1)$ and $C_P(L_2)$ coincide and the categories $K_P(L_1)$ and $K_P(L_2)$ are isomorphic.

The category $K_P(L)$ can be considered as a natural geometric invariant of the P-algebra L. It can also be viewed as a kind of characteristic which measures P-algebraic closeness of the field L. In other words, it characterizes possibilities to solve in L systems of equations with coefficients from P. We say that a field L is P-algebraically closed if, for any finite X and a proper ideal $U \subset P[X]$, the set U'_L is not empty. If the field P is algebraically closed, then each of its extensions L is P-algebraically closed. If L_1 and L_2 are P-algebraically closed, then they are geometrically equivalent. If L_1 is P-algebraically closed, and L_1 and L_2 are geometrically equivalent, then L_2 is P-algebraically closed. The definition above can be viewed as a definition of strict algebraic closeness of the field L over P. The weak algebraic closeness of the field L over P means that every polynomial with one variable and coefficients from P has roots in L. In fact, strict and weak closeness coincide. Indeed, algebraic (absolute) closure $\bar{P}$ of the field P coincides with the weak P-closure of the P. Therefore, $\bar{P}$ is contained in every P-closed extension of L. Then Hilbert's Nullstellensatz works.

In the definitions above, the classical Nullstellensatz can be formulated as follows: if the field L is P-algebraically closed, then for every ideal $U \in P[X]$ the equality

$$U''_L = \sqrt{U}$$

holds.

The equivalence pointed out above in the definitions plays a crucial role in algebraic geometry in different varieties Var-P. This equivalence fails in general varieties Θ and $H \in \Theta$.

Note that we distinguish algebraic sets in affine space and algebraic varieties. It is assumed that an algebraic variety is an algebraic set considered up to isomorphisms in the category $K_P(L)$.

The following problem is of special interest.

Let L_1 and L_2 be two extensions of the field P. When are the categories $K_P(L_1)$ and $K_P(L_2)$ isomorphic?

Let $\mu_1 : P \to L_1$ and $\mu_2 : P \to L_2$ be injections defining extensions.

An isomorphism of extensions means that there is a commutative diagram

$$\begin{array}{ccc} P & \xrightarrow{h_1} & L_1 \\ & \searrow^{h_2} & \downarrow \mu \\ & & L_2 \end{array}$$

where μ is an isomorphism of rings.

Two extensions $h_1 : P \to L_1$ and $h_2 : P \to L_2$ are called *semi-isomorphic* if there is a diagram

$$\begin{array}{ccc} P & \xrightarrow{h_1} & L_1 \\ \sigma \downarrow & & \downarrow \mu \\ P & \xrightarrow{h_1} & L_2 \end{array}$$

where μ is an isomorphism of rings and σ is an automorphism of the field P.

The semi-isomorphism does not imply the equivalence but if L_1 and L_2 are semi-isomorphic, then the categories $K_P(L_1)$ and $K_P(L_2)$ are isomorphic.

DEFINITION 1.2. Two extensions L_1 and L_2 are said to be geometrically equivalent up to a semi-isomorphism, if there exists a field L, such that L is semi-isomorphic to L_1 and geometrically equivalent to L_2.

THEOREM 1.2. *Two categories $K_P(L_1)$ and $K_P(L_2)$ are isomorphic if and only if the extensions L_1 and L_2 are geometrically equivalent up to a semi-isomorphism.*

The proof of this theorem is based on consideration of automorphisms of the category of polynomial algebras $P[X]$ with different X. It can be proved that every such automorphism is a semi-inner automorphism. Definitions of inner and semi-inner automorphisms are natural and will be given in Lecture 3.

In order to get further information on geometric equivalence of fields, we use the idea of a quasi-identity.

A quasi-identity u in the variety Var-P is a formula of the form

$$f_1 \equiv 0 \wedge f_2 \equiv 0 \wedge \cdots \wedge f_n \equiv 0 \Rightarrow f \equiv 0$$

where $f_i \in P[X]$, $i = 1, \ldots, n$.

THEOREM 1.3. *Two extensions L_1 and L_2 of a field P are geometrically equivalent if and only if they satisfy the same quasi-identities.*

Thus, in this case, geometric equivalence coincides with equivalence in logic of quasi-identities in the variety Var-P.

This theorem implies that any extension L of P is geometrically equivalent to each its ultrapower. Every L and all its ultrapowers have the same elementary theories.

The following problem looks quite natural:

PROBLEM 1.1. When do two geometrically equivalent extensions L_1 and L_2 of a field P have different elementary theories in the logic of Var-P?

There are examples that answer this question. Let, for example, P be an algebraically closed field and L be its non-algebraically closed extension. Then L

and P are geometrically equivalent, while they are not elementary equivalent. In particular, they are not elementary equivalent in the logic of Var-P.

Indeed, let $f(x) = \alpha_0 + \alpha_1 x \cdots + \alpha_n x^n$ be a polynomial with the coefficients in L and without roots in L. Take a polynomial over P

$$\varphi(x, y_0, \dots, y_n) = y_0 + y_1 x + \cdots + y_n x^n$$

and consider a P-formula

$$\forall y_0 \dots y_n \exists x (\varphi(x, y_0, \dots, y_n) = 0).$$

This formula holds in P and does not hold in L.

In Problem 1.1, one has to consider the general situation.

1.3. Some information from universal algebra. Almost all that has been determined for the classical case admits natural generalizations.

Let us recall some information from universal algebra. Fix a signature Ω, i.e., a set of symbols of operations of the arbitrary arity. Consider the class of all Ω-algebras. Every set of identities, satisfied by the operations from Ω, specifies some subclass in the class of all Ω-algebras. Every such subclass Θ is called *a variety of algebras.* Groups, rings, semigroups, associative and Lie rings, associative and Lie algebras over fields, Var-P are all examples of varieties.

Speaking of algebra, we think of universal algebra, i.e., algebra in an arbitrary but fixed variety Θ. One can assume that this is a group, or an associative or Lie algebra. Algebra can be viewed as universal or concrete.

In each variety Θ there are free algebras $W(X)$, where X is a set. This is a significant feature of varieties which plays a crucial role in the theory under consideration.

The characteristic property of $W(X)$ can be described by the commutativity of the following diagram

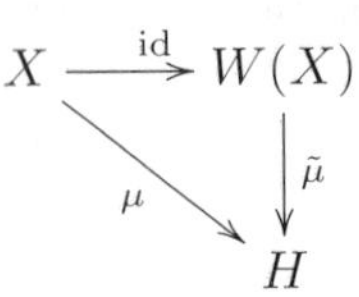

Here H is an arbitrary algebra in Θ, id is the identity map, μ is an arbitrary map, and $\overline{\mu}$ is the homomorphism in Θ uniquely determined by μ.

In other words, this diagram means that the set X freely generates $W(X)$ (or, X is a basis for W), the variables from X take values from H by the rule μ, and the homomorphism $\overline{\mu}$ computes in H the values of arbitrary elements $w \in W$.

Elements from W are used to call words, terms, polynomials, or Θ-polynomials. They are constructed from the variables from X by the rules of the variety Θ. Every $w \in W$ is uniquely presented as $w = w(x_1, \dots, x_n)$, $x_i \in X$, and uniqueness means that $w(x_1, \dots, x_n) = w'(x_1, \dots, x_n)$ holds if and only if it is an identity in Θ.

Every variety Θ can be regarded as a category whose morphisms are homomorphisms in Θ. To each homomorphism $\mu : G \to H$ corresponds its kernel $\operatorname{Ker} \mu$. This is a special binary relation, defined by the rule: $g_1 (\operatorname{Ker} \mu) g_2$ if and only if $g_1^\mu = g_2^\mu$. A relation T in G is called a *congruence* if $g_1 T g_1' \wedge \cdots \wedge g_n T g_n'$ implies $g_1 \dots g_n \omega T g_1' \dots g_n' \omega$ for every n-ary operation ω in Ω. The kernel $Ker \mu$ is always a congruence.

For every congruence T in G, one can take a factor algebra G/T with the natural homomorphism $G \to G/T$. The diagram

$$\begin{array}{ccc} G & \xrightarrow{\mu} & H \\ & \searrow^{\mu_0} & \uparrow \mu_1 \\ & & G/\operatorname{Ker}\mu \end{array}$$

gives a canonical decomposition of a homomorphism μ. The set of all homomorphisms $G \to H$ is denoted by $\operatorname{Hom}(G, H)$.

Every variety of algebras is closed under taking Cartesian products, subalgebras and homomorphic images. In particular, if I is a set and $H \in \Theta$, then the Cartesian power H^I also belongs to Θ. If $|I| = n$, then H^I is denoted by $H^{(n)}$. The algebra $H^{(n)}$ is viewed as an affine space of points $(a_1, \dots, a_n)$, $a_i \in H$. Let $X = \{x_1, \dots, x_n\}$. Then there is a bijection

$$\alpha_X : \operatorname{Hom}(W(X), H) \to H^{(n)},$$

where for every $\nu : W \to H$,

$$\alpha_X(\nu) = (\nu(x_1), \dots, \nu(x_n)).$$

This bijection induces a structure of algebras in Θ on $\operatorname{Hom}(W(X), H)$. By definition, if $\omega \in \Omega$ is an m-ary operation, $\nu_1, \dots, \nu_m \in \operatorname{Hom}(W(X), H)$, then

$$(\nu_1 \dots \nu_m\omega)(x) = \nu_1(x) \dots \nu_m(x)\omega,$$

for every $x \in X$.

We define the notion of a commutative algebra H. For a commutative algebra H, the equality

$$(\nu_1, \dots \nu_m\omega)(w) = \nu_1(w) \dots \nu_n(w)\omega$$

always holds for every $w \in W$.

Let $w_1, w_2 \in \Omega$ be two operations of arity m_1, m_2, respectively. Consider the matrix

$$\begin{pmatrix} x_{11} & \dots & x_{1m_1} \\ \dots & \dots & \dots \\ x_{m_21} & \dots & x_{m_2m_1} \end{pmatrix}.$$

Two operations w_1 and w_2 are said to be commutative if

$$\begin{aligned} &((x_{11} \dots x_{1m_1}\omega_1) \dots (x_{m_21} \dots x_{m_2m_1})\omega_1)\omega_2 \\ &= ((x_{11} \dots x_{m_21}\omega_2) \dots (x_{1m_1} \dots x_{m_2m_1}\omega_2)\omega_1. \end{aligned}$$

An algebra H is called commutative if every two operations, not necessarily different, commute. For nullary operations, this means that all of them coincide.

If H is a commutative algebra, then the algebra $\operatorname{Hom}(W(X), H)$ is also commutative and it is a subalgebra in the algebra of mappings $G^{W(X)}$.

In the general case, the set $\operatorname{Hom}(W(X), H)$, $|X| < \infty$ is considered as an affine space over an algebra H.

Let now Θ^0 denote the category of all free algebras $W(X)$ in Θ with finite X. The category Θ^0 is a full subcategory in the category Θ. Morphisms in Θ^0 are presented by arbitrary homomorphisms $s : W(X) \to W(Y)$.

Let us make remarks on equations. Let us fix a finite set $X = \{x_1, \dots, x_n\}$ and consider equations of the form $w = w'$, $w, w' \in W(X)$. Every such equation with the given X is considered also as a formula in the logic in Θ. In the later case, we write $w \equiv w'$.

A point $\nu : W(X) \to H$ is a root of the equation $w(x_1, \dots, x_n) = w'(x_1, \dots, x_n)$, if $w(x_1^\nu, \dots, x_n^\nu) = w'(x_1^\nu, \dots, x_n^\nu)$. This also means that the pair (w, w') belongs to $\operatorname{Ker} \nu$. We will identify the pair (w, w') and the equation $w = w'$.

In order to get a reasonable geometry in Θ, one has to consider the equations with constants. The next subsection deals with constants.

1.4. Algebras with a fixed algebra of constants. Let Θ be an arbitrary variety of algebras, G be a fixed algebra in Θ, $|G| > 1$. Consider the new variety, denoted by Θ^G. First we define the category Θ^G. The objects in Θ^G have the form $h : G \to H$, where h is a homomorphism in Θ, not necessarily injective. Morphisms in Θ^G are presented by the commutative diagrams

$$\begin{array}{ccc} G & \xrightarrow{h} & H \\ & \searrow^{h'} & \downarrow \mu \\ & & H \end{array}$$

where μ, h, h' are homomorphisms in Θ.

An algebra H, considered as a G-algebra, is denoted by (H, h). In this way, to elements $g \in G$ correspond constants, i.e., nullary operations in H. We add them to the signature Ω. This signature allows us to consider a variety of G-algebras. Identities of a variety of G-algebras are presented by identities of Θ and by the defining relations of the algebra G (see **[Pl7]** for details).

A free algebra $W = W(X)$ in Θ^G has the form $G * W_0(X)$, where $W_0(X)$ is the free algebra over X in Θ, $*$ is the free product in Θ and the embedding $i_G : G \to W(X) = G * W_0(X)$ follows from the definition of free product.

A G-algebra (H, h) is called a *faithful* G-algebra if $h : G \to H$ is an injection. A free algebra (W, i_G) is faithful. A G-algebra G with the identical $G \to G$ is also faithful. All other G-algebras G are isomorphic to this one. All of them are simple, i.e., they do not have faithful subalgebras and congruences. Let (H, h) be a G-algebra, and $\mu : H \to H'$ is a homomorphism in Θ. Then, by $h' = \mu h$, H' becomes a G-algebra, and μ is a homomorphism of G-algebras. Since one can start from an arbitrary congruence T in H and from the natural congruence H/T, we say that T is called *faithful* if the G algebra H/T is faithful. A congruence T is faithful if and only if $g_1^h = g_2^h$ is equivalent to $g_1 = g_2$.

Let a homomorphism

$$\begin{array}{ccc} G & \xrightarrow{h} & H \\ & \searrow^{h'} & \downarrow \mu \\ & & H' \end{array}$$

be given, and let (H', h') be a faithful G-algebra. Then (H, h) is a faithful G-algebra. If $T = \operatorname{Ker} \mu$, then T is a faithful congruence and H/T is also faithful.

Note that the inclusion $G \in \Theta$ means that all the constants of G-algebra H are covered by the elements from G.

EXAMPLES. A variety Var-P is a variety of the type Θ^G, where Θ is a variety of associative and commutative rings with 1, and G is a field P. In this example, elements of the field P are the constants in the P-algebras. They are considered as nullary operations, and, simultaneously, using multiplication, we can look at them as unary operations.

G-groups are another example of G-algebras. Here elements from G also can be viewed as unary operations.

We denote also a free G-algebra $W = W(X)$ by $G[X]$. Since Θ^G is a variety of algebras, all constructions such as Cartesian and free products, subalgebras and homomorphisms are naturally defined for Θ^G.

Consider, further, the special condition on Θ^G denoted by $(*)$. Namely, we assume that the algebra G generates the whole variety Θ^G, i.e., in G there are no non-trivial identities with the coefficients from G. This condition is fulfilled in Var-P if the field P is infinite and in Grp-$F = (\mathrm{Grp})^F$.

Every faithful G-algebra H contains G as a subalgebra. Thus, $(*)$ implies that every faithful G-algebra H generates the whole variety Θ^G, i.e., in Θ^G there are no proper subvarieties containing faithful algebras.

In the category Θ^G along with morphisms, one can consider also semimorphisms. They have the form

$$\begin{array}{ccc} G & \xrightarrow{h} & H \\ \sigma\downarrow & & \downarrow\mu \\ G & \xrightarrow[h']{} & H' \end{array}$$

where $\sigma \in \operatorname{End} G$. Then we can consider semi-isomorphic G-algebras.

Another possibility is to vary also the algebra of constants G. This leads to the diagram of the form

$$\begin{array}{ccc} G & \xrightarrow{h} & H \\ \sigma\downarrow & & \downarrow\mu \\ G' & \xrightarrow[h']{} & H' \end{array}$$

with component-wise multiplication.

Let us make a remark on equations. The equations of the form $w = w'$ with $w, w' \in W(X) = G * W_0(X)$ are equations with constants. Consider systems of such equations T. If T is a congruence, then T has a solution in a faithful G-algebra H if and only if T is a faithful congruence in $W = W(X)$. Thus, a system T has a solution if T is contained in a faithful congruence in W. Note that, by definition, all faithful congruences are proper.

1.5. Special categories. Let Θ be a variety of algebras and $H \in \Theta$. Consider the category of affine spaces $K_\Theta^0(H)$. Its objects are represented by affine spaces $\mathrm{Hom}(W(X), H)$. Morphisms have the form

$$\tilde{s} : \mathrm{Hom}(W(X), H) \to \mathrm{Hom}(W(Y), H),$$

where $s : W(Y) \to W(X)$ is a morphism in Θ^0. Here $\tilde{s}$ is defined by $\tilde{s}(\nu) = \nu s$ for every point $\nu : W(X) \to H$.

If an algebra H generates the variety Θ then $K_\Theta^0(H)$ is dual to the category Θ^0.

Another category is denoted by $\mathrm{Pol}_\Theta(H)$. Its objects have the form of affine spaces $H^{(n)}$, while morphisms are the special polynomial maps

$$s^\alpha : H^{(n)} \to H^{(m)}.$$

If, further, $s : W(Y) \to W(X)$ is a morphism in Θ and $|Y| = m$, $|X| = n$, then there is a commutative diagram

$$\begin{array}{ccc} \mathrm{Hom}(W(X), H) & \xrightarrow{\tilde{s}} & \mathrm{Hom}(W(Y), H) \\ {\scriptstyle \alpha_X}\downarrow & & \downarrow{\scriptstyle \alpha_Y} \\ H^{(n)} & \xrightarrow[s^{\alpha}]{} & H^{(m)} \end{array}$$

This yields a canonical duality between $K^0_\Theta(H)$ and Pol_Θ-H. In this diagram, all maps are algebra homomorphisms, if in Θ all operations commute.

Denote $\mathrm{Bool}_\Theta(H)$ to be a special category of Boolean algebras. Its objects $\mathrm{Bool}(W(X), H)$ are the Boolean algebras of subsets in $\mathrm{Hom}(W(X), H)$.

Morphisms

$$s_* : \mathrm{Bool}(W(X), H) \to \mathrm{Bool}(W(Y), H)$$

are defined for every $s : W(X) \to W(Y)$. Here $\tilde{s} : \mathrm{Hom}(W(Y), H) \to \mathrm{Hom}(W(X), H)$ is given, and for every $A \subset \mathrm{Hom}(W(X), H)$, $B = s_* A$ is the inverse image of the set A under the map $\tilde{s}$.

Lecture 2

Algebraic Sets and Algebraic Varieties

Contents

2.1. Systems of equations and algebraic sets. Let Θ be a variety of algebras, $W = W(X)$ be the free algebra in Θ with $|X| < \infty$.

Fix an algebra H from Θ. Consider equations of the form $w \equiv w'$, $w, w' \in W$. Denote the value of this equation in H by $\mathrm{Val}_H(w \equiv w')$. It is defined by

$$\mathrm{Val}_H(w \equiv w') = \{\mu : W \to H \mid w^\mu = w'^\mu\}.$$

Systems of equations in W are denoted by T. They can be viewed as binary relations in W. Some of these relations can be congruences in the algebra W.

Consider, also, sets of points $A \subset \mathrm{Hom}(W, H)$. The Galois correspondence between T and A is as follows:

$$\begin{cases} T' = A = \{\mu : W \to H \mid T \subset \mathrm{Ker}\,\mu\} = T'_H \\ A' = T = \bigcap\limits_{\mu \in A} \mathrm{Ker}\,\mu \end{cases}$$

The same can be written in the form

$$T' = \bigcap_{(w,w') \in T} \mathrm{Val}_H\ (w \equiv w')$$

$$A' = \{w \equiv w' \mid A \subset \mathrm{Val}_H\ (w \equiv w')\}.$$

DEFINITION 2.1. Every set A such that $A = T'$ for some T is called a *closed* or an *H-closed set.*

These are algebraic sets over the algebra H.

DEFINITION 2.2. Algebraic sets considered up to isomorphisms in the category of algebraic sets $K_\Theta(H)$ are called *algebraic varieties.*

The definition of the category $K_\Theta(H)$ is given in section 2.3 of this lecture.

If A is an arbitrary set of points, then its closure is defined to be $A'' = (A')'$.

DEFINITION 2.3. Congruences T in W of the form $T = A'$ are called *H-closed congruences.*

For an arbitrary system of equations T, its closure is defined by $T''_H = (T'_H)'$.

PROPOSITION 2.1. *A congruence T in W is H-closed if and only if for some set I there is an injection*

$$W/T \to H^I.$$

Let, further, T be a congruence in W, and $\mu_0 : W \to W/T$ be a natural homomorphism. Consider the sets $\mathrm{Hom}(W/T, H)$ and $\mathrm{Hom}(W/T, H)\mu_0$, where

$$\mathrm{Hom}(W/T, H)\mu_0 = \{\nu\mu_0 : W \to H\},\ \nu \in \mathrm{Hom}(W/T, H).$$

Denote $\tilde{\mu}_0(\nu) = \nu\mu_0$.

PROPOSITION 2.2. *The following formula*

$$T'_H = A = \mathrm{Hom}(W/T, H)\mu_0$$

holds. Moreover, $\tilde{\mu}_0 : \mathrm{Hom}(W/T, H) \to A$ *is a bijection.*

We consider, separately, the case where H is a commutative algebra. Then the sets $\mathrm{Hom}(W, H)$ and $\mathrm{Hom}(W/T, H)$ are also commutative algebras. Every algebraic set A is a subalgebra in $\mathrm{Hom}(W, H)$. The bijection $\tilde{\mu}_0 : \mathrm{Hom}(W/T, H) \to A$ is an isomorphism of algebras.

Now let us make some remarks on trivial cases.

A congruence T is called a *zero congruence* if wTw' means that w and w' coincide in W. Here T is a relation of equality. T is called a *unital* or a *non-proper congruence*, if wTw' is fulfilled for every $w, w' \in W$. These congruences are denoted $T = 0$, $T = 1$, respectively.

We have $0' = \mathrm{Hom}(W, H)$ and, therefore, $\mathrm{Hom}(W, H)$ is an algebraic set. As for $1'$, this is either an empty set, or a zero point in $\mathrm{Hom}(W, H)$, sending W to the zero subalgebra in H, if the latter exists. A zero subalgebra consists of one element and is defined by a unique nullary operation.

It is easy to see that if $A = \mathrm{Hom}(W, H)$, then $T = A'$ is the congruence of all identities of an algebra H in the free algebra W. This is the minimal closed congruence in W. If an algebra H generates the whole variety Θ, then $A' = 0$. If A is the empty set, then $A' = 1$. If A consists of one zero point, then $A' = 1$ as well. If T is the empty set, then assume that $T' = \mathrm{Hom}(W, H)$.

For arbitrary algebras H and G, denote

$$(H - \mathrm{Ker})(G) = \bigcap_{\nu : G \to H} \mathrm{Ker}\,\nu.$$

Let T be a congruence in W and take $\mu_0 : W \to W/T$. Then we have

PROPOSITION 2.3 (Hilbert's Nullstellensatz).

$$T''_H = \mu_0^{-1}((H - \mathrm{Ker}\,\mu)(W/T)).$$

This proposition can be considered as one of the forms of Hilbert's theorem. There are, also, others.

Let A be an algebraic set. To each element $w \in W$ corresponds a map

$$\overline{w} : A \to H$$

defined by $\overline{w}(\nu) = w^\nu$ for every $\nu \in A$. Such $\overline{w}$ are called *regular maps.* One can define the algebra of regular maps. This algebra is isomorphic to an algebra W/A', which is called the *coordinate algebra* of the algebraic set A.

From Proposition 2.1 follows that an algebra $G \in \Theta$ can be presented as a coordinate algebra of an algebraic set A over a given algebra $H \in \Theta$ if and only if G is finitely generated algebra and there is an injection $G \to H^I$ for some set I.

2.2. Lattices of algebraic sets. Given a set I, the following formulas hold:

1. $(\bigcup A_\alpha)' = \bigcap A'_\alpha$
2. $(\bigcup T_\alpha)' = \bigcap T'_\alpha$
3. $\bigcup T'_\alpha \subset (\bigcap T_\alpha)'$
4. $\bigcup A'_\alpha \subset (\bigcap A_\alpha)'$

where $\alpha \in I$.

If all A_α and T_α are H-closed sets, then

5. $(\bigcup T'_\alpha)'' = (\bigcap T_\alpha)'$
6. $(\bigcup A'_\alpha)'' = (\bigcap A_\alpha)'$.

Thus, one can state that the intersection of algebraic sets is an algebraic set, and the intersection of closed congruences is again a closed congruence.

It is evident that the union of two closed congruences is not a closed congruence, and the union of two algebraic sets is not necessarily an algebraic set (say, if H is a commutative group it is not an algebraic set).

For given $H \in \Theta$ and $W = W(X)$, $|X| < \infty$, denote by $\mathrm{Alv}_H(W)$ the set of all algebraic sets in $\mathrm{Hom}(W, H)$, $\mathrm{Cl}_H(W)$ the set of all H-closed congruences in W.

This gives rise to the functors

$$\mathrm{Alv}_H : \Theta^0 \to \mathrm{Set},$$

$$\mathrm{Cl}_H : \Theta^0 \to \mathrm{Set}.$$

The first functor is covariant, while the second one is contravariant. If $s : W(Y) \to W(X)$ is a morphism in Θ^0, then for every $B \in \mathrm{Alv}_H(W(Y))$ the corresponding $s_* B = A$ is contained in $\mathrm{Alv}_H(W(X))$.

There is a map

$$s_* : \mathrm{Alv}_H(W(Y)) \to \mathrm{Alv}_H(W(X)).$$

If $T \in Cl_H(W(X))$ and s_*T is the inverse image of T in $W(Y)$, then $s_*T \in \mathrm{Cl}_H(W(Y))$. This gives the map

$$s_* : \mathrm{Cl}_H(W(X)) \to \mathrm{Cl}_H(W(Y)).$$

The sets $\mathrm{Alv}_H(W)$ and $\mathrm{Cl}_H(W)$ constitute the lattices. If A and B are the elements from $\mathrm{Alv}_H(W)$, then we set

$$A \bar{\cup} B = (A \cup B)''.$$

If T_1 and T_2 belong to $\mathrm{Cl}_H(W)$, then

$$T_1 \bar{\cup} T_2 = (T_1 \cup T_2)''.$$

PROPOSITION 2.4. *Lattices* $\mathrm{Alv}_H(W)$ *and* $\mathrm{Cl}_H(W)$ *are dual.*

The duality is determined by $A \to A'$ and the properties 5,6 above.

In the general case, we cannot state that the functors Alv_H and Cl_H are coordinated with lattice operations.

DEFINITION 2.4. An algebra H is called *geometrically stable*, if for every $W(X) = W$ and any $A, B \in \mathrm{Alv}_H(W)$, the equality

$$A \bar{\cup} B = A \cup B$$

takes place.

This is equivalent to the fact that for closed systems of equations T_1 and T_2 in W, the equality

$$T_1' \cup T_2' = (T_1 \cap T_2)'$$

holds.

If H is a stable algebra, then every lattice $\mathrm{Alv}_H(W)$ is a distributive lattice in $\mathrm{Bool}_\Theta(W,H)$. A dual lattice $\mathrm{Cl}_H(W)$ is also distributive.

In this case, the functors Cl_H and Alv_H are the functors to the category of distributive lattices.

PROPOSITION 2.5. *If $s : W(Y) \to W(X)$ is surjective, then the maps*

$$s_* : \mathrm{Cl}_H(W(X)) \to \mathrm{Cl}_H(W(Y))$$

and

$$s^* : \mathrm{Alv}_H(W(Y)) \to \mathrm{Alv}_H(W(X))$$

are homomorphisms of lattices.

Here the map s^* is conjugated to the map s_*.

For every algebraic set $A \in \mathrm{Alv}_H(W)$, denote by $L(A)$ the lattice of all algebraic sets in A. This lattice is considered as a lattice invariant of A.

Another invariant is based on the semigroup $\mathrm{End}(H)$. For every $\delta : H \to H$, we have $\tilde{\delta} : \mathrm{Hom}(W,H) \to \mathrm{Hom}(W,H)$ by the rule $\tilde{\delta}(\nu) = \delta\nu$. Thus, the semigroup $\mathrm{End}(H)$ acts in the affine space $\mathrm{Hom}(W,H)$. It is clear that every algebraic set A in this space is invariant with respect to the action of $\mathrm{End}(H)$. Thus, we can look for the structure of A with respect to this action.

2.3. Categories of algebraic sets. Denote the category of algebraic sets in Θ over the given H by $K_\Theta(H)$. Its objects are denoted by

$$(X, A)$$

where A is an algebraic sets in the affine space $\mathrm{Hom}(W(X), H)$.

In order to define morphisms

$$(X, A) \to (Y, B),$$

we start from $s : W(Y) \to W(X)$. We say that s is admissible with respect to A and B, if $\tilde{s}(\nu) = \nu s \in B$ for every $\nu \in A$. The equivalent condition is $A \subset s_* B$. If s is admissible with respect to A and B, then the map $[s] \to [A, B]$ corresponds to s. We write $[s] : (X, A) \to (Y, B)$. The multiplication of such morphisms is defined in a natural way and the category $K_\Theta(H)$ is defined.

Let us define now the category $C_\Theta(H)$. Its objects are algebras from Θ of the form $W(X)/T$, where $|X| < \infty$, and T is an H-closed congruence. Morphisms are presented by homomorphisms

$$\sigma : W(Y)/T_2 \to W(X)/T_1.$$

Here $C_\Theta(H)$ is a full subcategory in the category Θ.

Every such σ can be represented as $\sigma = \bar{s}$, where $s : W(Y) \to W(X)$ is an admissible homomorphism of free algebras with respect to congruences T_2 and T_1. This means that wT_2w' implies $w^s T_1 w'^s$.

PROPOSITION 2.6. *A homomorphism $s : W(Y) \to W(X)$ is admissible with respect to A and B if and only if s is admissible with respect to $B' = T_2$ and $A' = T_1$.*

In addition, if $s_1 : W(Y) \to W(X)$ is admissible with respect to A and B, and, hence, with respect to T_2 and T_1, then the equality $[s_1] = [s]$ holds if and only if $\overline{s}_1 = \overline{s}$.

This leads to the following theorem.

THEOREM 2.1. *The transition*

$$(X, A) \to W(X)/A'$$

defines the duality of the categories $K_\Theta(H)$ and $C_\Theta(H)$.

This fact confirms the correctness of the definitions of $K_\Theta(H)$ and $C_\Theta(H)$.

Define now the categories K_Θ and C_Θ. In these categories, an algebra H is not fixed. The objects of K_Θ have the form

$$(X, A, H),$$

where A is an algebraic set in $\mathrm{Hom}(W(X), H)$. The objects in C_Θ have the form

$$(W(X)/T, H),$$

where T is an H-closed congruence in $W(X)$.

Consider morphisms

$$(X, A, H_1) \to (Y, B, H_2).$$

Take $s : W(Y) \to W(X)$ and a homomorphism $\delta : H_1 \to H_2$.

Consider a commutative diagram

$$\begin{CD} W(Y) @>{s}>> W(X) \\ @V{\nu'}VV @VV{\nu}V \\ H_2 @<<{\delta}< H_1 \end{CD}$$

For every point $\nu : W(X) \to H_1$ denote $(s, \delta)(\nu) = \nu' = \delta\nu s$.

A pair (s, δ) is said to be admissible with respect to A and B, if $\nu' = (s, \delta)(\nu) \in B$ for every $\nu \in A$. Let (s, δ) be an admissible pair with respect to A and B. Fix δ and consider the map $[s]_\delta : A \to B$. The pair $([s]_\delta, \delta)$ is considered to be a morphism

$$(X, A, H_1) \to (Y, B, H_2).$$

Let now

$$([s]_\delta, \delta) : (X, A, H_1) \to (Y, B, H_2)$$

and

$$([s']_{\delta'}, \delta') : (Y, B, H_2) \to (Z, C, H_3)$$

be two morphisms. Then set

$$([s']_{\delta'}, \delta')([s]_\delta, \delta) = ([ss']_{\delta'\delta}, \delta'\delta) : (X, A, H_1) \to (Z, C, H_3).$$

These definitions are correct and the category K_Θ is defined.

The categories K_Θ and C_Θ are dual. The categories $K_\Theta(H)$ and $C_\Theta(H)$ are subcategories in K_Θ and C_Θ, and duality between K_Θ and C_Θ induces duality between $K_\Theta(H)$ and $C_\Theta(H)$.

Let T be a system of equations in $W(X)$. Define a full subcategory $K_\Theta(T)$ in K_Θ.

The objects have the form

$$(X, A, H)$$

where X is fixed, $A = T'_H$, and H, A are changed. The category $K_\Theta(T)$ characterizes possibilities to solve the system T in different algebras $H \in \Theta$.

Consider separately the case when there is an injection $\delta : G \to H$. It can be proven that this injection induces in K_Θ an isomorphism between the category $K_\Theta(G)$ and a subcategory in $K_\Theta(H)$.

2.4. On the notion of algebraically closed algebras. As previously in the classical situation, we consider $K_\Theta(H)$ as a geometrical invariant of the algebra H. This category is viewed also as a measure of algebraic closeness of an algebra H. An algebra H can be not an algebraically closed algebra, but H is algebraically closed with some measure.

We consider the notion of algebraically closed algebras.

DEFINITION 2.5. An algebra $H \in \Theta$ is called *algebraically closed* in Θ, if for every finite X and every non-unital congruence T in $W(X)$ this T has a non-zero root $\mu : W(X) \to H$.

Here $\operatorname{Ker}\mu$ is a non-unital congruence and $T \subset \operatorname{Ker}\mu$. In particular, if T is a maximal congruence in $W = W(X)$, then $\operatorname{Ker}\mu = T$ and T is an H-closed congruence. Every maximal congruence T is H-closed.

It is clear that if an algebra H is contained in some other $H_1 \in \Theta$, and H is algebraically closed, then H_1 is also algebraically closed.

Consider this definition with respect to varieties of the kind Θ^G. An algebra $H \in \Theta$ is considered as a G-algebra for different G with an injection $h : G \to H$. For such (H, h) and different G, the definition of algebraic closeness with respect to different varieties Θ^G differ.

In Var-P, an algebra L over P can be P-closed but not absolutely algebraically closed. Here and in the general situation, absolutely algebraically closed means algebraic closeness in the variety Θ^L.

In the general case, all equations and systems of equations have coefficients in H. So, fix $G \in \Theta$ and consider a homomorphism in Θ^G

$$\begin{array}{ccc} G & \xrightarrow{i_G} & G * W_0 = W \\ & \searrow^{\varepsilon} & \downarrow \mu \\ & & G \end{array}$$

For every such μ, the congruence $T = \operatorname{Ker}\mu$ is faithful and maximal. The condition of algebraic closeness of G in Θ^G means that if T is a maximal faithful congruence in W, then $T = \operatorname{Ker}\mu$ for some μ.

This is well-coordinated with the situation of the field P in Var-P. Note that there are other approaches to the notion of algebraic closeness.

Consider systems of equalities T of the form $w \equiv w'$ and inequalities $w \not\equiv w'$. Such a system has a solution if there exists a proper congruence T_1 such that all $w \equiv w'$ from T belong to T_1 and all $w \not\equiv w'$ from T do not belong to T_1. Now one can say that an algebra H is algebraically closed if for any compatible system T of equalities and inequalities for every finite X, there is a point $\mu : W(X) \to H$, which is a root of such T.

Such a point of view on algebraic closeness of an algebra H is closer to algebraic geometry in logic (see Lecture 6) than to equational algebraic geometry.

2.5. Changing of the basic variety of algebras. Let Θ be a variety of algebras, $\Theta_0 \subset \Theta$ be a subvariety, and X be a finite set. Then $W = W(X)$ and $W_0 = W_0(X)$ are the free algebras in Θ and Θ_0, respectively. The homomorphism $\mu_0 : W(X) \to W_0(X)$ with the kernel $T = \operatorname{Ker} \mu$ consists of identities in W of the variety Θ_0. For every such H, there is a decomposition

$$\begin{array}{ccc} W & \xrightarrow{\mu_0} & W_0 \\ & {\scriptstyle\mu}\searrow & \downarrow{\scriptstyle\nu} \\ & & H \end{array}$$

To every point $\mu : W \to H$ uniquely corresponds $\nu : W_0 \to H$ with $\mu = \nu\mu_0 = \tilde{\mu}_0(\nu)$. We have the bijection $\mu_0 : \operatorname{Hom}(W_0, H) \to \operatorname{Hom}(W, H)$.

PROPOSITION 2.7. *The bijection $\tilde{\mu}_0$ induces the isomorphism of lattices of the algebraic sets $\mathrm{Alv}_H(W)$ and $\mathrm{Alv}_H(W_0)$; and the lattices $\mathrm{Cl}_H(W)$ and $\mathrm{Cl}_H(W_0)$.*

The same bijections μ_0 for different finite X_0 is used to prove

THEOREM 2.2. *The categories $K_\Theta(H)$ and $K_{\Theta_0}(H)$ are canonically isomorphic.*

For $\Theta_0 \subset \Theta$, one can consider the categories K_{Θ_0} and K_Θ. It can be proved that the first one is isomorphic to a subcategory of the second.

Let, again, $H \in \Theta_0 \subset \Theta$. Then we have the following

PROPOSITION 2.8. *If H is algebraically closed in Θ, then H is algebraically closed in Θ_0.*

The opposite statement is not true in general.

One can imagine the situation when a system of equations T in the given $W = W(X)$ in Θ contradicts to identities of the algebra H. Then T does not have a common root in H.

One can consider algebraic closeness of the given H with respect to a different finite X. Then the question is whether the absolute algebraic closeness follows from the algebraic closeness with respect to one-element set X. The answer depends on Θ and on the structure of algebraic closure of an arbitrary H in Θ.

2.6. Zariski topology. This is the minimal topology in the space $\operatorname{Hom}(W, H)$ in which all algebraic sets are closed.

If the algebra H is stable, then algebraic sets are, precisely, all closed sets. In the general case, the closed sets in Zariski topology are represented by finite unions of algebraic sets and their arbitrary intersections.

There is another approach which we now discuss.

Consider formulas of a more general form called *pseudoequalities*. They have the form

$$w_1 = w_1' \vee \cdots \vee w_n \equiv w_n', \quad w_i, w_i' \in W = W(X).$$

One can consider also pseudoequations. If u is the pseudoequality above, then its value $\mathrm{Val}_H(u)$ in the algebra H is defined by

$$\mathrm{Val}_H(u) = \mathrm{Val}_H(w_1 \equiv w_1') \vee \cdots \vee \mathrm{Val}_H(w_n \equiv w_n').$$

Similar to the usual equalities, for every system T of pseudoequalities, one can define

$$T'_H = A = \bigcap_{u \in T} \mathrm{Val}_H(u).$$

The sets A of such form are called *pseudoalgebraic sets.*

Every algebraic set is pseudoalgebraic. Conversely, if an algebra H is stable, then every pseudoalgebraic set is algebraic.

PROPOSITION 2.9. *Closed sets in Zariski topology coincide with pseudoalgebraic sets.*

One can consider the category of affine spaces $K^0_\Theta(H)$ and every category $K_\Theta(H)$ as categories with topological objects. The morphisms defined above turn out to be continuous maps.

Lecture 3

Geometric Properties and Geometric Relations of Algebras

Contents

3.1. Geometric equivalence of algebras.

Given a variety Θ.

DEFINITION 3.1. Algebras H_1 and H_2 in Θ are *geometrically equivalent* if for every finite X and every system of equations T in the free algebra $W = W(X)$, the equality

$$T''_{H_1} = T''_{H_2}$$

takes place.

This is equivalent to the fact that always $\mathrm{Cl}_{H_1}(W) = \mathrm{Cl}_{H_2}(W)$, i.e., the functors Cl_{H_1}, and Cl_{H_2} coincide.

Also the lattices $\mathrm{Cl}_{H_1}(W)$ and $\mathrm{Cl}_{H_2}(W)$ coincide and the lattices $\mathrm{Alv}_{H_1}(W)$ and $\mathrm{Alv}_{H_2}(W)$ are isomorphic.

PROPOSITION 3.1. *Every algebra H is geometrically equivalent to any of its cartesian powers H^I.*

This proposition follows from the fact that the algebras H_1 and H_2 are geometrically equivalent if and only if for any W and a congruence T in W the equality

$$(H_1\text{-}\mathrm{Ker})(W/T) = (H_2\text{-}\mathrm{Ker})(W/T)$$

takes place.

PROPOSITION 3.2. *Let Θ be a variety and Θ_0 a subvariety in Θ. The algebras H_1 and H_2 from Θ_0 are geometrically equivalent in Θ_0 if and only if they are equivalent in Θ.*

Consider one more point of view on the notion of geometric equivalence.

Let H_1 and H_2 from Θ be given. There exists a canonical functor

$$F = F(H_1, H_2) : K_\Theta(H_1) \to K_\Theta(H_2).$$

If (X, A) is an object in $K_\Theta(H_1)$, then

$$F((X, A)) = (X, (A')'_{H_2}).$$

The definition of the action of F on morphisms is more complicated and is omitted.

PROPOSITION 3.3. *The algebras H_1 and H_2 are geometrically equivalent if and only if the functors $F(H_1, H_2)$ and $F(H_2, H_1)$ are mutually inverse.*

In particular, if H_1 and H_2 are geometrically equivalent, then the categories $K_\Theta(H_1)$ and $K_\Theta(H_2)$ are isomorphic.

Isomorphism of categories can be established also from the following commutative diagram

$$\begin{array}{ccc} C_\Theta(H_1) & \longrightarrow & C_\Theta(H_2) \\ \uparrow & & \uparrow \\ K_\Theta(H_1) & \longrightarrow & K_\Theta(H_2) \end{array}$$

Here the upper arrow is an identity map, and the vertical arrows are dual.

As an example, let us consider the following result which is due to A. Berzins.

PROPOSITION 3.4. *Two abelian groups H_1 and H_2 are geometrically equivalent if and only if*

1. *they have the same exponents;*
2. *for every prime p the exponents of Sylow subgroups H_{1p} and H_{2p} coincide.*

3.2. Geometric equivalence and quasi-identities. A quasi-identity in Θ has the form

$$w_1 \equiv w_1' \wedge \cdots \wedge w_n \equiv w_n' \to w \equiv w'.$$

We can assume that all w_i, w_i', w, w' belong to the same $W(X)$ with finite X.

THEOREM 3.1. *If algebras H_1 and H_2 are geometrically equivalent, then they have the same quasi-identities.*

In particular, if algebras H_1 and H_2 are geometrically equivalent, then

$$\mathrm{Var}(H_1) = \mathrm{Var}(H_2).$$

COROLLARY 3.1. *If two groups G_1 and G_2 are geometrically equivalent and one of them is torsion-free, then the second one is also torsion-free.*

PROBLEM 3.1. Is it true that if H_1 and H_2 have the same quasi-identities then they are geometrically equivalent? Is it true that if H_1 and H_2 are elementary equivalent, then they are geometrically equivalent?

There is a negative answer to both of these questions. Namely, there exists a group H and its ultrapower $\tilde{H}$ such that H and $\tilde{H}$ are not geometrically equivalent. Similar fact is true also for G-groups. This very important expected result is obtained in [**MR2**] and uses ideas of the dissertation of V.A. Gorbunov (1996) [**Gor**]. For groups there are also beautiful solutions in [**GoSh**] and [**BlG**].

For associative and Lie algebras, this result is also valid (see [**Pl11**] for details. For associative algebras, it uses the result from [**Li**]).

On the other hand, recall that in the classical situation every extension L of the field P is geometrically equivalent to every its ultrapower.

We will consider also generalized (infinitary) quasi-identities. They have the form

$$\bigwedge_{\alpha \in I} (w_\alpha \equiv w_\alpha') \to w \equiv w'.$$

Here I is an arbitrary set, and all $w_\alpha, w_\alpha', w, w'$ belong to one and the same $W(X)$ with finite X.

THEOREM 3.2. *Algebras H_1 and H_2 are geometrically equivalent if and only if they have the same generalized quasi-identities.*

This theorem is based on the following form of the theorem about zeroes.

Let T be an arbitrary system of equations in $W(X)$ with finite X. Denote by I the set of all indices α of equations $w_\alpha \equiv w'_\alpha$ from T. Then we have

PROPOSITION 3.5. *Inclusion* $w \equiv w' \in T''_H$ *is satisfied if and only if*

$$\wedge_{\alpha \in I}(w_\alpha \equiv w'_\alpha) \to (w \equiv w')$$

is fulfilled in the algebra H.

If T is a finite set, then with T''_H the usual quasi-identity is associated. Thus, we have

THEOREM 3.3. *Algebras* H_1 *and* H_2 *are geometrically equivalent with respect to finite* T *if and only if* H_1 *and* H_2 *generate the same quasivariety.*

Note also the following

THEOREM 3.4. *Algebras* H_1 *and* H_2 *are geometrically equivalent if and only if*

$$LSC(H_1) = LSC(H_2).$$

Here L, S, and C are the standard closure operators on classes of algebras, used in the characterization of prevarieties.

For any class $\mathfrak{X}$, the class $LSC(\mathfrak{X})$ is a locally closed prevariety over $\mathfrak{X}$ which is contained in the quasivariety, generated by $\mathfrak{X}$ [**PPT**].

For every algebra $H \in \Theta$, finitely generated algebras in the prevariety $SC(H)$ are the algebras presented as coordinate algebras of algebraic sets over H.

It follows from [**MR2**] that the class $LSC(\mathfrak{X})$ is not a quasivariety and, moreover, not an axiomatized class. In this sense, the relation of geometric equivalence of algebras is not an axiomatizable relation. This relation is axiomatizable in terms of generalized quasi-identities [**Pl10**].

3.3. Geometric similarity of algebras. This notion generalizes the notion of geometric equivalence of algebras, and, like the notion of geometric equivalence, is associated with the problem of isomorphism of categories of algebraic sets. It is the main notion of Lecture 5. Geometric equivalence means that the functors Cl_{H_1} and Cl_{H_2} coincide. Geometric similarity assumes the more complicated connection between Cl_{H_1} and Cl_{H_2}.

If H_1 and H_2 are geometrically equivalent, then $\mathrm{Var}(H_1) = \mathrm{Var}(H_2)$ and the categories $\mathrm{Var}(H_1)^0$ and $\mathrm{Var}(H_2)^0$ also coincide.

Let us consider the functors

$$\mathrm{Cl}_{H_1} : \mathrm{Var}(H_1)^0 \to \mathrm{Set} \quad \text{and} \quad \mathrm{Cl}_{H_2} : \mathrm{Var}(H_2)^0 \to \mathrm{Set}.$$

Similarity of algebras means that there is an isomorphism $\varphi : \mathrm{Var}(H_1)^0 \to \mathrm{Var}(H_2)^0$ with the commutative diagram

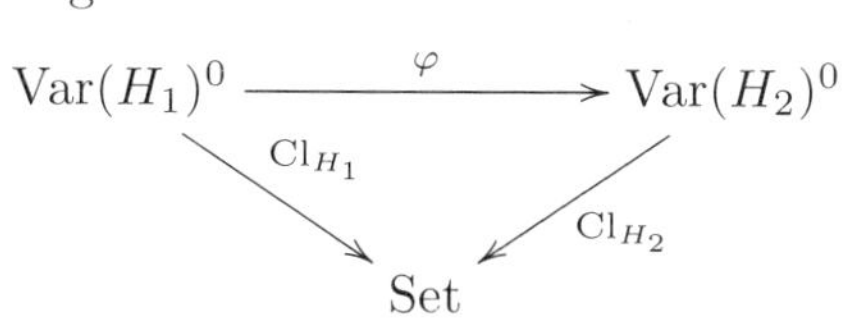

Commutativity of the diagram indicates an isomorphism (not necessarily equality) of the functors Cl_{H_1} and $\mathrm{Cl}_{H_2}\varphi$. This isomorphism $\alpha = \alpha(\varphi)$ depends on the isomorphism of categories φ and is constructed in a special way.

The notion of geometric similarity appears to be sophisticated, but in "good" cases it reduces to geometric equivalence.

We will use, further, some remarks on congruences in free algebras.

Let the algebra $W = W(X)$ be given. For a congruence T in W, consider a relation $\rho = \rho(T)$ in the semigroup $\mathrm{End}(W)$. Let us set $\nu\rho\nu', \nu, \nu' \in \mathrm{End}(W)$, if $\nu(w)T\nu'(w)$ holds for every $w \in W$.

Here ρ-equivalence on $\mathrm{End}(W)$, and $\nu\rho\nu'$ implies $(\nu_1\nu)\rho(\nu_1\nu')$ for every $\nu_1 \in \mathrm{End}(W)$.

Let now ρ be an arbitrary congruence on $\mathrm{End}(W)$. Define a relation $T = T(\rho)$ on the algebra W :

$$w_1 T w_2 \Leftrightarrow \exists w \in W, \nu, \nu' \in \mathrm{End}(W)$$

such that $w^\nu = w_1$, $w^{\nu'} = w_2$, and $\nu\rho\nu'$.

PROPOSITION 3.6. *If T is a congruence on the algebra W and $\rho = \rho(T)$, then $T(\rho) = T$.*

If T is a fully characteristic congruence on W, then $\rho = \rho(T)$ is a congruence in the semigroup $\mathrm{End}(W)$, and the semigroup $\mathrm{End}(W)/\rho$ is isomorphic to the semigroup $\mathrm{End}(W/T)$.

Now let an isomorphism of categories

$$\varphi : \mathrm{Var}(H_1)^0 \to \mathrm{Var}(H_2)^0$$

be given.

We want to define a function $\alpha = \alpha(\varphi) : \mathrm{Cl}_{H_1} \to \mathrm{Cl}_{H_2}\varphi$ which yields an isomorphism of functors. A function α for every W from $\mathrm{Var}(H_1)^0$ is given by the map $\alpha_W : \mathrm{Cl}_{H_1}(W) \to \mathrm{Cl}_{H_2}\varphi(W)$, and for every $s : W \to W'$, the commutative diagram

$$\begin{array}{ccc} \mathrm{Cl}_{H_1}(W') & \xrightarrow{\mathrm{Cl}_{H_1}(s)} & \mathrm{Cl}_{H_1}(W) \\ {\scriptstyle \alpha_{W'}}\downarrow & & \downarrow{\scriptstyle \alpha_W} \\ \mathrm{Cl}_{H_2}\varphi(W') & \xrightarrow[\mathrm{Cl}_{H_2}(s)]{} & \mathrm{Cl}_{H_2}\varphi(W) \end{array}$$

should be fulfilled.

This means that $\alpha = \alpha(\varphi)$ is an isomorphism of functors Cl_{H_1} and $\mathrm{Cl}_{H_2}\varphi$. Let us construct such an α.

First, for any Θ, define functions β and γ which for an object W from Θ^0 give the maps β_W and γ_W.

The function β_W is defined by $\beta_W(T) = \rho = \rho(T)$, where T is a congruence in W and $\rho(T)$ is defined above.

The function γ_W is given by $\gamma_W(\rho) = T = T(\rho)$, where ρ is an equivalence in $\mathrm{End}(W)$. We have

$$\gamma_W(\beta_W(T)) = T$$

if T is a congruence in W.

For every $T \in \mathrm{Cl}_{H_1}(W)$, define

$$\alpha(\varphi)_W(T) = \gamma_{\varphi(W)}(\varphi(\beta_W(T)).$$

Here if ρ is a relation on $\mathrm{End}(W)$, then $\varphi(\rho) = \rho^*$ is a relation on $\varphi(W)$ defined by $\mu\rho^*\mu'$ if and only if $\mu = \varphi(\nu)$, $\mu' = \varphi(\nu')$, $\nu\rho\nu'$. Here $\alpha(\varphi)_W(T)$ is a relation on the algebra $\varphi(W)$ which is not necessarily a congruence.

The meaning of the function $\alpha = \alpha(\varphi)$ is that it represents the action of the isomorphism φ on the congruences of free algebras.

Define now a function τ. Let W, W' be two objects in Θ^0. Then for any congruence T in W', define $\tau_{W,W'}(T) = \rho$ to be a relation over the set $\mathrm{Hom}(W, W')$ defined by

$$s\rho s' \;\Leftrightarrow\; w^s T w^{s'},\ \forall w \in W,$$

where s and s' are homomorphisms $W \to W'$. We have $\tau_{W,W} = \beta_W$. We say that α is compatible with τ if for every W and W', and a congruence $T \triangleleft W'$,

$$\varphi(\tau_{W,W'}(T)) = \tau_{\varphi(W),\varphi(W')}(\alpha(\varphi)_{W'}(T))$$

holds.

Now we state the main definition.

DEFINITION 3.2. Two algebras H_1 and H_2 are called *geometrically similar* if there exists an isomorphism

$$\varphi : \mathrm{Var}(H_1)^0 \to \mathrm{Var}(H_2)^0,$$

such that the function $\alpha = \alpha(\varphi) : \mathrm{Cl}_{H_1} \to \mathrm{Cl}_{H_2}\varphi$ is compatible with the function τ and for every object W in $\mathrm{Var}(H_1)^0$ there is a bijection

$$\alpha(\varphi)_W : \mathrm{Cl}_{H_1}(W) \to \mathrm{Cl}_{H_2}(\varphi(W)).$$

From this definition, it follows that a function $\alpha = \alpha(\varphi)$ defines an isomorphism of functors Cl_{H_1} and $\mathrm{Cl}_{H_2}\varphi$.

In addition, the same $\alpha(\varphi)$ defines an isomorphism of lattices

$$\alpha(\varphi)_W : \mathrm{Cl}_{H_1}(W) \to \mathrm{Cl}_{H_2}(\varphi(W)).$$

THEOREM 3.5. *The algebras H_1 and H_2 are geometrically similar if and only if there exists an isomorphism $\varphi : \mathrm{Var}(H_1)^0 \to \mathrm{Var}(H_2)^0$, and an isomorphism of functors $\alpha = \alpha(\varphi) : \mathrm{Cl}_{H_1} \to \mathrm{Cl}_{H_2}\varphi$ which is compatible with the function τ.*

Note the following:

1. If H_1 and H_2 are geometrically equivalent, then they are similar. Here $\mathrm{Var}(H_1) = \mathrm{Var}(H_2)$ and for φ take the identity map.
2. A relation of geometric similarity is symmetric, reflexive and transitive.
3. If Θ_0 is a subvariety in Θ, and $H_1, H_2 \in \Theta_0$, then H_1 and H_2 are similar in Θ if and only if they are similar in Θ_0.

3.4. Similarity and equivalence. Consider the case when $\mathrm{Var}(H_1) = \mathrm{Var}(H_2) = \Theta$. In this case, φ is an automorphism of the category Θ^0.

DEFINITION 3.3. An automorphism $\varphi : C \to C$ of an arbitrary category C is called *inner* if it corresponds to a function s which attaches an isomorphism $s_A : A \to \varphi(A)$ to every object A from C such that for any $\nu : A \to B$ there is a commutative diagram

$$\begin{array}{ccc} A & \xrightarrow{\nu} & B \\ {\scriptstyle s_A}\downarrow & & \downarrow{\scriptstyle s_B} \\ \varphi(A) & \xrightarrow{\varphi(\nu)} & \varphi(B) \end{array}$$

i.e.,

$$\varphi(\nu) = s_B \nu s_A^{-1}.$$

From this definition, it follows that an automorphism φ is inner if and only if the functor φ is isomorphic to the identity functor of the category C.

THEOREM 3.6. *If the algebras H_1 and H_2 are geometrically similar with respect to the inner automorphism φ, then they are equivalent.*

Let us make some remarks on inner automorphisms of categories.

If an inner automorphism φ is realized by a function s, then we write $\varphi = \hat{s}$. Let now φ be an arbitrary automorphism and $\hat{s}$ be an inner automorphism. Take an automorphism $\varphi\hat{s}\varphi^{-1}$ and consider a function $\varphi(s)$ defined by the rule

$$\varphi(s)_B = \varphi(s_{\varphi^{-1}B})$$

for any object B in the given category C.

Let us check that

$$\varphi\hat{s}\varphi^{-1} = \widehat{\varphi(s)}.$$

Thus if $\mathrm{Aut}(C)$ is a group of all automorphisms of the category C, then there is a normal subgroup $\mathrm{Int}(C)$ of all inner automorphisms.

Let now $\varphi \in \mathrm{Aut}(C)$ and A be an arbitrary object in C. Then φ induces an isomorphism of semigroups $\mathrm{End}(A)$ and $\mathrm{End}(\varphi(A))$ and the groups $\mathrm{Aut}(A)$ and $\mathrm{Aut}(\varphi(A))$. If $\varphi(A) = A$, then φ induces an automorphism of the semigroup $\mathrm{End}(A)$ and of the group $\mathrm{Aut}(A)$. If φ is an inner automorphism, then the corresponding induced automorphism of the semigroup $\mathrm{End}(A)$ and the group $\mathrm{Aut}(A)$ is also inner.

DEFINITION 3.4. A variety Θ is called *perfect* if every automorphism of the category Θ^0 is an inner automorphism.

We will consider the examples of perfect Θ. We can state that if Θ is a perfect variety and $\mathrm{Var}(H_1) = \mathrm{Var}(H_2) = \Theta$, then H_1 and H_2 are similar if and only if they are equivalent.

Let us define the weakly inner automorphisms $\varphi : C \to C$. An automorphism φ is called *weakly inner* if there exists a function s with isomorphisms $s_A : A \to \varphi(A)$, such that for every $\nu : A \to A$ holds

$$\varphi(\nu) = s_A \nu s_A^{-1} : \varphi(A) \to \varphi(A).$$

It can be proven that in Theorem 3.6 one can proceed from a weakly inner automorphism. Using this fact, one can prove

PROPOSITION 3.7. *Two abelian groups H_1 and H_2, each of which generates the whole variety of abelian groups, are similar if and only if they are equivalent.*

A reasonable conjecture states that the same is true for H_1 and H_2, generating the variety of all groups Grp.

The solution of this problem is closely connected with the following problem.

PROBLEM 3.2. Let $F = F(X)$ be the non-commutative free group with finite X. It is known that the group $\mathrm{Aut}(F)$ is a perfect group. Is the same true for the semigroup $\mathrm{End}(F)$?

Let φ be an automorphism of the semigroup $\mathrm{End}(F)$. This φ induces an automorphism of the group $\mathrm{Aut}(F)$. Thus, there exists an element $\sigma \in \mathrm{Aut}(F)$, such that $\varphi(x) = \sigma x \sigma^{-1}$ for every $x \in \mathrm{Aut}(F)$. Is it true that the same holds for every $x \in \mathrm{End}(F)$?[1]

3.5. Similarity and semiisomorphisms. Let Θ be an arbitrary variety, $G \in \Theta$, and consider the variety Θ^G. Assume that for $G \in \Theta^G$ the condition (*) (see Lecture 2) is fulfilled. This means that if (H_1, h_1) and (H_2, h_2) are two faithful G-algebras, then each of them generates the variety Θ^G.

In the category Θ^G, one can consider semiisomorphisms and semiisomorphic algebras.

THEOREM 3.7. *If algebras (H_1, h_1) and (H_2, h_2) are semiisomorphic, then they are similar.*

Thus, if there is a sequence of G-algebras

$$(H_1, h_1) = (H_1^0, h_1^0), (H_1^1, h_1^1), \dots (H_1^n, h_1^n) = (H_2, h_2),$$

such that the neighbors are geometrically equivalent or semiisomorphic, then (H_1, h_1) and (H_2, h_2) are similar.

DEFINITION 3.5. Algebras (H_1, h_1) and (H_2, h_2) are called *geometrically equivalent up to a semiisomorphism*, if there exists (H, h) such that (H_1, h_1) and (H, h) are semiisomorphic, and (H, h) is equivalent to (H_2, h_2).

We will consider semi-inner automorphisms of the category $(\Theta^G)^0$.

Let $\varphi : (\Theta^G)^0 \to (\Theta^G)^0$ be an automorphism of the category of free algebras in Θ^G. These algebras have the form

$$W(X) = G * W_0(X) = W = G * W_0.$$

Consider a pair (σ, s), where σ is an automorphism of the algebra G and $s_W : W \to \varphi(W)$ is an algebra isomorphism in Θ for every W. The commutative diagram

$$\begin{array}{ccc} G & \xrightarrow{i_G} & W \\ {\scriptstyle\sigma}\downarrow & & \downarrow{\scriptstyle s_W} \\ G & \xrightarrow[i'_G]{} & \varphi(W) \end{array}$$

defines a semiisomorphism

$$(\sigma, s_W) : (W, i_G) \to (\varphi(W), i'_G).$$

Suppose that with the automorphism φ one can associate such a pair (σ, s) and, besides, for every morphism $\nu : W \to W'$ there is a commutative diagram

$$\begin{array}{ccc} W & \xrightarrow{\nu} & W' \\ {\scriptstyle(\sigma, s_W)}\downarrow & & \downarrow{\scriptstyle(\sigma, s_{W'})} \\ \varphi(W) & \xrightarrow[\varphi(\nu)]{} & \varphi(W') \end{array}$$

[1]This question is solved positively in [**For**]. From this follows that the conjecture is also true. From the result of Formanek follows that the variety Grp is weekly perfect. Thus, arises the following problem.

Is it true that the variety Grp *is perfect? A positive answer can be obtained using the technique of Formanek. The same question can be asked for the variety of all abelian groups.*

Then

$$\varphi(\nu) = (\sigma, s_{W'})(1, \nu)(\sigma, s_W)^{-1} = (1, s_{W'}\nu s_W^{-1}) = s_{W'}\nu s_W^{-1},$$

and this is a morphism in $(\Theta^G)^0$.

DEFINITION 3.6. An automorphism φ is called *semi-inner*, if there exists a pair (σ, s) with the property above.

We denote $\varphi = (\widehat{\sigma, s})$.

It can be proved that semi-inner automorphisms constitute a subgroup in $\mathrm{Aut}((\Theta^G)^0)$ which contains the normal subgroup of inner automorphisms.

From Definition 3.6, it follows that an automorphism φ is semi-inner if and only if the functor φ is semiisomorphic to the identity functor of the category $(\Theta^G)^0$.

The main theorem here is as follows:

THEOREM 3.8. *If the similarity of algebras (H_1, h_1) and (H_2, h_2) is given by a semi-inner automorphism φ, then they are equivalent up to a semiisomorphism.*

3.6. Geometrically Noetherian algebras. We define a property of an algebra $H \in \Theta$, which always takes place in the classical situation and plays a crucial part.

DEFINITION 3.7. An algebra H in the variety Θ is called *geometrically Noetherian* if for every finite set X and every system of equations T in $W = W(X)$, there exists a finite subsystem $T_0 \subset T$, such that $T'_H = T'_{0H}$, or, in other words, $T''_H = T''_{0H}$.

We say that T and T_0 are an equivalent system of equations.

PROPOSITION 3.8. *An algebra H is geometrically Noetherian if and only if for every finite X the lattices $\mathrm{Cl}_H(W(X))$ and $\mathrm{Alv}_H(W(X))$ satisfy the maximal and minimal condition, respectively.*

PROPOSITION 3.9. *Let Θ_0 be a subvariety in Θ and $H \in \Theta^0$. Then H is geometrically Noetherian in Θ if and only if it is geometrically Noetherian in Θ_0.*

DEFINITION 3.8. A variety Θ is called *Noetherian* if every finitely generated algebra from Θ is Noetherian.

EXAMPLES.

1. The variety $Var\text{-}P$ is Noetherian.
2. Arbitrary variety of nilpotent groups is Noetherian.
3. In the variety of arbitrary associative algebras over the field P, every Noetherian subvariety can be distinguished by special known identities.

PROPOSITION 3.10. *If the variety Θ is Noetherian, then every algebra H in Θ is geometrically Noetherian. In particular, if the variety $\mathrm{Var}(H)$ is Noetherian, then the algebra H is geometrically Noetherian.*

THEOREM 3.9. *Let H_1 and H_2 be two geometrically Noetherian algebras. They are geometrically equivalent if they have the same quasi-identities.*

THEOREM 3.10. *Let H_1 and H_2 have the same quasi-identities. Then, if H_1 is geometrically Noetherian, then H_2 is also geometrically Noetherian.*

COROLLARY 3.2. *If an algebra H is geometrically Noetherian, then each of its cartesian power and ultrapower is geometrically Noetherian.*

The same is true for arbitrary filtered power.

PROPOSITION 3.11. *If algebras H_1 and H_2 are geometrically similar and one of them is geometrically Noetherian, then the second one is also geometrically Noetherian.*

It can be easily seen that every subalgebra of a geometrically Noetherian algebra is also geometrically Noetherian, and a finite cartesian product of geometrically Noetherian algebras is also a Noetherian algebra. Every finite algebra is geometrically Noetherian.

Geometrically Noetherian algebras admit a Noetherian Zariski topology in the corresponding affine spaces and the Lasker-Noether Theorem on the decomposition of algebraic varieties.

DEFINITION 3.9. A variety of the type Θ^G is called *faithfully Noetherian* if it is Noetherian with respect to faithful congruences.

If the variety Θ^G is faithfully Noetherian, then each of its faithful algebras (H, h) is geometrically Noetherian.

There are many important results on geometrically Noetherian groups (see [**Br**], [**Gu**], [**BMR1**], [**BMRo**] and others. There are also many interesting problems, especially for the case of G in Θ^G.

3.7. Geometric stability. This notion was already defined and will be used and discussed in the next lecture. Let us make now some remarks on G-algebras.

PROPOSITION 3.12. *Let the algebras (H_1, h_1) and (H_2, h_2) be semiisomorphic. Then*

1. *If one of them is stable, then the second one is also stable.*
2. *If one of them is geometrically Noetherian, then the second one is also geometrically Noetherian.*

In addition, if (H_1, h_1) and (H_2, h_2) are two arbitrary G-algebras, then their product $(H_1 \times H_2, h_1 \times h_2)$ cannot be a stable algebra [**BelP**]. From this follows that if (H_1, h_1) and (H_2, h_2) are equivalent and one of them is stable, then the second one is not necessarily stable.

DEFINITION 3.10. An algebra H in the variety Θ is called *geometrically distributive* if every lattice $\mathbb{C}_H(W)$ is distributive.

If H is stable, then H is geometrically distributive.

PROBLEM 3.3. Is it true that an algebra H is geometrically distributive if and only if it is similar to a stable algebra?

One can consider also geometrically modular algebras.

It can be proven that if an algebra $G \in \Theta$ admits faithful finite dimensional linearization over a field, then it is geometrically Noetherian in Θ and in Θ^G. This fact, in particular, relates to finite dimensional, associative and Lie algebras. (For groups, see [**BMR1**], the general case was done by A. Belov [**BelP**]). The proof uses the natural Zariski topology over the same field.

A. Miasnikov and V. Remeslennikov have noticed the following interesting question. Is it true that a free Lie algebra $F(X)$ with finite X is geometrically Noetherian?

Lecture 4

Algebraic Geometry in Group-based Algebras

Contents

4.1. Group-based algebras.

DEFINITION 4.1. A group-based algebra H is an additive group (not necessarily commutative) with some additional signature Ω. For every $\omega \in \Omega$ of the arity $n(\omega) = n > 0$, the condition $0 \cdots 0\omega = 0$ should be fulfilled.

Here 0 is the zero element in the additive group H and on the left side we have n times for 0.

Such group-based algebras are also called Ω-groups. They were introduced by P. Higgins [**Hi**] in 1956, see also [**Ku**], [**Pl2**].

A group is an Ω-group with empty set Ω, in rings the set Ω consists of a single multiplication, in modules over a ring R all elements of R belong to the set Ω. Groups over rings, considered by R. Lyndon [**L1**] are also Ω-groups.

In an arbitrary Ω-group G, we have the usual commutator $[a, b] = -a - b + a + b = -a + a^b$, and ω-commutators for ω with $n(\omega) > 0$. By definition we have:

$$[a_1, \cdots, a_n; b_1, \cdots, b_n; \omega] = -a_1 \cdots a_n\omega - b_1 \cdots b_n\omega + (a_1 + b_1) \cdots (a_n + b_n)\omega.$$

From this definition follows that if all $a_1, \cdots, a_n$ are zeros or all $b_1, \cdots, b_n$ are zeros, then the ω-commutator is a zero.

In rings we have:

$$[a_1, a_2; b_1, b_2; \cdot] = a_1b_2 + b_1a_2.$$

Now we consider ideals in Ω-groups. Let $\mu\colon H \to H_1$ be a homomorphism of two Ω-groups, U is the coimage of the zero element from H_1. Then

1. U is closed relative to all $\omega \in \Omega$ with $n(\omega) > 0$.
2. U is a normal subgroup in additive group H.
3. The ω-commutator $[a_1, \cdots, a_n; b_1, \cdots, b_n; \omega]$ belongs to U if $a_1, \cdots, a_n \in U$, and $b_1, \cdots, b_n$ are arbitrary elements of H.

DEFINITION 4.2. An ideal U in H is a set U with the three conditions above.

From the definition follows also that the commutator $[a_1, \cdots, a_n; b_1, \cdots, b_n; \omega]$ belongs to U if all $b_1, \cdots, b_n$ in U and $a_1, \cdots, a_n$ are arbitrary. The ideal in a group is a normal subgroup, the ideal in a ring is a usual ideal. It can be proved that if T is a congruence in an Ω-group H, then the class $[0] = U$ for T is an ideal and we have: $[a] = a + U$ for arbitrary $a \in H$. So we have one-to-one correspondence between congruences and ideals. We write now $\operatorname{Ker}\mu = U$ instead of $\operatorname{Ker}\mu = T$ and H/U instead of H/T.

Now let A and B be two sets in the Ω-group H, $\{A, B\}$ be an Ω-subgroup in H, generated by A and B.

DEFINITION 4.3. The mutual commutant $[A, B]$ is the ideal in $\{A, B\}$, generated by all commutators of the kind $[a, b]$, $a \in A, b \in B$ and all $[a_1, \cdots, a_n; b_1, \cdots, b_n; \omega]$, where $a_1, \cdots, a_n \in A$, $b_1, \cdots, b_n \in B$.

We have $[A, B] = [B, A]$ and for every two ideals U_1 and U_2 we have: $[U_1, U_2] \subset U_1 \cap U_2$. Let now U_1 and U_2 be ideals in the associative ring R. Then

$$[U_1, U_2] = U_1U_2 + U_2U_1.$$

In the Lie case, we have $[U_1, U_2] = U_1U_2$; in groups it is the usual mutual commutant.

DEFINITION 4.4. The group-based algebra H is abelian if $[H, H] = 0$.

This means that the group H is abelian and for every $\omega \in \Omega$, $n = n(\omega) > 0$, we have

$$(a_1 + b_1) \cdots (a_n + b_n)\omega = a_1 \cdots a_n\omega + b_1 \cdots b_n\omega$$

for every $a_1, \cdots, a_n$; $b_1, \cdots, b_n$. In rings this means that $ab = 0$ for every a and b.

DEFINITION 4.5. A group-based algebra H is antiabelian if:

1. U is an abelian ideal in H then $U = 0$.
2. U_1 and U_2 are two nontrivial ideals then the ideal $U_1 \cap U_2$ is also nontrivial.

4.2. Zero divisors and nilpotent elements. From now on, all algebras are group-based algebras. Let an algebra H be given, $a \in G$ and (a) be the ideal in H, generated by the element a.

DEFINITION 4.6. A non-zero element $a \in G$ is called a *zero divisor* if for some non-zero $b \in G$, we have

$$[(a), (b)] = 0.$$

Consider this notion for associative and Lie rings and groups. We study the case without zero divisors.

PROPOSITION 4.1. *A ring R is without zero divisors if and only if for every non-zero a and b, there exist c_1, c_2, c_1', c_2' such that*

$$c_1ac_2 \cdot c_1'bc_2' \neq 0 \text{ or } c_1'bc_2' \cdot c_1ac_2 \neq 0.$$

COROLLARY 4.1. *The class of all R without zero divisors is an axiomatizable class.*

PROPOSITION 4.2. *The Lie ring L is without zero divisors if and only if for every non-zero a and b in L, there exist $m \geq 0$, $n \geq 0$, and elements $c_1, \cdots, c_m, c_1', \cdots, c_n' \in L$ such that*

$$[[a, c_1, \cdots, c_m]; \ [b, c_1', \cdots, c_n']] \neq 0.$$

COROLLARY 4.2. *The class of all L without zero divisors, may not be an axiomatizable class.*

PROPOSITION 4.3. *The group G is a group without zero divisors iff for every non-unital a and b in G there exists $c \in G$ such that*

$$[a, b^c] \neq 1,$$

where $b^c = c^{-1}bc$.

COROLLARY 4.3. *The class of all groups without zero divisors is an axiomatizable class.*

Groups without zero divisors were considered in [**BMR1**] and in [**BPP**] where they were called antiabelian. In [**BMR1**] it was proved (in additional conditions) that the free product of two groups without zero divisors is a group without zero divisors as well.

From the corollary above, we can conclude that the ultraproduct of groups without zero divisors is also a group without zero divisors.

Now we have the following general result.

PROPOSITION 4.4. *An Ω-group H does not have zero divisors iff H is antiabelian.*

EXAMPLES. An algebra H is simple if H has no non-trivial ideals. Every non-abelian and simple algebra is antiabelian and so without zero-divisors. Every free group and every free Lie algebra is antiabelian. Every free associative or free associative and commutative algebras are without zero divisors, and thus such an algebras are antiabelian. It is easy to characterize all antiabelian finite dimensional associative and Lie algebras.

Note that using commutants in Ω-groups, the nilpotent and solvable Ω-groups are naturally defined.

DEFINITION 4.7. An element a in an Ω-group H is called *strictly nilpotent* if the ideal (a) is nilpotent. An element a in an Ω-group H is called *weakly nilpotent* if the ideal (a) is solvable.

It is easy to see that if a is a weakly nilpotent element, then in (a) there is a zero divisor. If a is a strictly nilpotent element, then a is a zero divisor.

4.3. Group-based algebras with an algebra of constants. For an arbitrary variety Θ and an arbitrary $G \in \Theta$, we consider a new variety of algebras with constants from G, which is denoted by Θ^G. We take for Θ a variety of Ω-groups and let G be a fixed Ω-group from Θ. As before, the key point for considering such a variety is the fact that groups and rings in their pure form are not so convenient for constructing an algebraic geometry. For algebraic geometry, we need some large set of constants in the signature Ω, and we need equations with constants.

For a given $h\colon G \to H$, the ideal U in H is the same as the ideal in G-algebra H. For such U, we have the embedding $\bar{h}\colon G \to H/U$, induced by h. If h is faithful and $Im(h) \cap U = 0$, then $\bar{h}$ is also faithful.

Let us define now relative ideals and relative zero divisors with respect to the group G.

Now we consider the notion of relative ideal or G-ideal. We connect the notion of ideal with the algebra of constants G.

DEFINITION 4.8. The set $U \subset H$ in the algebra $h\colon G \to H$ is a G-ideal if

1) U is closed with respect to all $\omega \in \Omega$, $n(\omega) > 0$, and U is a subgroup in the additive group H;
2) For every $a \in U$ and $g \in G$, we have $[a, h(g)] \in U$;
3) $[a_1, \cdots, a_n; h(g_1), \cdots, h(g_n); \omega] \in U$ if $\omega \in \Omega, n(\omega) > 0$; $a_1, \cdots, a_n \in U$, $g_1, \cdots, g_n \in G$.

If conditions 1,2,3 are fulfilled, the set U is called *invariant* with respect to constants from G.

From conditions 1), 2), 3), we have also

$$[h(g_1), \cdots, h(g_n); a_1, \cdots, a_n; \omega] \in U.$$

Every ideal at the same time is a G-ideal.

The algebra G can be considered as a G-algebra with the identical $G \to G$. In this case, the G-ideal in G is the same as ideal.

For every $a \in H$ by $(a)^G$, we denote the G-ideal, generated by a.

DEFINITION 4.9. A non-zero $a \in H$ is called a *G-zero divisor* if for some $b \neq 0$ we have

$$[(a)^G, (b)^G] = 0.$$

Every zero divisor in H is also a G-zero divisor. So if H is without G-zero divisors, then H is without (absolute) zero divisors. Such H is antiabelian.

PROPOSITION 4.5. *Let $h\colon G \to H$ be a G-group. Then H is without G-zero divisors if and only if for every non-unital a and b there exists $c \in G$ such that*

$$[a, b^{h(c)}] \neq 1.$$

We can consider now a set T of equations

$$[x, y^{h(c)}] = 1, \quad c \in G.$$

Let $T' = A$ be the set of all solutions in the given H. We can say that H is without G-zero divisors iff from $(a, b) \in A$ follows $a = 1$ or $b = 1$. In other words, the infinitary formula

$$\bigwedge_{c\in G}([x, y^{h(c)}] = 1) \to (x = 1) \vee (y = 1)$$

holds in H. A similar situation can be considered in Lie-L and in Assoc-R.

An algebra $G \to G$ is without G-zero divisors iff G is antiabelian.

EXAMPLES. As before, if $\Theta = \text{Grp}$, then we write $\Theta^G = \text{Grp-}G$. In the similar Lie case we have Lie-L, and in the associative case $\Theta^G = \text{Assoc-}R$.

4.4. Domains. A G-algebra H is called a *domain* if H is without G-zero divisors. It is trivial and very important that every G-subalgebra in the domain H is also a domain.

Now let $h_1\colon G \to H_1$ and $h_2\colon G \to H_2$ be two faithful G-algebras. We have in this case an isomorphism $h_0\colon Im(h_1) \to Im(h_2)$. Using this h_0, we can consider the amalgamated free product

$$h\colon G \to H_1 *_G H_2$$

that is a free product in $\Theta(G)$. In [**BMR1**], it is proven that if H_1 and H_2 from Grp-G are domains, then their free product is also a domain.

DEFINITION 4.10. An ideal U in G-algebra H is called *prime* if H/U is a domain.

For every G-algebra H, denote by $Spec(H)$ the set of all prime ideals in H.

If H/U is a domain then this algebra is antiabelian. In this case, the ideal U is irreducible and U cannot be non-trivially represented as $U = U_1 \cap U_2$.

Let now $\mu\colon H \to H_1$ be a homomorphism of G-algebras and H_1 be a domain. Then $U = \operatorname{Ker}\mu$ is prime.

PROPOSITION 4.6. *Let H and H_1 be two semiisomorphic G-algebras. Then H_1 is a domain if and only if H is a domain.*

PROPOSITION 4.7. *Let $\mu\colon H \to H_1$ be a homomorphism of G-algebras, $a \in H$. Then we have*

$$((a)^G)^\mu = (a^\mu)^G.$$

This Proposition plays an important role in the proofs of Theorems 4.1 and 4.2 below.

4.5. Algebraic geometry in group-based algebras. Let us repeat the basic notions for the case when the variety Θ is a variety of Ω-groups.

Let X be a finite set and $W = W(X)$ is a free algebra in Θ over X. For the given $H \in \Theta$, we consider the set of homomorphisms $\operatorname{Hom}(W, H)$ as an affine space.

For every point $\mu\colon W \to H$, we have the kernel $U = \operatorname{Ker}\mu$, and μ is a solution of some equation $w \equiv 0$ iff $w \in \operatorname{Ker}\mu$.

Now let U be a set of "polynomials" in W and A be a set of points in the space $\operatorname{Hom}(W, H)$. We establish the following Galois correspondence:

$$\begin{cases} U' = A = \{\mu | U \subset \operatorname{Ker}\mu\} \\ A' = U = \bigcap\limits_{\mu \in A} \operatorname{Ker}\mu \end{cases}$$

We call the set A such that $A = U'$ for some U an *algebraic set over the algebra H*. We call the ideal U of the form $U = A'$ for some A an *H-closed ideal.*

As usual we call algebraic sets also algebraic varieties.

Now let two algebraic varieties A and B in $\operatorname{Hom}(W, G)$ be given, $A' = U_1$ and $B' = U_2$. Then $(A \cap B) = (U_1 \cup U_2)'$ and so the intersection of two algebraic varieties is also an algebraic variety. But we cannot say the same for the union $A \cup B$. In general, it is not true that $A \cup B = (U_1 \cap U_2)'$. It depends on the choice of the algebra H.

4.6. Stability.

DEFINITION 4.11. The algebra H is said to be *stable* if for every $W = W(X)$ and every two algebraic sets A and B in the space $\operatorname{Hom}(W, H)$ the union $A \cup B$ is also an algebraic set.

If H is stable, then in the category $K_\Theta(H)$ every object A can be considered as a topological space. The morphisms in $K_\Theta(H)$ are well coordinated with the topology, and they are continuous maps.

In the classical case all fields and all domains are stable. But if $\Theta = \mathrm{Grp}$ is the variety of all groups and H is abelian then H is not stable.

Now we come to the main question: when is the algebra H stable? And the main idea: the notions to be stable and without zero divisors are close to each other.

Consider now the varieties of the type Θ^G.

THEOREM 4.1. *If a G-algebra H is a domain, then H is stable.*

Now we consider the following special identities, the so-called CD-identities (commutator distributivity). They are defined for all $\omega \in \Omega$, $n = n(\omega) > 0$ and have the form:

$$\begin{aligned}&[x_1 + y_1, \cdots, x_n + y_n; z_1, \cdots, z_n; \omega]\\&\quad = [x_1, \cdots, x_n; z_1, \cdots, z_n; \omega] + [y_1, \cdots, y_n; z_1, \cdots, z_n; \omega].\end{aligned}$$

Here all x_i, y_i, z_i in X.

DEFINITION 4.12. An algebra $H \in \Theta$ is called a CD-algebra if all CD-identities hold in H.

PROPOSITION 4.8. *If the algebra G is a CD-algebra and G has zero-divisors, then G is not stable in Θ^G.*

THEOREM 4.2. *Let G be a CD-algebra in Θ^G. Then G is stable if and only if G has no zero divisors.*

CD-conditions are fulfilled in groups, associative and Lie algebras. In these cases the theorem was proved by A. Berzins. The result is true also in arbitrary rings.

4.7. Theorems on zeroes. We consider the case when the algebra G is algebraically closed in the variety Θ^G. Let first the variety Θ be an arbitrary variety, i.e., not necessarily the variety of group-based algebras.

If T is a maximal faithful congruence in the algebra $W = G{*}W_0$, then $T = \operatorname{Ker} \mu$ for some point μ.

A G-algebra G is a simple algebra since it has no faithful congruences.

DEFINITION 4.13. G-algebra H is called *semisimple* if it is approximated by simple G-algebras. G-algebra H is called *locally semisimple*, if every finitely generated subalgebra in H is simple.

THEOREM 4.3. *Let algebra G be algebraically closed in the variety Θ^G and let H be a faithful locally semisimple G-algebra. Then for every faithful congruence T in $W = W(X)$ with finite X, the congruence T''_H is the intersection of all maximal faithful congruences in W which contain T.*

PROOF. By definition, T''_H is the intersection of all $\operatorname{Ker} \mu$ where $T \subset \operatorname{Ker} \mu$. For every such μ, take in H the image $Im\mu = H_1$. The algebras H_1 and $W/Ker\mu$ are faithful and semisimple. Therefore, the congruence $\operatorname{Ker} \mu$ is the intersection of some maximal faithful congruences. Then T''_H is an intersection of maximal faithful congruences.

Let now T_1 be a maximal faithful congruence in W containing T. Show that T_1 contains T''_H. We have $T_1 = \operatorname{Ker} \mu$ for some point $\mu : W \to G$. This point μ is a homomorphism $W \to H$. This means that $T_1 = T''_{1H}$. Since $T \subset T_1$, then $T''_H \subset T''_{1H}$. This implies the theorem. □

COROLLARY 4.4. *If H_1 and H_2 are faithful locally simple G-algebras, and G is algebraically closed, then H_1 and H_2 are geometrically equivalent.*

Let us come back to the situation of group-based algebras. Define relative nilpotents, i.e., G-nilpotents.

Let (H, h) be a $G - \Omega$-group and $a \in H$.

DEFINITION 4.14. An element a is called *G-nilpotent* if in an Ω-group $(a)^G$ there is a series of G-invariant ideals

$$0 = U_0 \subset U_1 \cdots \subset U_n = (a)^G$$

with the abelian factors U_{i+1}/U_i. If, moreover, $[(a)^G, U_{i+1}] \subset U_i$, then a is said to be a strictly G-nilpotent element.

In groups (Ω is empty), an element a is a nilpotent element if the group $(a)^G$ is solvable. If $(a)^G$ is nilpotent, then a is strictly nilpotent.

In associative rings both these notions coincide. Namely, an element $a \in H$ is a nilpotent if there exists n such that

$$b_0 a b_1 a b_2 a \cdots b_{n-1} a b_n = 0$$

for any $b_i = h(g_i)$, $g_i \in G$.

If the ring is commutative, then this condition is equivalent to $a^n = 0$. If the algebra H is a domain, then it has no non-zero nilpotent elements.

Given a G-algebra H, denote $N(H)$ the ideal in H generated by all its G-nilpotent elements. This is the N-radical of H. The function N can be iterated and this gives rise to the upper N-radical. Denote it by $\tilde{N}(H)$. There are no non-trivial nilpotent elements in $H/\tilde{N}(H)$.

If U is an ideal in H, then denote by $\sqrt{U}$ the inverse image in H of the radical $\tilde{N}(H/U)$.

It is clear that if a G-homomorphism $\mu : H \to H'$ is given and $U \subset \operatorname{Ker} \mu$ then $\sqrt{U}$ is also contained in $\operatorname{Ker} \mu$ in case H' is a domain. This implies

PROPOSITION 4.9. *Let U be an ideal in the free G-algebra $W = G * W_0$, and let H be a G-domain. Then*

$$\sqrt{U} \subset U''_H.$$

We are looking for conditions which imply the equality in the formula above.

DEFINITION 4.15. A variety is called *special* if for every finitely generated G-algebra H, its radical $\tilde{N}(H)$ is an intersection of maximal faithful congruences in H.

THEOREM 4.3′. *If an algebra G is algebraically closed and the variety Θ^G is special then for every faithful G-domain H the equality*

$$\sqrt{U} = U''_H$$

holds for every ideal U and every W.

In fact, this theorem, as well as Theorem 4.3, immediately follows from the definitions. The problem is to study the situations when all this is applicable.

Here the ideas of radical, semisimplicity and algebraic closure naturally intersect in Theorems 4.3 and 4.3′.

Lecture 5

Isomorphisms of Categories of Algebraic Sets

Contents

5.1. Correct isomorphism. Let H_1 and H_2 be two algebras in the variety Θ. We are looking for an isomorphism of categories

$$F : K_\Theta(H_1) \to K_\Theta(H_2).$$

Denote $\Theta_1 = \mathrm{Var}(H_1)$ and $\Theta_2 = \mathrm{Var}(H_2)$.

There is a one-to-one correspondence between the isomorphism F and the isomorphism

$$F : K_{\Theta_1}(H_1) \to K_{\Theta_2}(H_2)$$

which is also denoted by F.

Let $W^1 = W^1(X)$ and $W^2 = W^2(X)$ be the free algebras in Θ_1 and Θ_2 respectively, X is a finite set.

DEFINITION 5.1. An isomorphism F is called a *correct isomorphism* if

1. $F(\mathrm{Hom}(W^1(X), H_1) = \mathrm{Hom}(W^2(Y), H_2)$ for some Y such that $|Y| = |X|$.
2. If $[s] : A \to \mathrm{Hom}(W^1(X), H_1)$ is an identity embedding, then

$$F([s]); F(A) \to \mathrm{Hom}(W^2(Y), H_2)$$

is also an identity embedding.

Categories $K_\Theta(H_1)$ and $K_\Theta(H_2)$ are correctly isomorphic if there exists a correct isomorphism of categories $K_{\Theta_1}(H_1)$ and $K_{\Theta_2}(H_2)$.

Let us translate this definition into the language of dual categories. Consider the diagram

$$\begin{array}{ccc} K_{\Theta_1}(H_1) & \xrightarrow{F} & K_{\Theta_2}(H_2) \\ {\scriptstyle F_1}\uparrow & & \uparrow{\scriptstyle F_2} \\ C_{\Theta_1}(H_1) & \xrightarrow[\Phi]{} & C_{\Theta_2}(H_2) \end{array}$$

Here F_1 and F_2 are dual isomorphisms and Φ is an isomorphism of categories induced by F.

We consider the correctness condition in terms of the functor Φ. First, let us connect directly the functors Φ and F.

Let us take objects (X, A) in $K_{\Theta_1}(H_1)$ and (Y, B) in $K_{\Theta_2}(H_2)$.

Let $T = A'_{H_1}$ and $T^* = B'_{H_2}$, $A = T'_{H_1}$, and $B = T^{*\prime}_{H_2}$. Then $F(X, A) = (Y, B)$ if and only if $\Phi(W^1(X)/T) = W^2(Y)/T^*$.

Consider morphisms. Let $[s] : (X_1, A_1) \to (X_2, A_2)$ be a morphism in $K_{\Theta_1}(H_1)$ and

$$\bar{s} : W^1(X_2)/T_2 \to W^1(X_1)/T_1$$

be the corresponding morphism in $C_{\Theta_1}(H_1)$.

In the categories $K_{\Theta_2}(H_2)$ and $C_{\Theta_2}(H_2)$, we have

$$[s_1] : (Y_1, B_1) \to (Y_2, B_2)$$

and

$$\bar{s_1} : W^2(Y_2)/T_2^* \to W^2(Y_1)/T_1^*.$$

Now, note that

$$F[s] : F((X_1, A_1)) \to F((X_2, A_2)$$

is

$$[s_1] : (Y_1, B_1) \to (Y_2, B_2)$$

if and only if

$$\Phi(\bar{s}) : \Phi(W^1(X_2)/T_2) \to \Phi(W^1(X_1)/T_1)$$

equals to

$$\bar{s_1} : W^2(Y_2)/T_2^* \to W^2(Y_1)/T_1^*.$$

DEFINITION 5.2. An isomorphism $\Phi : C_{\Theta_1}(H_1) \to C_{\Theta_2}(H_2)$ is a correct isomorphism if

1. $\Phi(W^1(X)) = W^2(Y)$ for some Y, $|Y| = |X|$.
2. If $\mu_X : W^1(X) \to W^1(X)/T_1$ is the natural homomorphism in $C_{\Theta_1}(H_1)$, then

$$\Phi(\mu_X) : \Phi(W^1(X)) \to \Phi(W^1(X)/T_1)$$

is also a natural homomorphism,

$$\Phi(\mu_X) = \mu_Y : W^2(Y) \to W^2(Y)/T_2.$$

The main result here is as follows.

THEOREM 5.1. *An isomorphism $F : K_{\Theta_1}(H_1) \to K_{\Theta_2}(H_2)$ is a correct isomorphism if and only if $\Phi : C_{\Theta_1}(H_1) \to C_{\Theta_2}(H_2)$ is a correct isomorphism.*

5.2. Isomorphism, similarity, equivalence. In [**Pl3**], the following two theorems are proved.

THEOREM 5.2. *The categories $K_\Theta(H_1)$ and $K_\Theta(H_2)$ is correctly isomorphic if and only if the algebras H_1 and H_2 are geometrically similar.*

Let now H_1 and H_2 be abelian groups, each of which generates the variety of all abelian groups, and let Θ be the variety of all groups.

THEOREM 5.3. *Categories $K_\Theta(H_1)$ and $K_\Theta(H_2)$ are correctly isomorphic if and only if H_1 and H_2 are geometrically equivalent.*

Let H_1 and H_2 be groups such that $\mathrm{Var}(H_1) = \mathrm{Var}(H_2) = \Theta = \mathrm{Grp}$.

CONJECTURE. Categories $K_\Theta(H_1)$ and $K_\Theta(H_2)$ are correctly isomorphic if and only if the groups H_1 and H_2 are geometrically equivalent.

The crucial point here is Problem 5.3 below on Aut(End(F)), where F is free, is stated earlier.[2]

[2]Since Problem 5.3 is solved positively (E. Formanek [**For**]), the conjecture is true.

Problems

PROBLEM 5.1. What is the situation in semigroups, i.e., are H_1 and H_2 commutative semigroups?

PROBLEM 5.2. What is the situation in modules over commutative rings?

PROBLEM 5.3. What is the relation between arbitrary isomorphisms of the categories $K_\Theta(H_1)$ and $K_\Theta(H_2)$ and correct isomorphisms? This problem depends on the choice of Θ and $H \in \Theta$.

The following two problems are of a very general nature.

PROBLEM 5.4. Consider the question of equivalence of the categories $K_\Theta(H_1)$ and $K_\Theta(H_2)$.[3]

PROBLEM 5.5. When are the categories K_{Θ_1} and K_{Θ_2} isomorphic or equivalent? Consider, in particular, the case when Θ_1 and Θ_2 are equivalent categories. Consider also the cases $\Theta_1 = \Theta^{G_1}$, $\Theta_2 = \Theta^{G_2}$ for the different algebras G_1 and G_2 in the given Θ.

Note that in [**Pl4**] and [**Pl1**] there is an invariant approach (without equations) to the category of algebraic varieties and it is quite natural to proceed here from the idea of equivalence of categories. There are some results in this direction.

5.3. Perfect and semiperfect varieties of algebras. In Lecture 1, the theorem on the isomorphism of the categories of type $K_P(L)$ was formulated. The notion of semiisomorphism of P-algebras played key role in this theorem. All this was done for the variety Var-P.

Now we consider the general situation. We take varieties of the type $\Theta^G = \Theta$-G and consider the question on isomorphisms of the categories of the type $K_{\Theta^G}(H)$ where H is a faithful G-algebra. We are going to apply Theorem 5.1 in the G-algebra case.

Fix a variety Θ and $G \in \Theta$. Consider the variety Θ^G and the category $(\Theta^G)^0$ of free in Θ^G algebras of the form $W(X) = G * W_0(X)$ with different finite X.

We are interested in automorphisms of this category, and want to find out when all of them are inner or semi-inner.

We assume that the condition $(*)$ is fulfilled. This means that the algebra G generates the variety Θ^G. We apply results on automorphisms of the category $(\Theta^G)^0$ to the problem of isomorphism of categories of algebraic varieties with faithful G-algebras H.

With the given conditions, the category $(\Theta^G)^0$ is dual to the category of affine spaces $K^0_{\Theta^G}(G)$, while the last one is connected with the category of polynomial maps Pol-G.

Let $\varphi : (\Theta^G)^0 \to (\Theta^G)^0$ be an automorphism of the category $(\Theta^G)^0$. The automorphism τ of the category $K^0_{\Theta^G}(G)$ corresponds to φ. The automorphisms φ and τ are connected by the following rules.

First of all,

$$\tau(\mathrm{Hom}(W(X), G) = \mathrm{Hom}(\varphi(W(X)), G).$$

[3]This problem is considered in [**Pl11**].

Let, further, $s : W(Y) \to W(X)$ be a morphism in $(\Theta^G)^0$. In the category $K^0_{\Theta^G}(G)$,

$$\tilde{s} : (\mathrm{Hom}(W(X), G) \to \mathrm{Hom}(W(Y), G))$$

corresponds to s. For every $\nu : W(X) \to G$, we have $\tilde{s}(\nu) = \nu s$. Then

$$\tau(\tilde{s})(\nu) = \nu\varphi(s) = \tilde{\varphi(s)}(\nu)$$

for every point $\nu : \varphi(W(X)) \to G$, $\tau(\tilde{s}) = \tilde{\varphi(s)}$. This gives one-to-one correspondence between φ and τ.

5.3.1. We would like to explain that every automorphism τ of the category of affine spaces is, in a sense, a quasi-inner automorphism. Denote the category of affine spaces by $K^0_{\Theta^G}(G) = K^0_{\Theta^G}$.

Recall that every point $\nu \colon W \to G$ satisfies the commutative diagram

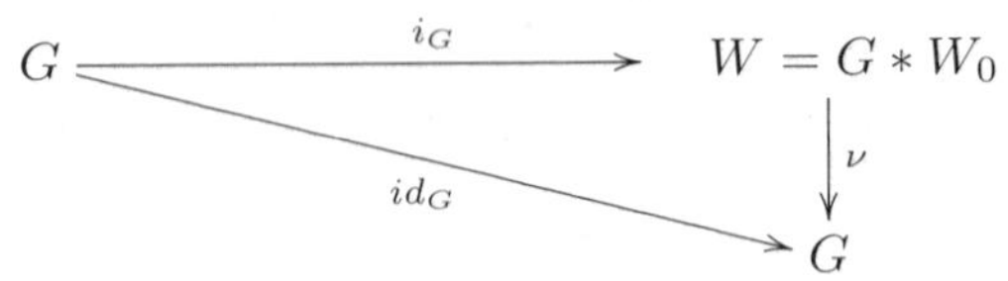

Let $\nu \colon W \to G$ be a point. Consider the homomorphism defined by

$$W \xrightarrow{\nu} G \xrightarrow{i_G} W.$$

Check that $i_G\nu \colon W \to W$ is an endomorphism of the algebra W in the variety Θ^G. We have to check that there is a commutative diagram

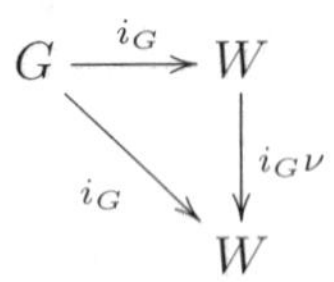

We have

$$(i_G\nu)i_G = i_G(\nu i_G) = i_G id_G = i_G.$$

Denote $i_G \cdot \nu = \bar{\nu}$. For every $w \in W$, the element $\bar{\nu}(w) = i_G(\nu(w))$ is a constant in W and, therefore, $\bar{\nu}$ is called a *constant endomorphism*. Every endomorphism $s \colon W \to W$ leaves constants and, therefore, $(s\bar{\nu})(w) = \bar{\nu}(w)$, $s\bar{\nu} = \bar{\nu}$. Endomorphism $\bar{\nu}$ defines the map

$$\tilde{\bar{\nu}} \colon \mathrm{Hom}(W, G) \longrightarrow \mathrm{Hom}(W, G).$$

Note that for every $\nu_0 \colon W \longrightarrow G$, we have $\tilde{\bar{\nu}}(\nu_0) = \nu$. Indeed, $\tilde{\bar{\nu}}(\nu_0) = \nu_0 \cdot \bar{\nu} = \nu_0(i_G\nu) = (\nu_0 i_G)\nu = id_G \cdot \nu = \nu$.

Thus, the map $\tilde{\bar{\nu}}$ takes an arbitrary ν_0 to one and the same element ν, and, therefore $\tilde{\bar{\nu}}$ is a constant map.

Consider an arbitrary $\sigma \colon W \to W$, such that $s\sigma = \sigma$ for every $s \colon W \to W$. Since one can take for s an endomorphism taking w to a constant, σ takes any w to a constant.

It can be shown that to every such σ there is a one-to-one correspondence $\nu : W \to G$ such that $\sigma = \bar{\nu}$.

Let now $\varphi \colon (\Theta^G)^0 \to (\Theta^G)^0$ be an automorphism of the category $(\Theta^G)^0$, and let $\tau : K^0_{\Theta^G} \to K^0_{\Theta^G}$ be the corresponding automorphism in the category of affine spaces.

In particular, the constant

$$\tilde{\bar{\nu}}\colon \operatorname{Hom}(W,G) \longrightarrow \operatorname{Hom}(W,G)$$

is characterized by the condition $s\bar{\nu} = \bar{\nu}$ for every s. This condition can be rewritten as $\tilde{\bar{\nu}}\tilde{s} = \tilde{\bar{\nu}}$. Apply τ to the equality $\tilde{\bar{\nu}}\tilde{s} = \tilde{\bar{\nu}}$. We have

$$(\tilde{\bar{\nu}}\tilde{s})^\tau = \tilde{\bar{\nu}}^\tau \cdot \tilde{s}^\tau = \tilde{\bar{\nu}}^\tau : \operatorname{Hom}(W^1,G) \longrightarrow \operatorname{Hom}(W^1,G).$$

Since $\tilde{s}^\tau$ is arbitrary, $\tilde{\bar{\nu}}^\tau$ is a constant, i.e., $\tilde{\bar{\nu}}^\tau = \tilde{\bar{\nu}}_1$, where $\nu_1\colon \varphi(W) = W^1 \to G$ is a point. Denote $\nu_1 = \mu(\nu)$. The map

$$\mu = \mu_W\colon \operatorname{Hom}(W,G) \to \operatorname{Hom}(W^1,G) = \operatorname{Hom}(\varphi(W),G)$$

is a bijection. For every point $\nu\colon W \to G$, there is $\tilde{\bar{\nu}}^\tau = \widetilde{\overline{\mu_W(\nu)}}$.

DEFINITION 5.3. An automorphism τ of the category $K^0_{\Theta(G)}$ is called *quasi-inner* if for an arbitrary $\tilde{s}\colon \operatorname{Hom}(W^2,G) \to \operatorname{Hom}(W^1,G)$, the formula

$$\tilde{s}^\tau = \mu_{W^1}\tilde{s}\mu^{-1}_{W^2}$$

takes place.

THEOREM 5.4 (see [**Be2**] for Var-P). *Every automorphism τ of the category $K^0_{\Theta G}$ is quasi-inner, i.e., for every $s : W(Y) \to W(X)$ we have*

$$\tilde{s}^\tau = \mu_{W^1}\tilde{s}\mu^{-1}_{W^2} : \operatorname{Hom}(\varphi(W(X),G) \to \operatorname{Hom}(\varphi(W(Y),G).$$

Note that the definitions of inner and semi-inner automorphism in the category $(\Theta^G))^0$ is well correlated with the definition of quasi-inner automorphism in the category $K^0_{\Theta G}$. Theorem 5.4 plays a key role in the proofs of the theorems for the conditions on an automorphism φ to be inner or semi-inner.

5.3.2. We define now a substitutional automorphism of the category $(\Theta^G))^0$. Let, first, φ be a substitution on the objects $W = W(X)$ of the category $(\Theta^G))^0$. Suppose that if $\varphi(W(X)) = W(Y)$ then $|X| = |Y|$. Let $X = \{x_1, \cdots, x_n\}$, $Y = \{y_1, \cdots, y_n\}$. Define an isomorphism $s_W\colon W(X) \to \varphi(W(Y))$ by $s_W(x_i) = y_i$, $i = 1, \cdots, n$. If, further, $\nu\colon W^1 \to W^2$ is a morphism, then set

$$\varphi(\nu) = s_{W^2}\nu s^{-1}_{W^1} \colon \varphi(W^1) \to \varphi(W^2).$$

From the substitution φ, we come to the automorphism $\bar{\varphi}$. We call $\bar{\varphi}$ a *substitutional automorphism.*

PROPOSITION 5.1. *Every automorphism φ in the category $(\Theta^G))^0$ can be presented in the form*

$$\varphi = \varphi^2\varphi^1$$

where φ^2 is a substitutional automorphism depending on φ, and $\varphi^1 = (\varphi^2)^{-1}\varphi$ does not change objects.

A similar decomposition can be considered for any variety, not necessarily of the form Θ^G. Such a decomposition of an automorphism of the category Θ^0 implies, under some conditions, the decomposition of the corresponding relation of similarity for algebras in Θ.

5.3.3. Let us pass to the category of polynomial maps Pol-G and let $s : W(Y) \to W(X)$, where $X = \{x_1, \cdots, x_n\}$, $Y = \{y_1, \cdots, y_n\}$, be given. Consider the commutative diagram (Lecture 1)

$$\begin{array}{ccc} \mathrm{Hom}(W(X), G) & \xrightarrow{\tilde{s}} & \mathrm{Hom}(W(Y), G) \\ \Big\downarrow{\scriptstyle \alpha_X} & & \Big\downarrow{\scriptstyle \alpha_Y} \\ G^{(n)} & \xrightarrow[s^\alpha]{} & G^{(m)} \end{array}$$

We have $s^\alpha = \alpha_Y \tilde{s} \alpha_X^{-1}$, $\tilde{s} = \alpha_Y^{-1} s^\alpha \alpha_X$ (Lecture 1). Let φ be an automorphism in Θ^{G^0} and φ does not change objects. The corresponding automorphism τ also acts identically on objects. It corresponds to an automorphism τ^α of the category Pol-G which also preserves the objects. Let $s^\alpha : G^{(n)} \to G^{(m)}$ be given. We set $\tau^\alpha(s^\alpha) = \alpha_Y \tau(\tilde{s}) \alpha_X^{-1}$, where $\tilde{s} = \alpha_Y^{-1} s^\alpha \alpha_X$. Here τ^α is an automorphism of the category Pol-G. To an automorphism τ corresponds a function μ, defining τ as a quasi-inner automorphism. Here

$$\mu_X = \mu_{W(X)} = \mu_W \colon \mathrm{Hom}(W, G) \to \mathrm{Hom}(W, G).$$

This is a bijection and $W = \varphi(W)$. A bijection

$$\mu_n \colon G^{(n)} \to G^{(n)}$$

corresponds to μ_X. Take $a = (a_1, \ldots, a_n) \in G^{(n)}$. Let

$$\mu_n(a) = \alpha_X(\mu_X(\alpha_X^{-1}(a)) = \alpha_X \mu_X \alpha_X^{-1}(a).$$

Then

$$\mu_n = \alpha_X \mu_X \alpha_X^{-1}, \ \mu_X = \alpha_X^{-1} \mu_n \alpha_X.$$

For the given homomorphism $s \colon W(Y) \to W(X)$, $|X| = n, |Y| = m$, we have

$$\tilde{s} \colon \mathrm{Hom}(W(X), G) \to \mathrm{Hom}(W(Y), G)$$

and

$$\tilde{s}^\tau = \mu_Y \tilde{s} \mu_X^{-1} = \tilde{s}_1 \colon \mathrm{Hom}(W(X), G) \to \mathrm{Hom}(W(Y), G).$$

Here $\mu_Y \tilde{s} = \tilde{s}_1 \mu_X$. In the category Pol-$G$, this equality has the form

$$\mu_m s^\alpha = s_1^\alpha \mu_n.$$

5.3.4. Let us make some general remarks on the transition $W \to \varphi(W)$ in arbitrary Θ. Here φ can change objects. Consider $s \colon W(X) \to W(X)$, $X = \{x_1, \ldots, x_n\}$ and present it as $s = (s_1, \ldots, s_n)$, where all s_i, $i = 1, \ldots, n$ are morphisms $W(x) \to W(X)$. Here s_i are defined by the condition

$$s_i(x) = s(x_i) = w_i(x_1, \ldots, x_n) = w_i.$$

The presentation $s = (s_1, \ldots, s_n)$ depends on the basis X. We write $s = (w_1, \ldots, w_n)$. Consider an automorphism $\varphi \colon \Theta^0 \to \Theta^0$. What can be said about the equality

$$\varphi(s) = (\varphi(s_1), \ldots, \varphi(s_n))?$$

We will see that application of φ preserves the corresponding presentation, but this is a presentation in some special base, connected with φ.

Consider a system of injections $(\varepsilon_1, \ldots, \varepsilon_n)$,

$$\varepsilon_i \colon W(x) \to W(X).$$

DEFINITION 5.4. We say that $(\varepsilon_1, \ldots, \varepsilon_n)$ freely defines an algebra W, if for arbitrary morphisms $f_1, \ldots, f_n$, $f_i \colon W(x) \to W(X)$, there exist a unique $s \colon W(X) \to W(X)$, such that $f_i = s\varepsilon_i$, $i = 1, 2, \ldots, n$.

PROPOSITION 5.2. *A collection $(\varepsilon_1, \ldots, \varepsilon_n)$ freely defines an algebra W if and only if the elements $\varepsilon_1(x), \ldots, \varepsilon_n(x)$ freely generate W.*

Consider, further, an automorphism φ of the category Θ^0 with the condition $\varphi(W(x)) = W(y)$.

PROPOSITION 5.3. *Let the set of morphisms $(\varepsilon_1, \ldots, \varepsilon_n)$, $\varepsilon_i \colon W(x) \to W(X)$ freely define $W = W(X), X = \{x_1, \ldots, x_n\}$. Then the set $(\varphi(\varepsilon_1), \ldots, \varphi(\varepsilon_n))$,*

$$\varphi(\varepsilon_i) \colon \varphi(W(x)) = W(y) \to \varphi(W(X)) = W(Y)$$

freely defines $W(Y)$.

DEFINITION 5.5. A variety Θ is called a *regular variety*, if for any free algebra $W = W(X), |X| = n$, every other system of free generators of W also consists of n elements.

Now we can state that if $W(Y) = \varphi(W(X))$, then $|Y| = |X|$ if Θ is regular. Fix $(\varepsilon_1, \ldots, \varepsilon_n), \varepsilon_i(x) = x_i$, take $(\varphi(\varepsilon_1), \ldots, \varphi(\varepsilon_n))$ and $y_1' = \varphi(\varepsilon_1)(y), \ldots, y_n' = \varphi(\varepsilon_n)(y), Y' = \{y_1', \ldots, y_n'\}$.

PROPOSITION 5.4. *If an endomorphism $s \colon W \to W$ in the basis $X = \{x_1, \ldots, x_n\}$ has a presentation $s = (s_1, \ldots, s_n)$, then in the base Y' we have the presentation $\varphi(s) = (\varphi(s_1), \ldots, \varphi(s_n))$.*

A transition from the base X to another base determines an automorphism used below in the proof of Theorem 5.5.

DEFINITION 5.6. A variety Θ^G is called *perfect* if every automorphism $\varphi : (\Theta^G)^0 \to (\Theta^G)^0$ is inner. If every φ is semi-inner, then Θ^G is called *semiperfect*.

We consider conditions which provide Θ^G to be perfect or semiperfect.

To every decomposition $\varphi = \varphi^2\varphi^1$ from Proposition 5.1 corresponds a decomposition $\tau = \tau^2\tau^1$. The function μ corresponds to τ and we have a pair (μ, φ). There are the corresponding pairs for τ^2 and τ^1. Then $(\mu, \varphi) = (\mu^2\mu^1, \varphi^2\varphi^1)$ and $(\mu^2\mu^1)_W = \mu^2_{\varphi^1(W)}\mu^1_W$. This decomposition rule is used in the proofs of the theorems.

Let an automorphism φ not change the objects. For every $W = W(X)$, the automorphism φ induces an automorphism of the semigroup $\operatorname{End} W$.

Take $W = W(x)$, $X = \{x\}$ and let φ_0 be the corresponding induced automorphism of the semigroup $\operatorname{End}(W)$.

DEFINITION 5.7. An automorphism ψ of the semigroup $\operatorname{End} W$ is called *semi-inner* if ψ is connected with the diagram

$$\begin{array}{ccc} G & \xrightarrow{i_G} & W = G * W_0(x) \\ {\scriptstyle\sigma}\big\downarrow & & \big\downarrow{\scriptstyle s} \\ G & \xrightarrow{i_G} & W = G * W_0(x) \end{array}$$

in such a way that for every endomorphism $\nu : W(x) \to W(x)$, the equality $\varphi(\nu) = s\nu s^{-1}$ holds.

Here (σ, s) is a semiautomorphism of the algebra $W = G * W_0$. The same ψ is called inner if one takes $\sigma = 1$.

DEFINITION 5.8. A semigroup $\operatorname{End} W$ is called perfect if each of its automorphism is inner. A semigroup $\operatorname{End} W$ is called semiperfect if each of its automorphism is semi-inner.

Let now ψ be an automorphism of the semigroup $\operatorname{End} W$ where $W = W(x)$. An automorphism $\bar{\psi} : (\Theta^G)^0 \to (\Theta^G)^0$ corresponds to ψ. This $\bar{\psi}$ is constructed in such a way that if ψ is inner or semi-inner, then $\bar{\psi}$ is also inner or semi-inner.

Return to the initial $\varphi : (\Theta^G)^0 \to (\Theta^G)^0$ and to the corresponding φ_0. Then $\bar{\varphi}_0$ as well as φ preserves the objects. This $\bar{\varphi}_0$ also induces φ_0 in $\operatorname{End}(W)$.

Let $\varphi = \varphi_1\bar{\varphi}_0$, $\varphi_1 = \varphi \cdot \bar{\varphi}_0^{-1}$, φ_1 preserves the objects and, besides, does not change constant endomorphisms of the algebra $W(x)$. Decomposition of φ gives rise to decomposition of τ, $\tau = \tau_1\tau_2$, where τ_2 corresponds to the automorphism $\bar{\varphi}_0$. If now μ is a function for τ, then $\mu = \mu^1\mu^2$, $\mu_W = \mu^1_W \cdot \mu^2_W$. Let now $W_0 = W(x)$. Then $\mu_{W_0} = \mu^1_{W_0} \cdot \mu^2_{W_0}$. But $\mu_{W_0} = \mu^2_{W_0}$ since φ and $\bar{\varphi}_0$, τ_1 and τ_2 coincide with W_0. Therefore $\mu^1_{W_0} = 1$. This precisely means that φ_1 does not change constant endomorphisms of the algebra $W(x)$. Applying this fact and Proposition 5.4, it can be proved that φ_1 is an inner automorphism.

The main theorem here is the following.

THEOREM 5.5. *If the semigroup* $\operatorname{End} W(x)$ *is perfect in* Θ^G*, then the variety* Θ^G *is perfect too. If the semigroup* $\operatorname{End} W(x)$ *is semiperfect then the variety* Θ^G *is semiperfect.*

The proof of this theorem uses various reductions, based on transitions from φ to τ, from τ to τ^α, and to the corresponding functions μ and bijections of the form μ_X and μ_n.

Using Theorem 5.2 and Lecture 3, we have also

THEOREM 5.5$'$. *Let* H_1 *and* H_2 *be faithful* G*-algebras. Consider the categories* $K_{\Theta^G}(H_1)$ *and* $K_{\Theta^G}(H_2)$*. Then*

1. *If the semigroup* $\operatorname{End} W(x)$ *is semiperfect, then the categories are isomorphic if and only if* H_1 *and* H_2 *are geometrically equivalent up to a semiisomorphism.*
2. *If the semigroup* $\operatorname{End} W(x)$ *is perfect, then the isomorphism of categories coincides with geometrical equivalence of* H_1 *and* H_2*.*

Automorphisms of categories of free algebras of varieties are studied in [**MPP**].

5.4. Other results. A. Berzins proved [**Be2**] that:

1. If P is an infinite field, and $P[x]$ is analgebra of polynomials with one variable x, then the semigroup $\operatorname{End} P[x]$ is semiperfect, i.e., every its automorphism is semi-inner.
1. Let F be a free non-commutative group. The semigroup $\operatorname{End}(F * \{x\})$ is semiperfect.

So we have

THEOREM 5.6. *The variety* Var-P *is semiperfect. If the field* P *does not have automorphisms, then* Var-P *is perfect.*

THEOREM 5.7. *The variety* Grp-F *is semiperfect.*

And, further, we have:

1. If L_1 and L_2 are two extensions of the field P, then the categories $K_P(L_1)$ and $K_P(L_2)$ are correctly isomorphic if and only if L_1 and L_2 are geometrically equivalent up to a semiisomorphism.
2. If H_1 and H_2 are two faithful F-groups, then the categories $K_{\text{Grp-}F}(H_1)$ and $K_{\text{Grp-}F}(H_2)$ are isomorphic if and only if H_1 and H_2 are geometrically equivalent up to a semiisomorphism. It can be shown that in this case H_1 and H_2 are, in fact, equivalent.

5.5. Problems.

PROBLEM 5.6. What is the situation for Lie F-algebras, where F is a free Lie algebra?

PROBLEM 5.7. What is the situation for associative F-algebras, where F is a free associative algebra or an infinite dimensional over its center skew field [**Pl11**]?

It should be noted that Theorem 5.5 cannot be applied to associative algebras over a field, since a field P does not generate the whole variety of associative algebras over P and the condition $(*)$ is not fulfilled.

This lecture concludes the part devoted to equational algebraic geometry. We present now a general view of equational algebraic geometry.

First of all, note that there are problems which relate to the universal theory. Some of them have been mentioned. However, the principal idea is to consider situations in the various special Θ and special $H \in \Theta$.

The subject of algebraic geometry in groups is an area of growing interest, see [**BMR1**], [**BMR2**], [**KM**] [**MR1**], [**MR2**], [**BMRo**], [**Se**] and others. It is quite reasonable to expect a similar breakthrough in Lie algebras and semigroups.

For the case of associative algebras over a field or over some other algebra of constants, it is necessary to clarify how all this is connected with the theory which used to be called non-commutative algebraic geometry. In particular, it would be quite reasonable to compare the notions of Noetherian variety of algebras and geometrically Noetherian algebras with the notion of Noetherian scheme in the non-commutative algebraic geometry.

One has to distinguish also the cases of Noetherian and non-Noetherian non-commutative geometry. An algebraic set $A \subset Hom(W, H)$ is called *correct* if for every system of equations T in W such that $T' = A$ there exists a finite $T_0 \subset T$, such that $T_0' = A$. In the opposite case, the set A is called *non-correct.* A set A is called *almost correct* if A is non-correct but all its proper algebraic subsets are correct. If an algebra H is not geometrically Noetherian, then there exist non-correct algebraic sets over H. Each non-correct algebraic set contains an almost correct subset. This follows from the following observation. Take in the lattice $\mathrm{Alv}_H(W)$ an infinite descending system A_α, $\alpha \in I$, consisting of non-correct algebraic sets. Denote $A = \bigcap_{\alpha \in I} A_\alpha$. Then A is also non-correct algebraic set. Indeed, take $T_\alpha = A'_\alpha$, $T'_\alpha = A_\alpha$, and let T be $\bigcup T_\alpha$. Then $T' = A$ and $A' = T''$. Suppose that the set A is correct and $A = T_0'$ where T_0 is a finite subset in T. Since T_0 is a finite set, T_0 is contained in some T_α. Therefore, $T_0'' = T'' \subset T_\alpha$. This is impossible. Therefore, the set A is non-correct.

We have the following general problem. Which almost correct sets A arise for the given non-geometrically Noetherian algebra H? This question, first of all, is related to associative non-commutative algebras over a field.

The next natural object is modules over rings.

Note that there are some results for algebraic geometry for group representations [**KP**].

There are many problems associated with the solution of equations in specific groups, say in $GL_2(p)$, and with the investigation of the corresponding algebraic sets.

Note one more problem.

Let K be a ring with unity and Mod-K be the category of K-modules which is considered as a variety.

The notion of semiisomorphism makes sense also for this category.

PROBLEM 5.8. For which K can it be stated that all automorphisms of the category of free K-modules are semi-inner? [4]

Is the theory similar to the constructed one possible in this case?

The solution of this problem is connected with the solution of the problem of similarity for K-modules.

[4]This is true if a ring K is left Noetherian [**MPP**].

Lecture 6

Algebraic Geometry in First Order Logic

Contents

6.1. Introduction.

6.1.1. *The initial idea.* Equational algebraic geometry is a geometry whose algebraic sets are determined by the systems of equations of a special type $w \equiv w'$. These are equalities in logic.

Here we proceed from arbitrary formulas of elementary Θ-logic. A formula u is considered as an equation, and systems T of such generalized equations determine generalized algebraic sets. We call them *elementary sets*. The viewpoint of the Zariski topology, which is a main topology in such a geometry, is changed correspondingly.

For this new algebraic geometry, we need a special category of formulas which takes the role that the category Θ^0 plays in the equational theory.

This category of formulas assumes the transition from logic to algebraic logic. The logic is built according to some variety of algebras Θ.

6.1.2. *Algebra.* Fix a variety Θ. Now, keeping in mind applications from Lecture 7, we proceed from the situation when algebras are multi-sorted (not necessarily one-sorted). Fix for Θ a set of sorts Γ which is now finite, but in general it may be infinite. Every algebra $G \in \Theta$ is recorded as $G = (G_i, i \in \Gamma)$. Operation in the signature Ω is a Γ-sorted one. For every $\omega \in \Omega$, we have its type $\tau = \tau(\omega) = (i_1, \dots, i_n; j)$, $i, j \in \Gamma$. An operation ω of the type τ is a mapping $\omega : G_{i_1} \times \cdots \times G_{i_n} \to G_j$. All operations of the signature Ω satisfy some set of identities. This fixes the variety Θ of Γ-sorted Ω-algebras. Let us switch to homomorphisms in Θ and to free algebras. A homomorphism of algebras in Θ has the form

$$\mu = (\mu_i, i \in \Gamma)\colon G = (G_i, i \in \Gamma) \to G' = (G'_i, i \in \Gamma).$$

Here $\mu_i\colon G_i \to G'_i$ are mappings of sets, coordinated with operations in Ω. A congruence $\operatorname{Ker}\mu = (\operatorname{Ker}\mu_i, i \in \Gamma)$ is the kernel of a homomorphism μ.

We consider multi-sorted sets $X = (X_i, i \in \Gamma)$ and the corresponding free algebras in Θ

$$W = W(X) = (W_i, i \in \Gamma).$$

A set X and a free algebra W can be presented as a free union of all X_i and all W_i, respectively.

Every (multi-sorted) mapping $\mu : X \to G$ is extended up to a homomorphism $\mu : W \to G$. Denote the set of all such μ by $\operatorname{Hom}(W, G)$. If all X_i are finite, we treat this set as an affine space. The homomorphisms $\mu\colon W \to G$ are points of this space.

For the given $G = (G_i, i \in \Gamma)$ and $X = (X_i, i \in \Gamma)$, we can consider the set

$$G^X = (G_i^{X_i}, i \in \Gamma).$$

It is the set of mappings

$$\mu = (\mu_i, i \in \Gamma)\colon X \to G.$$

Then we have the natural bijection $\mathrm{Hom}(W, G) \to G^X$. More information about multi-sorted algebras can be found in [**Pl8**].

Now let us pass to the models. Fix some set of symbols of relations Φ. Every $\varphi \in \Phi$ has its type $\tau = \tau(\varphi) = (i_1, \dots, i_n)$. A relation, corresponding to φ, is a subset in the Cartesian product $G_{i_1} \times \dots \times G_{i_n}$. Now $\Phi\Theta$ denotes the class of all models (G, Φ, f), where $G \in \Theta$, and f is a realization of the set Φ in G. As for homomorphisms of models, they are homomorphisms of the corresponding algebras which are coordinated with relations.

6.1.3. *Logic.* We consider logic in the given variety Θ. For every finite X, it is determined a logical signature

$$L = L_X = \{\vee, \wedge, \neg, \exists x, \ \ x \in X\},$$

where X is $\bigcup_{i \in \Gamma} X_i$ for a finite Γ. We consider a set (more precisely, an L-algebra) of formulas $L\Phi W$ over the free algebra $W = W(X)$. This algebra is an L-algebra of formulas of FOL over the given Θ and Φ and for the given X.

First we define the atomic formulas. They are equalities of the form $w \equiv w'$ with $w, w' \in W$ of the same sort and the formulas $\varphi(w_1, \dots, w_n)$, where $w_i \in W$ and all w_i are positioned according to the type $\tau = \tau(\varphi)$ of the relations φ and to the sorts. The set of all atomic formulas we denote by $M = M_X$. Define $L\Phi W$ as the absolutely free L_X-algebra over M_X.

Let us consider another example of an L_X-algebra.

Given $W = W(X)$ and $G \in \Theta$, as before, we denote by $\mathrm{Bool}(W, G)$ the Boolean algebra $\mathrm{Sub}(\mathrm{Hom}(W, G))$ of all subsets in $\mathrm{Hom}(W, G)$. We define also the action of quantifiers in $\mathrm{Bool}(W, G)$. Let A be a subset in $\mathrm{Hom}(W, G)$ and $x \in X_i$ be a variable of the sort i. Then $\mu\colon W \to G$ belongs to $\exists x A$ if there exists $\nu\colon W \to G$ in A such that $\mu(y) = \nu(y)$ for every $y \in X$ of the sort j, $j \neq i$, and for every $y \in X_i$, $y \neq x$. Thus we get an L-algebra $\mathrm{Bool}(W, G)$.

Now let us define a mapping

$$\mathrm{Val}_f^X : M_X \to \mathrm{Bool}(W, G),$$

where f is a model which realizes the set Φ in the given G. If $w \equiv w'$ is an equality of the sort i, then we set:

$$\mu : W \to G \in \mathrm{Val}_f^X(w \equiv w') = \mathrm{Val}^X(w \equiv w')$$

if $\mu_i(w) = \mu_i(w')$ in G. Here the point μ is a solution of the equation $w \equiv w'$. If the formula is of the form $\varphi(w_1, \dots, w_n)$, then

$$\mu \in \mathrm{Val}_f^X(\varphi(w_1, \dots, w_n))$$

if $\varphi(\mu(w_1), \dots, \mu(w_n))$ is valid in the model (G, Φ, f). Here $\mu(w_j) = \mu_{i_j}(w_j)$, i_j is the sort of w_j. The mapping Val_f^X is uniquely extended up to the L-homomorphism

$$\mathrm{Val}_f^X : L\Phi W \to \mathrm{Bool}(W, G).$$

Thus, for every formula $u \in L\Phi W$ we have its value $\mathrm{Val}_f(u)$ in the model (G, Φ, f), which is an element in $\mathrm{Bool}(W, G)$.

Every formula $u \in L\Phi W$ can be viewed as an equation in the given model. The point $\mu\colon W \to G$ is the solution of the "equation" u if $\mu \in \mathrm{Val}_f(u)$.

6.1.4. *Geometrical Aspect.* In the L-algebra of formulas $L\Phi W$, $W = W(X)$, we consider its various subsets T, i.e., sets of formulas. We regard T also as a system of equations. On the other hand, we consider subsets A in the affine space $\mathrm{Hom}(W, G)$, i.e., elements of the L-algebra $\mathrm{Bool}(W, G)$. For each given model (G, Φ, f) and for these T and A, we establish the following Galois correspondence:

$$T^f = A = \bigcap_{u \in T} \mathrm{Val}_f(u)$$

$$A^f = T = \{u | A \subset \mathrm{Val}_f(u)\}.$$

Here $A = T^f$ is a locus of all points satisfying the system of equations $u \in T$. Every set A of such kind is said to be an *algebraic set* (or closed set, or elementary set), determined for the given model.

The set A can be treated also as a relation between elements of G, derived from equalities and relations of the basic set Φ. The relation $A = T^f$ belongs to the multi-sorted set

$$G^X = \{G_i^{X_i}, \ i \in \Gamma\}.$$

The set T of the form $T = A^f$ for some A is an f-closed set. For an arbitrary T, we have its closure $T^{ff} = (T^f)^f$ and for every $A \subset \mathrm{Hom}(W, G)$ we have the closure $A^{ff} = (A^f)^f$.

It is easy to understand that the following rule takes place:

The formula v belongs to the set T^{ff} if and only if the formula

$$\Big(\bigwedge_{u \in T} u\Big) \to v$$

holds in the model (G, Φ, f).

If the set T is infinite then the corresponding formula is infinitary.

Free algebras $W(X)$ in Θ with finite X are the objects of the category, denoted by Θ^0. Morphisms of this category $s\colon W(X) \to W(Y)$ are arbitrary homomorphisms of algebras. The category Θ^0 is a full subcategory in the category Θ.

Based on the first order logic in the given Θ, we intend to build a category which is similar, in a sense, to the category Θ^0 in the equational logic. Thus we pass from pure logic to algebraic logic. The sets of the type $T = A^f$ appear here more attractive.

6.2. Algebraic logic.

6.2.1. *The main idea.* Algebraic logic deals with algebraic structures, related to various logical structures, i.e., with logical calculi. Boolean algebras relate to classical propositional logic, Heyting algebras relate to non-classical propositional logic, Tarski cylindric algebras and Halmos polyadic algebras relate to FOL.

Every logical calculus assumes a set of formulas of the calculus, axioms of logic and rules of inference. On this basis the syntactical equivalence of formulas, well correlated with their semantical equivalence, is defined. The transition from pure logic to algebraic logic is based on treating logical formulas up to a certain equivalence, i.e., squeezed formulas. Such a transition leads to various special algebraic structures, in particular to the structures mentioned above.

As for logical calculi, usually they are associated with some infinite sets of variables. Denote such a set by X^0. In our situation, it is a multi-sorted set $X^0 = (X_i^0, i \in \Gamma)$. Keeping in mind algebraic geometry in logic, knowledge theory and its geometrical aspect, we will use a system of all finite subsets $X = (X_i,\ i \in \Gamma)$ of X^0 instead of this infinite universum. This leads to multi-sorted logic and multi-sorted algebraic logic. Every formula has a definite type (sort) X. Denote the new set of sorts by Γ^0. It is a set of all finite subsets of the initial set X^0.

6.2.2. *Halmos Categories.* Fix some variety of algebras Θ. This means that a finite set of sorts Γ, a signature $\Omega = \Omega(\Theta)$ connected with Γ, and a system of identities $Id(\Theta)$ are given.

Define Halmos categories for the given Θ.

First, for the given Boolean algebra B we define its existential quantifiers [**HMT**]. Existential quantifiers are the mappings $\exists\colon B \to B$ with the conditions:

1) $\exists 0 = 0$,
2) $a < \exists a$,
3) $\exists(a \wedge \exists b) = \exists a \wedge \exists b$, $0, a, b \in B$.

The universal quantifier $\forall\colon B \to B$ is defined dually:

1) $\forall 1 = 1$,
2) $a > \forall a$,
3) $\forall(a \vee \forall b) = \forall a \vee \forall b$.

Let B be a Boolean algebra and X a set. We say that B is a *quantifier X-algebra* if a quantifier $\exists x\colon B \to B$ is defined for every $x \in X$ and for every two elements $x, y \in X$ the equality $\exists x \exists y = \exists y \exists x$ holds true.

One may consider also *quantifier X-algebras B with equalities* over $W(X)$. In such algebras, to each pair of elements $w, w' \in W(X)$ of the same sort corresponds an element $w \equiv w' \in B$ satisfying the following conditions:

1) $w \equiv w$ is the unit in B;
2) $(w_1 \equiv w'_1 \wedge \ldots \wedge w_n \equiv w'_n) < (w_1 \ldots w_n \omega \equiv w'_1 \ldots w'_n \omega)$ where ω is an operation in Ω and everything is coordinated with the type of operation.

Now we will give a general definition of the Halmos category for the given Θ, which will be followed by examples.

A Halmos category H for an arbitrary finite $X = (X_i, i \in \Gamma)$ fixes some quantifier X-algebra $H(X)$ with X-equalities. $H(X)$ is an object in H.

The morphisms in H correspond to morphisms in the category Θ^0. Every morphism s_* in H has the form

$$s_*\colon H(X) \to H(Y),$$

where $s\colon W(X) \to W(Y)$ is a morphism in Θ^0. We identify s_* and s.

We assume that

1) The transitions $W(X) \to H(X)$ and $s \to s_*$ constitute a (covariant) functor $\Theta^0 \to H$.
2) Every $s_*\colon H(X) \to H(Y)$ is a Boolean homomorphism.
3) The coordination with the quantifiers is as follows:
   3.1) $s_1 \exists x a = s_2 \exists x a$, $\quad a \in H(X)$, if $s_1 y = s_2 y$ for every $y \in X$, $y \neq x$.
   3.2) $s \exists x a = \exists(sx)(sa)$ if $sx = y \in Y$ and $y = sx$ not in the support of sx', $x' \in X$, $x' \neq x$.
4) The following conditions describe coordination with equalities:

4.1) $s_*(w \equiv w') = (sw \equiv sw')$ for $s\colon W(X) \to W(Y)$, $w, w' \in W(X)$ are of the same sort.

4.2) $s^x_w a \wedge (w \equiv w') < s^x_{w'} a$ for an arbitrary $a \in H(X), x \in X, w, w'$ of the same sort with x in $W(X)$, and $s^x_w\colon W(X) \to W(X)$ is defined by the rule: $s^x_w(x) = w, sy = y, y \in X,\ \ y \neq x$.

So, the definition of Halmos category is given.

6.2.3. *The example* $\mathrm{Hal}_\Theta(G)$. Fix an algebra G in the variety Θ. Define the Halmos category $\mathrm{Hal}_\Theta(G)$ for the given G. Take a finite set X and consider the space $\mathrm{Hom}(W(X), G)$. We have defined the action of quantifiers $\exists x$ for all $x \in X$ in the Boolean algebra $\mathrm{Bool}(W(X), G)$. The equality $w \equiv w'$ in $\mathrm{Bool}(W(X), G)$ is defined as a diagonal, coinciding with the set of all $\mu : W(X) \to G$ for which $w^\mu = w'^\mu$ holds true. It is easy to check that in this case the algebra $\mathrm{Bool}(W(X), G)$ turns to be a quantifier X-algebra with equalities. We set

$$\mathrm{Hal}_\Theta(G)(X) = \mathrm{Bool}(W(X), G).$$

Let now $s\colon W(X) \to W(Y)$ be given in Θ^0. We have

$$\tilde{s}\colon \mathrm{Hom}(W(Y), G) \to \mathrm{Hom}(W(X), G)$$

defined by $\tilde{s}(\nu) = \nu s$ for any arbitrary $\nu\colon W(Y) \to G$.

Now if A is a subset in $\mathrm{Hom}(W(X), G)$, then $\nu \in s_*A = sA$ if and only if $\tilde{s}(\nu) = \nu s \in A$.

We have a mapping

$$s_*\colon \mathrm{Bool}(W(X), G) \to \mathrm{Bool}(W(Y), G)$$

which is a Boolean homomorphism. One can also check that s_* satisfies the conditions 3)–4). Thus, the Halmos category $\mathrm{Hal}_\Theta(G)$ is defined.

Note that to each s_* corresponds a conjugate mapping

$$s^*\colon \mathrm{Bool}(W(Y), G) \to \mathrm{Bool}(W(X), G),$$

where the set s^*B is the $\tilde{s}$-image of the set B for every $B \subset \mathrm{Hom}(W(Y), G)$.

Here s^* is not a boolean homomorphism, but it preserves sums and zero.

It may be seen [**Pl8**] that such a conjugate mapping can be defined in any Halmos category. Note the obvious relation between the categories $\mathrm{Hal}_\Theta(G)$ and $\mathrm{Bool}_\Theta(G)$.

6.2.4. *Multi-sorted Halmos algebras.* Fix some infinite set $X^0 = (X^0_i, i \in \Gamma)$ and let Γ^0 be the set of all finite subsets $X = (X_i, i \in \Gamma)$ in X^0. In this section multi-sorted algebra means Γ^0-sorted. Every such algebra is of the form $H = (H(X), X \in \Gamma^0)$.

A few words about the signature of the algebras to be constructed. First, the signature includes L_X for every X together with equalities ($w \equiv w', w, w'$ of one sort in $W(X)$) as nullary operations. This is the signature in $H(X)$. Second, we consider symbols of operations of the type $s\colon W(X) \to W(Y)$. To each such symbol corresponds an unary operation $s\colon H(X) \to H(Y)$. Denote this signature of all L_X, all equalities, and all $s\colon W(X) \to W(Y)$ by L_Θ. This is the signature of FOL in Θ in the multi-sorted variant.

Consider further the variety of Γ^0-sorted L_Θ-algebras, denoted by Hal_Θ. The identities of this variety exactly copy the definition of Halmos category. We call algebras from Hal_Θ *multi-sorted Halmos algebras.*

Every such algebra can be considered as a small Halmos category.

6.2.5. *Algebras of formulas.* First consider a multi-sorted set of atomic formulas $M = (M(X), X \in \Gamma^0)$, with $M(X) = M_X$ defined as above. All $w \equiv w'$ are viewed as symbols of nullary operations-equalities. The set of symbols of relations Φ is fixed.

Denote by $H_{\Phi\Theta} = (H_{\Phi\Theta}(X), X \in \Gamma^0)$ an absolutely free L_Θ-algebra over the set M. This is the algebra of formulas of pure FOL in the given Θ.

Now denote by $\tilde{H}_{\Phi\Theta}$ the result of factorization of the algebra $H_{\Phi\Theta}$ by the identities of the variety Hal_Θ. It is a free Halmos algebra over the set of atomic formulas M.

Let us introduce the following defining relations:

$$(*) \qquad s_*\varphi(w_1, \dots, w_n) = \varphi(sw_1, \dots, sw_n)$$

for all $s\colon W(X) \to W(Y)$ and all formulas of the type $\varphi(w_1, \dots, w_n)$ in $M(X)$.

In the sequel, the Halmos algebra $\mathrm{Hal}_\Theta(\Phi) = \mathrm{Hal}_{\Phi\Theta}$ will play a principal role, defined as a factor algebra of the free algebra $\tilde{H}_{\Phi\Theta}$ by the relations of (*) type. Elements of this algebra are defined as squeezed formulas.

Consider now values of formulas. First of all, consider a mapping

$$\mathrm{Val}_f = (\mathrm{Val}_f^X, X \in \Gamma^0)\colon\ M \to \mathrm{Hal}_\Theta(G)$$

for the model (G, Φ, f). Here the mappings $\mathrm{Val}_f^X : M_X \to \mathrm{Bool}(W(X), G) = \mathrm{Hal}_\Theta(G)(X)$ have been defined.

This mapping is uniquely extended up to the homomorphisms

$$\mathrm{Val}_f\colon H_{\Phi\Theta} \to \mathrm{Hal}_\Theta(G),$$
$$\mathrm{Val}_f\colon \tilde{H}_{\Phi\Theta} \to \mathrm{Hal}_\Theta(G).$$

Note that the relations $(*)$ hold in every algebra $\mathrm{Hal}_\Theta(G)$ and this gives a canonical homomorphism of Halmos algebras

$$\mathrm{Val}_f : \mathrm{Hal}_\Theta(\Phi) \to \mathrm{Hal}_\Theta(G).$$

It determines the value of the formulas $\mathrm{Val}_f(u)$ (pure and squeezed) in the given model (G, Φ, f).

It is easy to see that the kernel $\mathrm{Ker}(\mathrm{Val}_f)$ is precisely the elementary theory of the model (G, Φ, f) in the logic of the variety Θ.

In fact, elementary theory of the algebra G or the model (G, Φ, f) is considered also in the logic of the variety Θ^G. This logic is richer with respect to the given G.

We call two pure formulas u and v of the given type X *semantically equivalent*, if $\mathrm{Val}_f(u) = \mathrm{Val}_f(v)$ for every model (G, Φ, f).

We have the following main theorem.

THEOREM 6.1. *Two formulas u and v are semantically equivalent if and only if the corresponding squeezed formulas $\overline{u}$ and $\overline{v}$ coincide in the algebra* $\mathrm{Hal}_\Theta(\Phi)$.

This theorem explains the role of the algebra $\mathrm{Hal}_\Theta(\Phi)$ as a main structure of the multi-sorted algebraic logic for FOL in the given Θ. The same algebra plays the essential part in the algebraic geometry in the FOL in Θ. In addition, the role of the algebras $\mathrm{Hal}_\Theta(G)$ is underlined by the following theorem:

THEOREM 6.2. *All algebras* $\mathrm{Hal}_\Theta(G)$ *over different $G \in \Theta$ generate the variety of Halmos algebras* Hal_Θ.

Define the notion of the logical kernel of a homomorphism.

Let the homomorphism $\mu\colon W(X) \to G$ be given. One can view its kernel $\operatorname{Ker}\mu$ as a system of all formulas $w \equiv w'$ with w, w' of the same sort in $W(X)$, for which $\mu \in \operatorname{Val}(w \equiv w')$.

The logical kernel $\operatorname{Log}\operatorname{Ker}\mu$ naturally generates the standard $\operatorname{Ker}\mu$. We set: the formula $u \in \operatorname{Hal}_{\Phi\Theta}(X)$ belongs to $\operatorname{Log}\operatorname{Ker}(\mu)$ if the point μ lies in $\operatorname{Val}_f(u)$, i.e., μ is a solution of the "equation" u in the given model (G, Φ, f). It is easy to understand that for every point μ its logical kernel is an ultrafilter of the Boolean algebra $\operatorname{Hal}_{\Phi\Theta}(X)$. It is also clear that the kernel $\operatorname{Ker}\mu$ is the set of all equalities in the logical kernel.

6.3. Elementary (algebraic) sets.

6.3.1. *Preliminary remarks.* In the sequel, we call the sets algebraic sets although it would be more sensible to call them elementary sets.

Algebraic sets are the sets determined by FOL formulas. We work with squeezed formulas, i.e., formulas of the algebra $\operatorname{Hal}_\Theta(\Phi) = \operatorname{Hal}_{\Phi\Theta}$.

For the given place X, consider sets of formulas T in $\operatorname{Hal}_{\Phi\Theta}(X)$ and the sets of points A in the space $\operatorname{Hom}(W(X), G)$. We establish a Galois correspondence determined by the given model (G, Φ, f). It looks like

$$T^f = A = \bigcap_{u \in T} \operatorname{Val}_f(u) = \{\mu | T \subset \operatorname{Log}\operatorname{Ker}(\mu)\}$$

$$A^f = T = \{u | A \subset \operatorname{Val}_f(u)\} = \bigcap_{\mu \in A} \operatorname{Log}\operatorname{Ker}(\mu).$$

As above, we call a set A represented as $A = T^f$ an *algebraic set* or an *elementary set* for the given model (G, Φ, f).

The set T, represented as $A^f = T$, is always a filter of a Boolean algebra $\operatorname{Hal}_{\Phi\Theta}(X)$, since by the definition it is an intersection of ultrafilters. We call it an *f-closed filter*. If A is an algebraic set then $T = A^f$ can be considered also as the elementary theory of the given A. One can consider here the Boolean algebra $\operatorname{Hal}_{\Phi\Theta}(X)/T$ for this T. If $T^f = A$ and $A^f = T$, then the algebra $\operatorname{Hal}_{\Phi\Theta}(X)/T$ is considered as an invariant of the algebraic set A. This invariant is a *coordinate algebra* of the set A. It can be represented as an algebra of regular functions determined on A (see [**Pl8**]).

Every algebraic set, defined in Section 6.1, is also an algebraic set according to this new definition. The opposite is not true, because in the new variant additional operations of the type $s\colon W(X) \to W(Y)$ are involved in the formulas.

Consider now the relations between Galois correspondence and morphisms of Halmos categories.

For every $s\colon W(X) \to W(Y)$ and every A of the type X, we consider a set $B = s_*A$ of the type Y. If B is of the type Y, then $A = s^*B$ is of the type X. Define the operations s_* and s^* on the sets of formulas.

If T is a set of formulas in $\operatorname{Hal}_{\Phi\Theta}(Y)$, then s_*T is a set of formulas in $\operatorname{Hal}_{\Phi\Theta}(X)$ defined by the rule

$$u \in s_*T \Leftrightarrow su \in T.$$

If T is a set of formulas in $\operatorname{Hal}_{\Phi\Theta}(X)$, then s^*T is contained in $\operatorname{Hal}_{\Phi\Theta}(Y)$ and it is defined by

$$u \in s^*T \text{ if } \ u = sv, \quad v \in T.$$

The following theorem takes place:

THEOREM 6.3.

1. *If T lies in* $\mathrm{Hal}_{\Phi\Theta}(X)$*, then*

$$(s^*T)^f = s_*T^f = sT^f.$$

2. *If $B \subset \mathrm{Hom}(W(Y), G)$, then*

$$(s^*B)^f = s_*B^f.$$

3. *If $A \subset \mathrm{Hom}(W(X), G)$, then $s^*A^f \subset (s_*A)^f$.*

It follows from these rules that

1. If $A = T^f$ is an algebraic set, then sA is also an algebraic set.
2. If $T = B^f$ is f-closed, then $sT = s_*T$ is f-closed.

6.3.2. *Categories $K_{\Phi\Theta}(f)$ and $C_{\Phi\Theta}(f)$.* Fix a model (G, Φ, f) and define a category of algebraic sets $K_{\Phi\Theta}(f)$ for this model. Objects of this category have the form (X, A), where $A = T^f$ for some T. X is the place for both A and T.

We now define morphisms $(X, A) \to (Y, B)$. Proceed from $s\colon W(Y) \to W(X)$. We say that s is admissible for A and B if $\tilde{s}(\nu) = \nu s \in B$ for any $\nu \in A$. It is clear that s is admissible for A and B if $A \subset sB$. A mapping $[s] : A \to B$ corresponds to each s admissible for A and B.

We consider *weak* and *exact* categories $K_{\Phi\Theta}(f)$. In the first one, the morphisms look like $s : (X, A) \to (Y, B)$, while in the second one like $[s] : (X, A) \to (Y, B)$.

If s_1 is admissible for A and B and s_2 for B and C, then $A \subset s_1B$, $B \subset s_2C$, $s_1B \subset s_1s_2C$, and s_1s_2 is admissible for A and C.

Define now a category $C_{\Phi\Theta}(f)$. Its objects are Boolean algebras of the type $\mathrm{Hal}_{\Phi\Theta}(X)/T$, where $T = A^f$ for some A.

Consider morphisms

$$\mathrm{Hal}_{\Phi\Theta}(Y)/T_2 \xrightarrow{\overline{s}} \mathrm{Hal}_{\Phi\Theta}(X)/T_1.$$

We proceed here from $s\colon W(Y) \to W(X)$ and pass to the new $s : \mathrm{Hal}_{\Phi\Theta}(Y) \to \mathrm{Hal}_{\Phi\Theta}(X)$. Assume that $su \in T_1$ for every $u \in T_2$. The homomorphism s is admissible for T_2 and T_1 in this sense. Define homomorphisms $\overline{s}$ for such s. This defines morphisms in $C_{\Phi\Theta}(f)$.

The next two propositions determine the correspondence between the categories $K_{\Phi\Theta}(f)$ and $C_{\Phi\Theta}(f)$.

PROPOSITION 6.1. *Homomorphism $s\colon W(Y) \to W(X)$ is admissible for varieties (X, A) and (Y, B) if and only if this s is admissible for $T_2 = B^f$ and $T_1 = A^f$.*

PROPOSITION 6.2. *If $s_1, s_2\colon W(Y) \to W(X)$ are admissible for A and B, then $[s_1] = [s_2]$ implies $\overline{s_1} = \overline{s_2}$.*

It follows from these two propositions that the transition

$$(X, A) \to \mathrm{Hal}_{\Phi\Theta}(X)/A^f$$

determines a contravariant functor

$$K_{\Phi\Theta}(f) \to C_{\Phi\Theta}(f)$$

for weak and exact categories $K_{\Phi\Theta}(f)$. Duality takes place under some additional conditions. In particular, this is the case when we proceed from the variety Θ^G for the given G in Θ.

6.4. Categories $K_{\Phi\Theta}$ and $C_{\Phi\Theta}$. In the categories $K_{\Phi\Theta}$ and $C_{\Phi\Theta}$, the model (G, Φ, f) is not fixed. Objects of $K_{\Phi\Theta}$ have the form $(X, A; G, f)$. Here f is a realization of the set Φ, fixed for the category $K_{\Phi\Theta}$, in the algebra G, and $A = T^f$ for some $T \subset \mathrm{Hal}_{\Phi\Theta}(X)$.

Define morphisms

$$(X, A; G_1, f_1) \to (Y, B; G_2, f_2).$$

They act on all components of the objects. Proceed from the commutative diagram

$$\begin{array}{ccc} W(Y) & \xrightarrow{\ s\ } & W(X) \\ {\scriptstyle \nu'}\downarrow & & \downarrow{\scriptstyle \nu} \\ G_2 & \xleftarrow{\ \delta\ } & G_1 \end{array}$$

Consider a pair (s, δ) and write $(s, \delta)(\nu) = \nu' = \delta\nu s$.

Let now $A = T_1^{f_1}$ be of the type X and $B = T_2^{f_2}$ of the type Y. We say that the pair (s, δ) is *admissible* for A and B if $(s, \delta)(\nu) \in B$ for every $\nu \in A$.

We need some further auxiliary remarks. For every $\delta : G_1 \to G_2$ and every X, we have a mapping

$$\tilde{\delta} : \mathrm{Hom}(W(X), G_1) \to \mathrm{Hom}(W(X), G_2)$$

defined by the rule

$$\tilde{\delta}(\nu) = \delta\nu, \nu \in \mathrm{Hom}(W(X), G_1).$$

Determine $\delta_* A \subset \mathrm{Hom}(W(X), G_1)$ for every $A \subset \mathrm{Hom}(W(X), G_2)$, setting

$$\nu \in \delta_* A \ \text{ if } \ \delta\nu = \tilde{\delta}(\nu) \in A.$$

We change notations to $\delta_* A = \delta A$.

If $A \subset \mathrm{Hom}(W(X), G_1)$, then $\delta^* A \subset \mathrm{Hom}(W(X), G_2)$ and $\nu \in \delta^* A$ if $\nu = \delta\nu_1$, $\nu_1 \in A$.

Now we can say that the pair (s, δ) is admissible for A and B if $\delta^* A \subset sB$, or, the same, $A \subset \delta sB = s\delta B$.

We have morphisms

$$(s, \delta) : (X, A; G_1, f_1) \to (Y, B; G_2, f_2)$$

and

$$([s], \delta) \colon (X, A; G_1, f_1) \to (Y, B; G_2, f_2)$$

for the admissible (s, δ). Here $[s] : A \to B$ is a mapping, induced by the pair (s, δ). We get weak and exact categories $K_{\Phi\Theta}$. It can be proven that the pair (s, δ) is admissible for A and B if and only if the homomorphism $s\colon \mathrm{Hal}_{\Phi\Theta}(Y) \to Hal_{\Phi\Theta}(X)$ is admissible with respect to $T_2 = B^{f_2}$ and $T_1 = (\delta^* A)^{f_2}$. This leads to a natural definition of the category $C_{\Phi\Theta}$ with contravariant functor $K_{\Phi\Theta} \to C_{\Phi\Theta}$.

Let us define the categories $K_{\Phi\Theta}(G)$ and $C_{\Phi\Theta}(G)$. Here G is a fixed algebra in Θ, while the realizations f of the set Φ in G change.

The objects in $K_{\Phi\Theta}(G)$ have the form

$$(X, A; f).$$

The morphisms

$$(X, A, f_1) \to (Y, B, f_2)$$

are defined according to the general definition of the morphisms in $K_{\Phi\Theta}$ with identical $\delta = \varepsilon\colon G \to G$.

Objects in $C_{\Phi\Theta}(G)$ have the form

$$(\mathrm{Hal}_{\Phi\Theta}(X)/T, f), \quad \text{where } T = A^f$$

for some A of the type X.

The transition

$$(X, A; f) \to (\mathrm{Hal}_{\Phi\Theta}(X)/A^f, f)$$

determines the functor $K_{\Phi\Theta}(G) \to C_{\Phi\Theta}(G)$. Here $K_{\Phi\Theta}(G)$ is a subcategory in $K_{\Phi\Theta}$ and every $K_{\Phi\Theta}(f)$ is a subcategory in $K_{\Phi\Theta}(G)$. The same is true for C.

6.5. Galois theory in logic. The Galois theory that we are talking about is tied in with the considered algebraic geometry and, besides, it will be used in the next lecture.

6.5.1. *Automorphisms.* Let $\delta\colon G \to G$ be an automorphism of the algebra G, $\delta \in \mathrm{Aut}\, G$. Then for every X we have a substitution

$$\tilde{\delta} : \mathrm{Hom}(W(X), G) \to \mathrm{Hom}(W(X), G).$$

This substitution induces an automorphism of the Boolean algebra $\mathrm{Bool}(W(X), G)$. For each element A, from this Boolean algebra we have $\delta_* A = \delta A$.

Actually, it is an automorphism of the quantifier X-algebra with equalities. This leads to the automorphism $\sigma = \delta_*$ of the Halmos algebra $\mathrm{Hal}_\Theta(G)$. We have a representation

$$\mathrm{Aut}(G) \to \mathrm{Aut}(\mathrm{Hal}_\Theta(G)).$$

THEOREM 6.4. *The given representation is an isomorphism of the groups of automorphisms* $\mathrm{Aut}(G)$ *and* $\mathrm{Aut}(\mathrm{Hal}_\Theta(G))$.

Thus, the group of automorphisms $\mathrm{Aut}\, G$ *can be considered as a group of automorphisms of the algebra* $\mathrm{Hal}_\Theta(G)$.

Let, further, $\delta\colon G_2 \to G_1$ be an isomorphism of algebras in Θ. As above, we may take the isomorphism of Halmos algebras

$$\delta_*\colon \mathrm{Hal}_\Theta(G_1) \to \mathrm{Hal}_\Theta(G_2).$$

This isomorphism is well correlated with the homomorphisms of the type Val_f.

Let the model (G_2, Φ, f) be given. The isomorphism $\delta : G_2 \to G_1$ uniquely determines the model (G_1, Φ, f^δ), isomorphic to the initial model.

We have the commutative diagram

$$\begin{array}{ccc} & \mathrm{Hal}_\Theta(\Phi) & \\ {}^{\mathrm{Val}_{f\delta}}\swarrow & & \searrow^{\mathrm{Val}_f} \\ \mathrm{Hal}_\Theta(G_1) & \xrightarrow[\delta_*]{} & \mathrm{Hal}_\Theta(G_2) \end{array}$$

The arrow $\mathrm{Val}_{f\delta}$ is uniquely determined by the other two arrows due to the commutativity of the diagram.

6.6. The main theorems of Galois theory. Consider the algebra $\mathrm{Hal}_\Theta(G)$ together with its group of automorphisms $\mathrm{Aut}(G)$.

Let us define the standard Galois correspondence.

Let H be a subset of $\mathrm{Aut}(G)$. Then $H' = R$ is a subalgebra of $\mathrm{Hal}_\Theta(G)$ (with equalities), consisting of all A with $\delta A = A$ for every $\delta \in H$.

Let R be a subset in $\mathrm{Hal}_\Theta(G)$. Then $R' = H$ is a subgroup in $\mathrm{Aut}(G)$ consisting of all δ with $\delta A = A$ for every $A \in R$.

Define closures R'' and H''.

THEOREM 6.5. *If the algebra G is finite, then every subalgebra in* $\mathrm{Hal}_\Theta(G)$ *and every subgroup in* $\mathrm{Aut}(G)$ *are closed. There is a one-to-one correspondence between them.*

Let now two finite algebras G_1 and G_2 be given. Take a subalgebra R_1 in $\mathrm{Hal}_\Theta(G_1)$ and a subalgebra R_2 in $\mathrm{Hal}_\Theta(G_2)$.

THEOREM 6.6. *If there is an isomorphism $\gamma\colon R_1 \to R_2$, then there exists an isomorphism $\delta : G_2 \to G_1$, such that $\delta_* : \mathrm{Hal}_\Theta(G_1) \to \mathrm{Hal}_\Theta(G_2)$ induces the given $\gamma : R_1 \to R_2$.*

Here

1. $R_2 = \delta_*(R_1)$,
2. $R_1' = \delta R_2' \delta^{-1}$.

R_1' and R_2' determine R_1 and R_2 with $R_2 = \delta_*(R_1)$.

Note that the considered Galois theory goes back to the paper by M.I. Krasner [**KR**] and it would be reasonable to call it the Galois-Krasner theory.

6.6.1. *Automorphisms and algebraic sets.* Denote a group of automorphisms of the model (G, Φ, f) by $\mathrm{Aut}(f)$. It is a subgroup of $\mathrm{Aut}(G)$.

THEOREM 6.7. *Every algebraic set $A = T^f$ is invariant under the action of the group* $\mathrm{Aut}(f)$. *If $A = T^f$ and $\delta \in \mathrm{Aut}(f)$, then $\delta A = A$.*

This theorem to some extent determines the group structure of algebraic sets.

Consider $\mathrm{Val}_f\colon \mathrm{Hal}_\Theta(\Phi) \to \mathrm{Hal}_\Theta(G)$ for the model (G, Φ, f) and let again R_f be the image of this homomorphism. R_f is a subalgebra in $\mathrm{Hal}_\Theta(G)$ consisting of algebraic sets of the form $A = \mathrm{Val}_f(u)$. Here T consists of one element u. We call A a simple algebraic set.

THEOREM 6.8. $R_f' = \mathrm{Aut}(f)$ *holds for every model (G, Φ, f). If the algebra G is finite, then* $\mathrm{Aut}(f)' = R_f$.

In particular, it follows that if the algebra G is finite, then every algebraic set $A = T^f$ is a simple algebraic set, $A = \mathrm{Val}_f(u)$ for some $u \in \mathrm{Hal}_{\Phi\Theta} = \mathrm{Hal}_\Theta(\Phi)$.

Note in addition to Theorem 6.7 that the set $A \subset \mathrm{Hom}(W, G)$ is an elementary (algebraic) set over a model with the algebra G, if and only if $\delta A = A$, $\delta \neq 1$ holds for some $\delta \in \mathrm{Aut}(G)$. We can select a set Φ and its realization f in G by δ in such a way that $\delta \in \mathrm{Aut}(f)$ and, simultaneously, we have $\delta A = A$ for every $\delta \in \mathrm{Aut}(f)$.

6.7. Lattices of elementary sets. For every model (G, Φ, f) and every finite set X, denote $\mathrm{Alv}_f(X)$ to be the set of all elementary sets for the model (G, Φ, f) in the space $\mathrm{Hom}(W, G)$; $\mathrm{Cl}_f(X)$ is the set of all closed sets T in $\mathrm{Hal}_{\Phi\Theta}(X)$.

We consider both these sets as lattices as well. The intersection operation is defined here as set theory intersection, while the union is defined by the rules:

$$A \bigtriangledown B = (A \cup B)^{ff};\ A, B \in \mathrm{Alv}_f(X)$$

$$T_1 \bigtriangledown T_2 = (T_1 \cup T_2)^{ff};\ T_1, T_2 \in \mathrm{Cl}_f(X).$$

The transitions $A \to A^f = T$ and $T \to T^f = A$ give the duality of lattices.

Further, we can specify various levels of logic, all of which are the parts of the algebra $\mathrm{Hal}_\Theta(\Phi)$. Levels of algebraic geometry are connected with them. In particular, equational geometry is connected with equational logic. The attitude to the pointed lattices changes. Let us point out some interesting levels.

- L_0 – equational logic. Its formulas have the form $w \equiv w'$. They are equalities.
- L_1 – pseudoequational logic. Its formulas are pseudoequalities of the form $w_1 \equiv w_1' \vee \cdots \vee w_n = w_n'$.
- L_2 – universal logic over equalities. The formulas have the form $w_1 \equiv w_1' \vee \cdots \vee w_n \equiv w_n' \vee v_1 \not\equiv v_1' \vee \ldots v_m \not\equiv v_m'$.
- L_3 – we mean here positive logic whose formulas are built without negations.
- L_4 – universal logic (without quantifiers).
- L_5 – all that is built from atoms without quantifiers and negations (only $\vee$ and $\wedge$).
- L_6 – the whole algebra $\mathrm{Hal}_\Theta(\Phi)$.

Finally, we say that the logic L is $\vee$-closed, if $u, v \in L$ implies $u \vee v \in L$. With each of these logics L, a definite level ℓ of geometry is associated. A notion of logical kernel of a homomorphism relates to the level ℓ. If the homomorphism $\mu : W(X) \to G$ in Θ is given, then we set $\ell - \log\ker(\mu)$ is the set of formulas u of the level ℓ in $\mathrm{Hal}_{\Theta\Phi}(X)$ for which $\mu \in \mathrm{Val}_f(u)$. We assume that the algebra G is included in the model (G, Φ, f).

According to these considerations, we localize Galois connections, all sets of formulas T are on a definite level ℓ in a definite logic L. Let us rewrite these connections

For every $A \subset \mathrm{Hom}(W, G)$, we have

$$T = A^{f,\ell} = \bigcap_{\mu \in A} (\ell - \log\ker(\mu)) = \{u \mid u \in \ell \quad \text{and} \quad A \subset \mathrm{Val}_f(u)\}.$$

If, further, T is of the level ℓ, then

$$T^f = A = \bigcap_{u \in T} \mathrm{Val}_f(u) = \{\mu : W \to G \mid T \subset \ell - \log\ker(\mu)\}.$$

Consider the corresponding Galois closures. For the given $A \subset \mathrm{Hom}(W(X), G)$, we have $A^{ff,\ell} = (A^{f,\ell})^f$. For the given T of the level ℓ, we have $T^{ff,\ell} = (T^f)^{f,\ell}$. Consider also ℓ, f-closed A and T. We consider the lattices $\mathrm{Alv}_f(X)$ and $\mathrm{Cl}_f(X)$ on the level ℓ as well. We denote them $\mathrm{Alv}_{f,\ell}(X)$ and $\mathrm{Cl}_{f,\ell}(X)$, respectively.

6.8. Zariski topology. We speak of topology in the affine space $\mathrm{Hom}(W, G)$, $W = W(X)$, for the given model (G, Φ, f) on the given level of logic ℓ. We consider here only positive formulas.

For the given level ℓ, a Zariski topology in $\mathrm{Hom}(W, G)$ over the model (G, Φ, f) is a topology generated by all ℓ-algebraic sets A in $\mathrm{Hom}(W, G)$ as closed sets.

THEOREM 6.9. *If the level ℓ is a $\vee$-closed logic, then all closed sets in the Zariski topology are exactly all ℓ-algebraic sets.*

In this case, the lattice $\mathrm{Alv}_{f,\ell}(X)$ is a sublattice of the lattice $\mathrm{Bool}(W, G)$. It is distributive, as is the lattice $\mathrm{Cl}_{f,\ell}(X)$.

An innteresting case of Zariski topology is provided by the empty set Φ. In the special situation when $\Theta = \mathrm{Var}\text{-}R$ and the field R is the field of real numbers, the case when Φ is an one-element relation φ which is the order relation is of special interest.

It is natural to treat all the considerations of this lecture in the classical situation of the variety Var-P.

6.9. Geometrical properties. We work first in the largest FOL-L_6.

DEFINITION 6.1. The model (G, Φ, f) is called *geometrically Noetherian* if for every finite set X and every set of formulas T in $\mathrm{Hal}_{\Phi\Theta}(X)$ in T there is some finite part T_0 with $T_0^f = T^f$. This means that $T_0^{ff} = T^{ff}$.

THEOREM 6.10. *The model (G, Φ, f) is geometrically Noetherian if and only if the minimality condition holds in the lattice $\mathrm{Alv}_f(X)$. Correspondingly, in the lattice $\mathrm{Cl}_f(X)$ we have the maximality condition.*

Now let $T_0 = \{u_1, \ldots, u_n\}$ and u is $u_1 \wedge \cdots \wedge u_n$. Then $T_0^f = \mathrm{Val}_f(u) = \{u\}^f$. Thus, we may claim that if the model (G, Φ, f) is geometrically Noetherian, then every algebraic set over this model is a simple algebraic set.

However, the corresponding element $u = u_1 \wedge \cdots \wedge u_n$ does not necessarily belong to the initial set T. We call a model (G, Φ, f) weak geometrically Noetherian if every algebraic set over it is a simple algebraic set. A weak Noetherian model is not necessarily geometrically Noetherian.

However, we may claim that every finite model is geometrically Noetherian (compare Theorem 6.8 in Galois theory). We may also claim that any finite cartesian product of geometrically Noetherian models is also a geometrically Noetherian model. A submodel of a geometrically Noetherian model is also geometrically Noetherian. Similar properties are not true in general in the transition to cartesian powers.

The notion of the model to be geometrically Noetherian may be also defined on different special levels ℓ. It is easy to understand that if the model (G, Φ, f) is Noetherian on the absolute level ℓ_6, then it is Noetherian on each other level ℓ.

We may assume that by lowering the level ℓ, one may obtain good properties. For example, for which level ℓ does the property of the model (G, Φ, f) to be Noetherian imply that each of its cartesian powers is Noetherian as well. It is valid for L_0 and for the logic of all atomic formulas. It is not clear when else this is true. It is most likely never true (see also Theorem 6.11 below).

Let us pass to the notion of geometrical equivalence of two models. Let two models (G_1, Φ, f_1) and (G_2, Φ, f_2) with the same Φ be given.

DEFINITION 6.2. The models (G_1, Φ, f_1) and (G_2, Φ, f_2) are geometrically equivalent if $T^{f_1 f_1} = T^{f_2 f_2}$ holds for every finite X and every T in $\mathrm{Hal}_{\Phi\Theta}(X)$.

If the models (G_1, Φ, f_1) and (G_2, Φ, f_2) are geometrically equivalent, then

1. the lattices $\mathrm{Cl}_{f_1}(X)$ and $\mathrm{Cl}_{f_2}(X)$ coincide, while the lattices $\mathrm{Alv}_{f_1}(X)$ and $\mathrm{Alv}_{f_2}(X)$ are isomorphic;
2. the categories $C_{\Phi\Theta}(f_1)$ and $C_{\Phi\Theta}(f_2)$ coincide, while the categories $K_{\Phi\Theta}(f_1)$ and $K_{\Phi\Theta}(f_2)$ are isomorphic;
3. these models are elementary equivalent.

It follows from the first claim that if the models are geometrically equivalent and one of them is geometrically Noetherian, then the second one is also geometrically Noetherian.

THEOREM 6.11. *Let the model* (G, Φ, f) *be geometrically Noetherian. Then each of its ultrapowers is also geometrically Noetherian, and all these ultrapowers are geometrically equivalent to the initial* (G, Φ, f).

All the described notions are naturally connected with the logic of generalized (infinitary) formulas of the kind $(\wedge_{u \in T}) \to v$, or, what is the same $T \to v$. For the geometrically Noetherian models, it is sufficient to proceed from the usual finite formulas.

Note that the notion of geometrical equivalence of models can be also considered on different logical levels. Here if the models are equivalent on the absolute level ℓ_6, then they are equivalent on every other level.

Elementary equivalence of the models does not generally imply their geometrical equivalence. However, we may claim the following

THEOREM 6.12. *If two models are elementary equivalent and one of them is geometrically Noetherian, then the second one is aso geometrically Noetherian and these models are geometrically equivalent.*

Regarding the notion of a geometrically Noetherian model, let us return to the notion of the logical kernel of a homomorphism of the form $\mu : W(X) \to G$. It is easy to see that if $\mathrm{Log}_f \mathrm{Ker}(\mu)$ is such a kernel and $\mathrm{Val}_f(\mathrm{Log}_f \mathrm{Ker}(\mu))$ is its image in the algebra $R_f(X)$, then this image is a principal ultrafilter if the model (G, Φ, f) is geometrically Noetherian. This ultrafilter is generated by the algebraic set $\{\mu\}^{ff}$.

Lecture 7

Databases and Knowledge Bases: Geometrical Aspect

Contents

7.1. Introduction. We use here the material from Lecture 6, mainly concerning knowledge theory.

Elementary knowledge is considered to be a first order knowledge, i.e., knowledge that can be represented by the means of the First Order Logic (FOL). The corresponding applied field (field of knowledge) is based on some variety of algebras Θ, which is arbitrary but fixed and can be multi-sorted. This variety Θ is considered as a knowledge type, like in database theory, databases of a data type Θ are considered. We also fix a set of symbols of relations Φ. Finally, the subject of knowledge is a model (G, Φ, f), where G is an algebra in Θ and f is a realization of the set Φ in G. It is a model in the ordinary mathematical meaning. Similar to the previous lecture, we write f instead of (G, Φ, f) for short. Given Φ, we denote the corresponding applied field by $\Phi\Theta$. FOL is also oriented on the variety Θ.

We assume that every knowledge under consideration is represented by three components:

1) The description of the knowledge.

   It is a syntactical part of knowledge, written out in the language of the given logic. The description reflects what we do want to know.
2) The subject of the knowledge which is an object in the given applied field, i.e., an object for which we determine knowledge.
3) Content of the knowledge (its semantics).

It is assumed also that there is a finite type of the description of knowledge, which is denoted by X. This X determines the space where the content of knowledge is calculated.

We consider knowledge as a triple of the form $(Description, Subject, Content)$. In the knowledge base a query is the description of the knowledge. The reply to a query is content of knowledge. Subject of the knowledge is fixed for the whole knowledge base.

The first two components of knowledge are relatively independent, while the third one is uniquely determined by the previous two components. In the theory under consideration, this third component has a geometrical nature. In some sense, it is an algebraic variety in an affine space. If T is a description of knowledge and (G, Φ, f) is a subject, then T^f denotes the content of knowledge. We would like to fill the content with its own structure, algebraic or geometric, and we consider some elements of such structure.

On the one hand, the language is constructed according to primary notions of algebraic geometry. On the other hand, it uses algebraic logic. We want to underline that there are three aspects in our approach to knowledge representation: logical (for knowledge description), algebraic (for the subject of knowledge) and geometric (in the content of knowledge).

The core point of the Lecture is elementary knowledge, i.e., First Order Knowledge. The main goal is to construct a model to represent some non-elementary knowledge about elementary knowledge using the universal algebra approach. For the solution of this problem, we combine the methods of algebraic logic and universal algebraic geometry in logic, both defined over an arbitrary variety of algebras Θ. Let us stress also that we make emphasis on the geometric nature of knowledge.

We consider categories of elementary knowledge. Language of categories in the knowledge theory is a good way to organize and systematize primary elementary knowledge. Morphisms in the knowledge category give links between knowledge. In particular, one can speak of isomorphic knowledge. The categorical approach also allows us to use ideas of monada and comonada [**ML**]. It turns out that this leads to some general views on enrichment and computation of knowledge. Enrichment of a structure can be associated with a suitable monada over a category, while the corresponding computation is organized by comonada.

Let us make one more remark. In every well described field of knowledge one can study the category of elementary knowledge, belonging to this field. Consideration of such categories might be of special interest.

7.2. Categories of elementary knowledge.

7.2.1. *The category* $\mathrm{Know}_{\Phi\Theta}(f)$. Fix a model (subject of knowledge) (G, Φ, f). Let us define a category of knowledge for this model and denote it by $\mathrm{Know}_{\Phi\Theta}(f)$. It is a category for the given subject of knowledge. The objects of the category $\mathrm{Know}_{\Phi\Theta}(f)$ have the form (X, T, A). Their meaning is knowledge. We do not fix the subject of knowledge, although it occurs here implicitly, since it is fixed once and for all. The set X is multi-sorted in general. It marks the "place" where knowledge is situated. The set X points also to the "place of the knowledge" – the space of the knowledge $\mathrm{Hom}(W(X), G)$ while the subject of the knowledge (G, Φ, f) is given. The set T is the description of the knowledge in the algebra $\mathrm{Hal}_{\Phi\Theta}(X)$, and $A = T^f$, $A \subset \mathrm{Hom}(W(X), G)$, is the content of knowledge, depending on T and f. The set $T^{ff} = A^f$ is the full description of the knowledge (X, T, A) which is a Boolean filter in $\mathrm{Hal}_{\Phi\Theta}(X)$.

Now concerning morphisms $(X, T_1, A) \to (Y, T_2, B)$. Take $s\colon W(Y) \to W(X)$. We have also $s\colon \mathrm{Hal}_{\Phi\Theta}(Y) \to \mathrm{Hal}_{\Phi\Theta}(X)$. This is a homomorphism of Boolean algebras. The first s gives also

$$\tilde{s}\colon \mathrm{Hom}(W(X), G) \to \mathrm{Hom}(W(Y), G).$$

As above, the first s is admissible for A and B if $\tilde{s}(\nu) = \nu s \in B$ for every point $\nu\colon W(X) \to G$ in A.

As we know, s is admissible for A and B if and only if for every $u \in B^f$ we have $su \in A^f$. This is for the second s, for which we have also a homomorphism $\overline{s} : \mathrm{Hal}_{\Phi\Theta}(Y)/B^f \to \mathrm{Hal}_{\Phi\Theta}(X)/A^f$. It is easy to prove that s is admissible for A and B if and only if $su \in A^f$ holds for every $u \in T_2$. We consider such s as a morphism

$$s\colon (X, T_1, A) \to (Y, T_2, B)$$

in the weak category $\mathrm{Know}_{\Phi\Theta}(f)$. We have $\tilde{s}(\nu) = \nu s \in B$ if $\nu \in A$, and s induces a mapping $[s]\colon A \to B$. Simultaneously, a mapping $s\colon T_2 \to A^f$ arises and we have a homomorphism

$$\overline{s}\colon \mathrm{Hal}_{\Phi\Theta}(Y)/B^f \to \mathrm{Hal}_{\Phi\Theta}(X)/A^f.$$

We have already mentioned that $\overline{s}_1 = \overline{s}_2$ follows from $[s_1] = [s_2]$. Thus, we can consider the morphisms of the form

$$[s]\colon (X, T_1, A) \to (Y, T_2, B)$$

in the exact category $\mathrm{Know}_{\Phi\Theta}(f)$.

The canonical functors

$$\mathrm{Know}_{\Phi\Theta}(f) \to K_{\Phi\Theta}(f)$$

for weak and exact categories are given by the transition $(X, T, A) \to (X, A)$. In this transition, we "forget" to fix the description of knowledge.

7.2.2. *The category $Know_{\Phi\Theta}$.* Let us define the category of elementary knowledge for the whole applied field $\Phi\Theta$; the subject of the knowledge (G, Φ, f) is not fixed. As earlier, we proceed from the category $\Phi\Theta$ whose morphisms are homomorphisms in Θ. They ignore the relations from Φ.

Objects of the category $\mathrm{Know}_{\Phi\Theta}$ are knowledge, represented by

$$(X, T, A; (G, \Phi, f)),$$

and we write $(X, T, A; G, f)$, because Φ is fixed for the category. Here X denotes the place of knowledge; $A = T^f$, G and f may change.

Consider morphisms

$$(X, T_1, A; G_1, f_1) \to (Y, T_2, B; G_2, f_2).$$

We apply the same approach as before with some extensions.

Start from $s : W(Y) \to W(X)$ and $\delta : G_1 \to G_2$. These s and δ should correlate. Let us explain the correlation condition. Take a set $A_1 = \{\delta\nu, \nu \in A\} = \delta^* A$ and take further $T_1^\delta = A_1^{f_2}$. Correlation of s and δ means that $su \in T_1^\delta$ holds for any $u \in T_2$. The same for every $u \in B^{f_2}$. The latter also says that there is a homomorphism

$$\overline{s}\colon \mathrm{Hal}_{\Phi\Theta}(Y)/B^{f_2} \to \mathrm{Hal}_{\Phi\Theta}(X)/A_1^{f_2}.$$

The first of the two mappings $(s, \delta)\colon A \to B$ and $s\colon T_2 \to T_1^\delta$ transforms the content of knowledge, while the second one acts on the description. Here T_2 and T_1^δ describe knowledge, associated with the same subject (G_2, Φ, f_2).

With fixed δ, there is also an exact mapping $([s], \delta) : A \to B$. We come to weak and exact categories $\mathrm{Know}_{\Phi\Theta}$. The morphisms of the first one are (s, δ) and in the second one they are $([s], \delta)$ for $(X, T_1, A; G_2, f_1) \to (Y, T_2, B; G_2, f_2)$.

The canonical functor $\mathrm{Know}_{\Phi\Theta} \to K_{\Phi\Theta}$ is defined by the transition

$$(X, T, A; G, f) \to (X, A; G, f).$$

As above, we remove description of knowledge from the information about it.

7.2.3. *Categories $K_{\Phi\Theta}(G)$ and* $\mathrm{Know}_{\Phi\Theta}(G)$. An algebra $G \in \Theta$ is fixed in the categories $K_{\Phi\Theta}(G)$ and $\mathrm{Know}_{\Phi\Theta}(G)$. A set of symbols of relations Φ is fixed as usual, but realizations f of Φ in G may change. Thus, $K_{\Phi\Theta}(G)$ is a subcategory in $K_{\Phi\Theta}$ and $\mathrm{Know}_{\Phi\Theta}(G)$ is a subcategory in $\mathrm{Know}_{\Phi\Theta}$. Here the corresponding $\delta : G \to G$ are identical homomorphisms. Objects of the category $K_{\Phi\Theta}(G)$ are now recorded as (X, A, f), and those of the category $\mathrm{Know}_{\Phi\Theta}(G)$ as (X, T, A, f). There is a canonical functor $\mathrm{Know}_{\Phi\Theta}(G) \to K_{\Phi\Theta}(G)$. As for morphisms

$$(X, A, f_1) \to (Y, B, f_2)$$

and

$$(X, T_1, A, f_1) \to (Y, T_2, B, f_2),$$

we note that $A = A_1, A_1^{f_2} = T_1^{\delta}$ and $A^{f_2} = T_1^{f_1 f_2}$. Hence, the corresponding admissible $s : W(Y) \to W(X)$ transfers each $u \in T_2$ into $su \in T_1^{f_1 f_2}$ and it induces a homomorphism

$$\overline{s} : \mathrm{Hal}_{\Phi\Theta}(Y)/B^{f_2} \to \mathrm{Hal}_{\Phi\Theta}(X)/A^{f_2}.$$

Every s gives a mapping $[s] : A \to B$.

7.3. Databases and Knowledge bases.

7.3.1. *Databases.* The proposed model of a database differs from that of [**Pl2**]. We want to compare databases and knowledge bases. The geometrical aspect in databases reflects the fact that the reply to the query can be considered as an algebraic variety. It is a simple algebraic variety. In knowledge bases, we deal with arbitrary algebraic varieties. But this is not the only difference.

A database is represented as a category.

Let us fix an algebra $G \in \Theta$ and consider an (admissible) set F of realizations f of the set of symbols of relations Φ in G. These f are instances of a database. For every instance $f \in F$, we have a model (G, Φ, f), all of which form a system of models, denoted by (G, Φ, F). We call this system a *multimodel.*

Consider a DB as a category whose objects have the form

$$\mathrm{Val}_f \colon \mathrm{Hal}_{\Phi\Theta} \to R_f, \quad f \in F.$$

Here, as above, R_f is a subalgebra in the algebra $\mathrm{Hal}_\Theta(G)$, coinciding with the image of the homomorphism Val_f. If u is a query to a database, then the reply to this query in the instance f is $u * f = \mathrm{Val}_f(u)$.

The morphisms are homomorphisms $\gamma : R_{f_1} \to R_{f_2}$ with the commutative diagram

$$\begin{array}{ccc} \mathrm{Hal}_{\Phi\Theta} & \xrightarrow{\mathrm{Val}_{f_1}} & R_{f_1} \\ & \searrow^{\mathrm{Val}_{f_2}} & \downarrow \gamma \\ & & R_{f_2} \end{array}$$

We call these γ *correct homomorphisms.* The diagram associates replies to the same query in different instances and for different γ.

Note that all algebras R_f are simple algebras and hence all $\gamma : R_{f_1} \to R_{f_2}$ are injective.

Denote the database by $DB(G, \Phi, F)$.

7.3.2. *Knowledge bases.* We fix again a multimodel (G, Φ, F) and consider a knowledge base $KB(G, \Phi, F)$. This knowledge base is a category, whose objects are

$$\mathrm{Val}_f \colon \mathrm{Hal}_{\Phi\Theta} \to K_{\Phi\Theta}(f), \quad f \in F.$$

The mapping Val_f transforms formulas of the algebra $\mathrm{Hal}_{\Phi\Theta}$ into the objects of the category $K_{\Phi\Theta}(f)$, which is a subcategory in $K_{\Phi\Theta}(G)$. Denote by $R_{\Phi\Theta}(f)$ a full subcategory in $K_{\Phi\Theta}(f)$, whose objects form a subalgebra R_f in $\mathrm{Hal}_\Theta(G)$.

In each object, for every description of knowledge T, the content of knowledge $A = T^f$ is calculated. This A is considered as an object of the category $K_{\Phi\Theta}(f)$ with all its internal and external ties in this category.

Morphisms of the category $KB(G, \Phi, F)$ are represented as follows:

$$
\begin{array}{ccc}
\mathrm{Hal}_{\Phi\Theta} & \xrightarrow{\mathrm{Val}_{f_1}} & K_{\Phi\Theta}(f_1) \\
& \searrow^{\mathrm{Val}_{f_2}} & \downarrow \gamma \\
& & K_{\Phi\Theta}(f_2)
\end{array}
$$

This diagram needs explanation. Here γ is a functor of categories and commutativity of the diagram is supposed on the level of the objects of categories who are elements of the algebra $\mathrm{Hal}_\Theta(G)$. This γ induces the diagram

$$
\begin{array}{ccc}
\mathrm{Hal}_{\Phi\Theta} & \xrightarrow{\mathrm{Val}_{f_1}} & R_{f_1} \\
& \searrow^{\mathrm{Val}_{f_2}} & \downarrow \gamma \\
& & R_{f_2}
\end{array}
$$

Thus, there is a canonical functor

$$KB(G, \Phi, f) \to DB(G, \Phi, f).$$

This functor shows what databases and knowledge bases have in common. A homomorphism of algebras $R_{f_1} \to R_{f_2}$ corresponds to the functor $\gamma : K_{\Phi\Theta}(f_1) \to K_{\Phi\Theta}(f_2)$. Furthermore, we assume that if T is a set of formulas in $\mathrm{Hal}_{\Phi\Theta}(X)$, then

$$\gamma(T^{f_1}) = T^{f_2}.$$

This is a strengthened commutativity of the diagram. It connects knowledge content for the same description in different instances and for different γ.

In the knowledge category $\mathrm{Know}_{\Phi\Theta}(G)$, we can distinguish a subcategory for the given set of instances F.

Note also the ties between knowledge base and knowledge category.

We consider the following commutative diagram:

$$
\begin{array}{ccc}
\mathrm{Hal}_\Theta \Phi & \xrightarrow{\quad \mathrm{Val}_f \quad} & K_{\Phi\Theta}(f) \\
\nwarrow & & \nearrow \\
& \mathrm{Know}_{\Phi\Theta}(f) &
\end{array}
$$

for every $f \in F$.

The right arrow in the object (X, T, A) "forgets" the component T, while the left one "forgets" the component A. Such diagrams associate the category of knowledge with the knowledge base.

7.4. Equivalence of databases and knowledge bases.

7.4.1. *Equivalence of databases.* Let two databases with different (G_1, Φ_1, F_1) and (G_2, Φ_2, F_2) be given. We are interested in informational equivalence of these databases; for another approach see [**PlP**].

Consider pairs (α, γ) where $\alpha : F_1 \to F_2$ is a mapping of sets and γ is a function, defining a homomorphism $\gamma_f : R_f \to R_{f^\alpha}$ for every $f \in F_1$.

The pair (α, γ) is called an *equivalence* of the corresponding databases if α is a bijection and every $\gamma_f, f \in F_1$, is an isomorphism of algebras. Databases are equivalent, if there exists an equivalence (α, γ) between them.

Let us expound on this definition. Take first a kernel $\mathrm{Ker}(\mathrm{Val}_f)$ for every $\mathrm{Val}_f\colon \mathrm{Hal}_{\Phi\Theta} \to R_f$ and pass to a factor algebra $Q_f = \mathrm{Hal}_{\Phi\Theta}/\mathrm{Ker}(\mathrm{Val}_f)$. Let here $\delta_f\colon \mathrm{Hal}_{\Phi\Theta} \to Q_f$ be a natural homomorphism. Represent a homomorphism Val_f as $\mathrm{Val}_f = \mathrm{Val}^\circ_f \delta_f$ where $\mathrm{Val}^\circ_f\colon Q_f \to R_f$ is an isomorphism. We have a diagram of isomorphisms for $f \in F_1$

$$\begin{array}{ccc} Q_f & \xrightarrow{\beta^\circ_f} & Q_{f^\alpha} \\ {\scriptstyle \mathrm{Val}^\circ_f}\downarrow & & \downarrow{\scriptstyle \mathrm{Val}^\circ_{f^\alpha}} \\ R_f & \xrightarrow[\gamma_f]{} & R_{f^\alpha} \end{array}$$

Here $\beta^\circ_f = (\mathrm{Val}^\circ_{f^\alpha})^{-1}\gamma_f \mathrm{Val}^\circ_f$.

Along with the natural homomorphism δ_f, we fix also some function of choice $\delta_f^{-1}\colon Q_f \to \mathrm{Hal}_{\Phi\Theta}$ which chooses a definite $u \in \mathrm{Hal}_{\Phi\Theta}$ with $\delta_f(u) = q$ for every $q \in Q_f$.

Consider special functions β and β'. The function β gives a mapping $\beta_f\colon \mathrm{Hal}_{\Phi_1\Theta} \to \mathrm{Hal}_{\Phi_2\Theta}$ for every $f \in F_1$. First we define a homomorphism $\beta^1_f\colon \mathrm{Hal}_{\Phi_1\Theta} \to Q_{f^\alpha}$ by $\beta^1_f = \beta^\circ_f\delta_f$. Now $\beta_f = \delta^{-1}_{f^\alpha}\beta^1_f = \delta^{-1}_{f^\alpha}\beta^\circ_f\delta_f$. Here β_f is a multi-sorted mapping of algebras, which is not, obviously, a homomorphism.

The function β' chooses a mapping $\beta'_f\colon \mathrm{Hal}_{\Phi_2\Theta} \to \mathrm{Hal}_{\Phi_1\Theta}$ for every $f^\alpha \in F_2$. It is constructed similarly:

$$\beta'_f = \delta_f^{-1}(\beta^\circ_f)^{-1}\delta_{f^\alpha}.$$

Let now $u \in \mathrm{Hal}_{\Phi_1\Theta}$ and $f \in F_1$. Then

$$\begin{aligned} (u * f)^{\gamma_f} &= \gamma_f(\mathrm{Val}_f(u)) = \gamma_f(\mathrm{Val}^\circ_f\, \delta_f(u)) \\ &= \mathrm{Val}^\circ_{f^\alpha}(\mathrm{Val}^\circ_{f^\alpha})^{-1}\gamma_f \mathrm{Val}^\circ_f\, \delta_f(u) = \mathrm{Val}^\circ_{f^\alpha}\, \beta^\circ_f\delta_f(u) \\ &= \mathrm{Val}^\circ_{f^\alpha}\, \delta_{f^\alpha}\delta^{-1}_{f^\alpha}\beta^\circ_f\delta_f(u) = \mathrm{Val}_{f^\alpha}\, \beta_f(u) = u^{\beta_f} * f^\alpha. \end{aligned}$$

Analogously, if $u \in \mathrm{Hal}_{\Phi_2\Theta}$ and $f_1 = f^\alpha \in F_2$, then

$$(u * f^\alpha)^{\gamma_f^{-1}} = u^{\beta'_f} * f.$$

Thus, the answer to the query in the first database can be obtained via the second one and vice versa.

Consider separately a case when $\Phi_1 = \Phi_2 = \Phi$ and every isomorphism γ_f is correct. The latter means that

$$\begin{array}{ccc} \mathrm{Hal}_{\Phi\Theta} & \xrightarrow{\mathrm{Val}_f} & R_f \\ & {\scriptstyle \mathrm{Val}_{f^\alpha}}\searrow & \downarrow{\scriptstyle \gamma_f} \\ & & R_{f^\alpha} \end{array}$$

holds for every $f \in F_1$. Also, for every query $u \in \mathrm{Hal}_{\Phi\Theta}$ and every $f \in F_1$ we have $(f * u)^{\gamma_f} = u * f^\alpha$.

Take a morphism in the first database:

$$\begin{array}{ccc} \mathrm{Hal}_{\Phi\Theta} & \xrightarrow{\mathrm{Val}_{f_1}} & R_{f_1} \\ & \searrow^{\mathrm{Val}_{f_2}} & \downarrow \gamma \\ & & R_{f^\alpha} \end{array}$$

and construct the morphism in the second database:

$$\begin{array}{ccc} \mathrm{Hal}_{\Phi\Theta} & \xrightarrow{\mathrm{Val}_{f_1^\alpha}} & R_{f_1} \\ & \searrow^{\mathrm{Val}_{f_2^\alpha}} & \downarrow \gamma^\alpha \\ & & R_{f_2^\alpha} \end{array}$$

Here $\gamma^\alpha = \gamma_{f_2}\gamma\gamma_{f_1}^{-1}$ can be found from the diagram

$$\begin{array}{ccc} R_{f_1} & \xrightarrow{\gamma_{f_1}} & R_{f_1^\alpha} \\ \gamma \downarrow & & \downarrow \gamma^\alpha \\ R_{f_2} & \xrightarrow{\gamma_{f_2}} & R_{f_2^\alpha} \end{array}$$

We need to check that γ^α is correct: $\gamma^\alpha \mathrm{Val}_{f_1^\alpha} = \mathrm{Val}_{f_2^\alpha}$. We have: $\gamma^\alpha \mathrm{Val}_{f_1^\alpha} = \gamma_{f_2}\gamma\gamma_{f_1}^{-1}\gamma_{f_1} \mathrm{Val}_{f_1} = \gamma_{f_2}\gamma \mathrm{Val}_{f_1} = \gamma_{f_2} \mathrm{Val}_{f_2} = \mathrm{Val}_{f_2^\alpha}$. Hence, the transition $\gamma \to \gamma^\alpha$ gives the functor which is an isomorphism of databases, and in this specific situation equivalence of databases turns out to be their isomorphism.

Let us return to the general case. The relations with the special functions of choice β and β' for the sets F_1 and F_2 can be now represented as commutative diagrams:

$$\begin{array}{ccc} \mathrm{Hal}_{\Phi_1\Theta} & \xrightarrow{\mathrm{Val}_f} & R_f \\ \beta_f \downarrow & & \downarrow \gamma_f \\ \mathrm{Hal}_{\Phi_2\Theta} & \xrightarrow{\mathrm{Val}_{f^\alpha}} & R_{f^\alpha} \end{array} \qquad \begin{array}{ccc} \mathrm{Hal}_{\Phi_1\Theta} & \xrightarrow{\mathrm{Val}_f} & R_f \\ \beta'_f \uparrow & & \uparrow \gamma_f^{-1} \\ \mathrm{Hal}_{\Phi_2\Theta} & \xrightarrow{\mathrm{Val}_{f^\alpha}} & R_{f^\alpha} \end{array}$$

These diagrams along with further remarks mean that equivalence of databases in general can be treated as some semi-isomorphism or skew isomorphism [**Pl8**]. The remarks are as follows.

Let $\gamma\colon R_{f_1} \to R_{f_2}$ be a morphism in the first database. Take $\gamma^\alpha = \gamma_{f_2}\gamma\gamma_{f_1}^{-1}$. Check that for every $u \in \mathrm{Hal}_{\Phi_1\Theta}$ it holds

$$\gamma^\alpha \mathrm{Val}_{f_1^\alpha}(\beta_{f_1}(u)) = \mathrm{Val}_{f_2^\alpha}(\beta_{f_2}(u)).$$

Indeed,

$$\begin{aligned} \gamma^\alpha \mathrm{Val}_{f_1^\alpha}(\beta_{f_1}(u)) &= \gamma^\alpha(\beta_{f_1}(u) * f_1^\alpha) \\ &= \gamma^\alpha\gamma_{f_1}(u * f_1) = \gamma_{f_2}\gamma\gamma_{f_1}^{-1}\gamma_{f_1}(u * f_1) \\ &= \gamma_{f_2}\gamma(u * f_1) = \gamma_{f_2}(u * f_2) = \beta_{f_2}(u) * f_2^\alpha \\ &= \mathrm{Val}_{f_2^\alpha}(\beta_{f_2}(u)). \end{aligned}$$

Thus, γ^α is no longer a morphism in the second database, but rather some "skew" morphism. The same for the transition from the second database to the first one with the mapping β'.

7.5. Equivalence of knowledge bases. Again we regard multimodels (G_1, Φ_1, F_1) and (G_2, Φ_2, F_2) and the related knowledgebases, denoted by KB_1 and KB_2.

Consider pairs (α, γ) where $\alpha : F_1 \to F_2$ is a bijection of sets and γ is a function, determining an isomorphism of weak categories:

$$\gamma_f \colon K_{\Phi_1\Theta}(f) \to K_{\Phi_2\Theta}(f^\alpha).$$

We assume that the isomorphism γ_f induces an isomorphism of algebras $\gamma_f : R_f \to R_{f^\alpha}$. It preserves the type X and is coordinated with the inclusion of sets of points in affine spaces.

Consider every such pair (α, γ) as an equivalence of knowledge bases: knowledge bases are equivalent if there is some equivalence (α, γ).

As was done for databases, let us pass to motivations and distinguish the case when $\Phi_1 = \Phi_2 = \Phi$ and isomorphism γ_f is correct. The latter means that

$$\begin{array}{ccc} \mathrm{Hal}_{\Phi\Theta} & \xrightarrow{\mathrm{Val}_f} & R_f \\ & \searrow^{\mathrm{Val}_{f^\alpha}} & \downarrow{\scriptstyle \gamma_f} \\ & & K_{\Phi\Theta}(f^\alpha) \end{array}$$

for every $f \in F_1$.

Take an arbitrary set of formulas T in the definite $\mathrm{Hal}_{\Phi\Theta}(X)$. Then

$$\begin{aligned} T^{f^\alpha} &= \bigcap_{u \in T} \mathrm{Val}_{f^\alpha}(u) = \bigcap_{u \in T} \gamma_f \,\mathrm{Val}_f(u) \\ &= \gamma_f\Big(\bigcap_{u \in T} \mathrm{Val}_f(u)\Big) = \gamma_f T^f. \end{aligned}$$

We have used here the correlation of γ_f with inclusions which brings correlation with intersections. Thus, all T^{f^α} and T^f are well correlated by the isomorphism γ_f.

Check now that the pair (α, γ) gives an isomorphism of categories KB_1 and KB_2. If $\mathrm{Val}_f : \mathrm{Hal}_{\Phi\Theta} \to K_{\Phi\Theta}(f)$ is an object in the first category, then the corresponding object of the second category is $\mathrm{Val}_{f^\alpha} : \mathrm{Hal}_{\Phi\Theta} \to K_{\Phi\Theta}(f^\alpha)$, and vice versa. Consider a morphism

$$\begin{array}{ccc} \mathrm{Hal}_{\Phi\Theta} & \xrightarrow{\mathrm{Val}_{f_1}} & K_{\Phi\Theta}(f_1) \\ & \searrow^{\mathrm{Val}_{f_2}} & \downarrow{\scriptstyle \gamma} \\ & & K_{\Phi\Theta}(f_2) \end{array}$$

The corresponding diagram is

$$\begin{array}{ccc} \mathrm{Hal}_{\Phi\Theta} & \xrightarrow{\mathrm{Val}^\alpha_{f_1}} & K_{\Phi\Theta}(f_1^\alpha) \\ & \searrow^{\mathrm{Val}^\alpha_{f_2}} & \downarrow{\scriptstyle \gamma^\alpha = \gamma_{f_2}\gamma\gamma_{f_1}^{-1}} \\ & & K_{\Phi\Theta}(f_2^\alpha) \end{array}$$

This second diagram is actually a morphism in the second category.

Thus, in this special situation, equivalence of knowledge bases is reduced to their isomorphism. The same fact has been established for the databases.

Now let us make some remarks on the general case. For the given set of formulas T in $\mathrm{Hal}_{\Phi_1\Theta}(X)$ and the mapping $\beta_f : \mathrm{Hal}_{\Phi_1\Theta} \to \mathrm{Hal}_{\Phi_2\Theta}$, consider a set T^{β_f} in $\mathrm{Hal}_{\Phi_2\Theta}(X)$, defined by

$$T^{\beta_f} = \{\beta_f(u) | u \in T\}.$$

Check that $T^{\beta_f f^\alpha} = \gamma_f T^f$ for every $f \in F_1$. We have

$$\begin{aligned} T^{\beta_f f^\alpha} &= \bigcap_{u\in T} \mathrm{Val}_{f^\alpha}(\beta_f(u)) = \bigcap_{u\in T} (\beta_f(u) * f^\alpha) \\ &= \bigcap_{u\in T} \gamma_f(u * f) = \gamma_f(\bigcap_{u\in T} \mathrm{Val}_f(u)) = \gamma_f T^f. \end{aligned}$$

Similarly, if T is a set of formulas in $\mathrm{Hal}_{\Phi_2\Theta}(X)$, then

$$T^{\beta_f' f} = \gamma_f^{-1} T^{f^\alpha}.$$

The pointed relations mean that equivalence of knowledge bases KB_1 and KB_2 provides the corresponding informational equivalence on the level of the transition from the description of knowledge to its content.

Let now $\gamma\colon K_{\Phi_1\Theta}(f_1) \to K_{\Phi_1\Theta}(f_2)$ be a morphism in KB_1. Take $\gamma^\alpha = \gamma_{f_2}\gamma\gamma_{f_1}^{-1}$. One can check that $\gamma^\alpha T^{\beta_{f^1} f_1^\alpha} = T^{\beta_{f_2} f_2^\alpha}$ for every T in $\mathrm{Hal}_{\Phi_1\Theta}(X)$. In particular, it means that $\gamma^\alpha : K_{\Phi_2\Theta}(f_1^\alpha) \to K_{\Phi_2\Theta}(f_2^\alpha)$ is not a morphism in KB_2. Also here we have some "skew" property which needs additional motivation [**Pl8**].

7.5.1. *Main results.* Two *multimodels* (G_1, Φ, F_1) and (G_2, Φ, F_2) with the same Φ are called *geometrically equivalent* if there is a bijection $\alpha : F_1 \to F_2$ such that for every $f \in F_1$ models (G_1, Φ, f) and (G_2, Φ, f^α) are geometrically equivalent.

THEOREM 7.1. *If multimodels (G_1, Φ, F_1) and (G_2, Φ, F_2) are geometrically equivalent, then the knowledge bases KB_1 and KB_2 are isomorphic and, hence, equivalent.*

Let us say that two *multimodels* (G_1, Φ, F_1) and (G_2, Φ, F_2) are *isomorphic* if there is a bijection $\alpha : F_1 \to F_2$ such that for every $f \in F_1$ the models (G_1, Φ, f) and (G_2, Φ, f^α) are isomorphic. If multimodels are isomorphic, then they are geometrically equivalent and the corresponding bases KB_1 and KB_2 are isomorphic. If here the algebras G_1 and G_2 are finite, then the opposite is true as well.

THEOREM 7.2. *If algebras G_1 and G_2 in (G_1, Φ, F_1) and (G_2, Φ, F_2) are finite, then the knowledge bases KB_1 and KB_2 are isomorphic if and only if the multimodels are isomorphic.*

This theorem gives the algorithm of verification of isomorphism of two finite knowledge bases.

Consider now the question of equivalence of two finite knowledge bases. Here we need additional definitions.

DEFINITION 7.1. Let (G_1, Φ_1, f_1) and (G_2, Φ_2, f_2) be two models (Φ_1 and Φ_2 may be different). We call them automorphic equivalent, if

1) algebras G_1 and G_2 are isomorphic;

2) groups $\mathrm{Aut}(f_1)$ and $\mathrm{Aut}(f_2)$ are conjugated by some isomorphism of algebras G_1 and G_2.

In other words, there exists an isomorphism $\delta : G_2 \to G_1$ such that

$$\mathrm{Aut}(f_2) = \delta^{-1}\,\mathrm{Aut}(f_1)\delta.$$

DEFINITION 7.2. Two multimodels (G_1, Φ_1, F_1) and (G_2, Φ_2, F_2) are automorphic equivalent, if for some bijection $\alpha : F_1 \to F_2$ the models (G_1, Φ_1, f) and (G_2, Φ_2, f^α) are automorphic equivalent for every $f \in F_1$.

PROPOSITION 7.1. *If the multimodels (G_1, Φ_1, F_1) and (G_2, Φ_2, F_2) are automorphic equivalent then the corresponding knowledge bases KB_1 and KB_2 are equivalent.*

Let now the multimodels (G_1, Φ_1, F_1) and (G_2, Φ_2, F_2) with finite $G_1, G_2 \in \Theta$ be given; KB_1 and KB_2 are the corresponding knowledge bases.

THEOREM 7.3. *Knowledge bases KB_1 and KB_2 are equivalent if and only if their multimodels are automorphic equivalent.*

The proof of the theorem is based on Galois theory from the previous lecture.

It also gives the algorithm of verification of two finite knowledge bases (compare [**PlP**].

Let us note that it is natural to consider semi-isomorphisms along with the isomorphisms of knowledge bases with the same Φ. Semi-isomorphisms are described by the diagrams of the form

$$\begin{array}{ccc}
\mathrm{Hal}_{\Phi\Theta} & \xrightarrow{\mathrm{Val}_f} & K_{\Phi\Theta}(f) \\
{\scriptstyle\sigma}\downarrow & & \downarrow{\scriptstyle\gamma_f} \\
\mathrm{Hal}_{\Phi\Theta} & \xrightarrow{\mathrm{Val}_{f^\alpha}} & K_{\Phi\Theta}(f^\alpha)
\end{array}$$

where σ is an automorphism of the algebra $\mathrm{Hal}_{\Phi\Theta}$ and γ_f is an isomorphism of the categories. According to such a diagram, we have

$$\gamma_f\,\mathrm{Val}_f(u) = (u * f)^{\gamma_f} = \mathrm{Val}_{f^\alpha}(\sigma u) = \sigma u * f^\alpha$$

for every $u \in \mathrm{Hal}_{\Phi\Theta}$. Thus, we may replace the mappings β_f and β'_f in the general situation by universal mappings σ and σ^{-1}. In this case, all the pictures with "skew" become more visible. We see that semi-isomorphism implies equivalence of knowledge bases, as does isomorphism. We can also consider a category of knowledge bases for fixed Φ and Θ. These are categories with usual (special) morphisms, and categories with semimorphisms. We can consider monads and comonads in them.

7.6. Conclusion. The main problem in the computation of knowledge is to find the content of knowledge $A = T^f$ by the given description of knowledge T. Since $A = \bigcap_{u\in T} \mathrm{Val}_f(u)$, we need to compute the sets $\mathrm{Val}_f(u)$ for various $u \in \mathrm{Hal}_{\Phi\Theta}$.

We pass from the algebra $\mathrm{Hal}_{\Phi\Theta}$ to the corresponding free Halmos algebra $\tilde{H} = \tilde{H}_{\Phi\Theta}$ (see Lecture 6). We treat every u as an element in $\tilde{H}$. We consider an algebra $\tilde{H}$ to be the constructive one. In other words, every element u is well represented by atoms. Thus, the computation of the sets $\mathrm{Val}_f(u)$ with arbitrary u is reduced to that for atoms. This computation is somehow made for constructive models (G, Φ, f). All the above can be realized via some comonada in the category of knowledge bases. One may also consider monads in these categories for enrichment of knowledge.

Let us note the following general problem: what is the inference of knowledge in the categories of knowledge under consideration? The problem is to formalize well the corresponding idea.

References

[B] H. Bauer, *About Hilbert and Ruckkert Nullstellensatz*, preprint.

[BMR1] G. Baumslag, A. Myasnikov, and V. Remeslennikov, *Algebraic geometry over groups. I. Algebraic sets and ideal theory*, J. Algebra **219** (1999), no. 1, 16–79, DOI 10.1006/jabr.1999.7881. MR1707663

[BMR2] G. Baumslag, A. Myasnikov, and V. Remeslennikov, *Algebraic geometry over groups*, Algorithmic problems in groups and semigroups (Lincoln, NE, 1998), Trends Math., Birkhäuser Boston, Boston, MA, 2000, pp. 35–50. MR1750491

[BMRo] G. Baumslag, A. Myasnikov, and V. Roman′kov, *Two theorems about equationally Noetherian groups*, J. Algebra **194** (1997), no. 2, 654–664, DOI 10.1006/jabr.1997.7025. MR1467171

[BelP] A. Belov and B. Plotkin, *Abstract and natural Zariski topology*, preprint.

[Be1] A. Berzins, *Geometric equivalence of algebras*, Internat. J. Algebra Comput. **11** (2001), no. 4, 447–456, DOI 10.1142/S0218196701000668. MR1850211

[Be2] A. Berzins, *Variety Grp-F is semiperfect*, preprint, Riga, 1998.

[BPP] A. Berzins, B. Plotkin, and E. Plotkin, *Algebraic geometry in varieties of algebras with the given algebra of constants*, J. Math. Sci. (New York) **102** (2000), no. 3, 4039–4070, DOI 10.1007/BF02673880. MR1797978

[BlG] V. V. Bludov and B. V. Gusev, *Geometric equivalence of groups*, Proc. Steklov Inst. Math. **257** (2007), no. suppl. 1, S61–S82, DOI 10.1134/S0081543807050045. MR2338240

[Br] R. M. Bryant, *The verbal topology of a group*, J. Algebra **48** (1977), no. 2, 340–346, DOI 10.1016/0021-8693(77)90312-X. MR0453878

[Co] P. M. Cohn, *Universal algebra*, Harper & Row, Publishers, New York-London, 1965. MR0175948

[D] Y. Diers, *Categories of algebraic sets*, Appl. Categ. Structures **4** (1996), no. 2-3, 329–341, DOI 10.1007/BF00122260. MR1406106

[DF] J. L. Dyer and E. Formanek, *The automorphism group of a free group is complete*, J. London Math. Soc. (2) **11** (1975), no. 2, 181–190, DOI 10.1112/jlms/s2-11.2.181. MR0379683

[For] E. Formanek, *A question of B. Plotkin about the semigroup of endomorphisms of a free group*, Proc. Amer. Math. Soc. **130** (2002), no. 4, 935–937, DOI 10.1090/S0002-9939-01-06155-X. MR1873764

[F] L. Fuchs, *Infinite Abelian Groups*, Academic Press, New York, London, 1970.

[GoSh] R. Göbel and S. Shelah, *Radicals and Plotkin's problem concerning geometrically equivalent groups*, Proc. Amer. Math. Soc. **130** (2002), no. 3, 673–674, DOI 10.1090/S0002-9939-01-06108-1. MR1866018

[Gor] V.A. Gorbunov, *Algebraic theory of quasivarieties*, Doctoral Thesis, Novosibirsk, 1996, Plenum Publ. Co., 1998.

[GK] R. I. Grigorchuk and P. F. Kurchanov, *Some questions of group theory related to geometry*, Algebra, VII, Encyclopaedia Math. Sci., vol. 58, Springer, Berlin, 1993, pp. 167–232, 233–240. Translated from the Russian by P. M. Cohn. MR1265271

[Gro] A. Grothendieck, *The cohomology theory of abstract algebraic varieties*, Proc. Internat. Congress Math. (Edinburgh, 1958), Cambridge Univ. Press, New York, 1960, pp. 103–118. MR0130879

[Gu] V. S. Guba, *Equivalence of infinite systems of equations in free groups and semigroups to finite subsystems* (Russian), Mat. Zametki **40** (1986), no. 3, 321–324, 428. MR869922

[Hal] P. R. Halmos, *Algebraic logic*, Chelsea Publishing Co., New York, 1962. MR0131961

[Har] R. Hartshorne, *Algebraic geometry*, Graduate Texts in Mathematics, No. 52, Springer-Verlag, New York-Heidelberg, 1977. MR0463157

[HMT] L. Henkin, J. Monk, A. Tarski, *Cylindric Algebras*, North Holland, 1985.

[Hi] P. J. Higgins, *Groups with multiple operators*, Proc. London Math. Soc. (3) **6** (1956), 366–416, DOI 10.1112/plms/s3-6.3.366. MR0082492

[HZ] E. Hrushovski and B. Zilber, *Zariski geometries*, J. Amer. Math. Soc. **9** (1996), no. 1, 1–56, DOI 10.1090/S0894-0347-96-00180-4. MR1311822

[KM] O. Kharlampovich and A. Myasnikov, *Irreducible affine varieties over a free group. I. Irreducibility of quadratic equations and Nullstellensatz*, J. Algebra **200** (1998), no. 2, 472–516, DOI 10.1006/jabr.1997.7183. MR1610660

[KM1] O. Kharlampovich and A. Myasnikov, *Irreducible affine varieties over a free group. II. Systems in triangular quasi-quadratic form and description of residually free groups*, J. Algebra **200** (1998), no. 2, 517–570, DOI 10.1006/jabr.1997.7184. MR1610664

[KP] E. Koublanova and B. Plotkin, *Algebraic geometry in group representations*, preprint, Jerusalem, 1999.

[KR] M.I. Krasner, *Generalization et analogues de la theorie de Galois*, Comptes Rendus de congress de la Victorie de l'Ass. Frac. pour I'Avancem. Sci., 1945, 54–58.

[Ku] A. Kurosh, *Lectures on general algebra*, Moscow, Nauka, 1973.

[L] S. Lang, *Algebra*, Addison-Wesley Publishing Co., Inc., Reading, Mass., 1965. MR0197234

[Lav] F. W. Lawvere, *Continuously variable sets; algebraic geometry=geometric logic*, Logic Colloquium '73 (Bristol, 1973), North-Holland, Amsterdam, 1975, pp. 135–156. MR0392567

[Li] A. Lichtman, *Two-generator division algebras and Plotkin's question*, preprint.

[L1] R. C. Lyndon, *Groups with parametric exponents*, Trans. Amer. Math. Soc. **96** (1960), 518–533, DOI 10.2307/1993539. MR0151502

[L2] R. C. Lyndon, *Equations in groups*, Bol. Soc. Brasil. Mat. **11** (1980), no. 1, 79–102, DOI 10.1007/BF02584882. MR607019

[LS] R. C. Lyndon and P. E. Schupp, *Combinatorial group theory*, Springer-Verlag, Berlin-New York, 1977. MR0577064

[ML] S. Mac Lane, *Categories for the working mathematician*, Graduate Texts in Mathematics, Vol. 5, Springer-Verlag, New York-Berlin, 1971. MR0354798

[Ma] G. Makanin, *Decidability of the universal and positive theories of a free group*, Izv. Math. USSR. **25** (1985), no. 1, 75–88.

[Mal1] A.I. Mal′cev, *Algebraic Systems*, North Holland, 1973.

[Mal2] A. I. Mal′cev, *Several remarks on quasivarieties of algebraic systems* (Russian), Algebra i Logika Sem. **5** (1966), no. 3, 3–9. MR0205902

[Man1] Yu. I. Manin, *Quantum groups and noncommutative geometry*, Université de Montréal, Centre de Recherches Mathématiques, Montreal, QC, 1988. MR1016381

[Man2] Y. I. Manin, *Topics in noncommutative geometry*, M. B. Porter Lectures, Princeton University Press, Princeton, NJ, 1991. MR1095783

[MPP] G. Mashevitzky, B. Plotkin, and E. Plotkin, *Automorphisms of categories of free algebras of varieties*, Electron. Res. Announc. Amer. Math. Soc. **8** (2002), 1–10, DOI 10.1090/S1079-6762-02-00099-9. MR1911740

[Me] Ju. I. Merzljakov, *Positive formulae on free groups* (Russian), Algebra i Logika Sem. **5** (1966), no. 4, 25–42. MR0222149

[MR1] A. Myasnikov and V. Remeslennikov, *Algebraic geometry over groups I*, J. Algebra, **219:1** (1999) 16–79.

[MR2] A. Myasnikov and V. Remeslennikov, *Algebraic geometry over groups. II. Logical foundations*, J. Algebra **234** (2000), no. 1, 225–276, DOI 10.1006/jabr.2000.8414. MR1799485

[Ne1] B. H. Neumann, *Adjunction of elements to groups*, J. London Math. Soc. **18** (1943), 4–11, DOI 10.1112/jlms/s1-18.1.4. MR0008808

[Ne2] B. H. Neumann, *A note on algebraically closed groups*, J. London Math. Soc. **27** (1952), 247–249, DOI 10.1112/jlms/s1-27.2.247. MR0046363

[NP] D. B. Nikolova and B. I. Plotkin, *Some notes on universal algebraic geometry*, Algebra (Moscow, 1998), de Gruyter, Berlin, 2000, pp. 237–261. MR1754673

[OMe] O. T. O'Meara, *Lectures on linear groups*, American Mathematical Society, Providence, R.I., 1974. Expository Lectures from the CBMS Regional Conference held at Arizona State University, Tempe, Ariz., March 26–30, 1973; Conference Board of the Mathematical Sciences Regional Conference Series in Mathematics, No. 22. MR0349862

[Pl1] B. I. Plotkin, *Groups of automorphisms of algebraic systems*, Wolters-Noordhoff Publishing, Groningen, 1972. Translated from the Russian by K. A. Hirsch. MR0344322

[Pl2] B. Plotkin, *Universal algebra, algebraic logic, and databases*, Mathematics and its Applications, vol. 272, Kluwer Academic Publishers Group, Dordrecht, 1994. Translated from the 1991 Russian original by J. Cīrulis, A. Nenashev and V. Pototsky and revised by the author. MR1273136

[Pl3] B. Plotkin, *Varieties of algebras and algebraic varieties.* part B, Israel J. Math. **96** (1996), 511–522, DOI 10.1007/BF02937320. MR1433704

[Pl4] B. Plotkin, *Varieties of algebras and algebraic varieties. Categories of algebraic varieties*, Siberian Adv. Math. **7** (1997), no. 2, 64–97. MR1481222

[Pl5] B. I. Plotkin, *Some concepts of algebraic geometry in universal algebra* (Russian, with Russian summary), Algebra i Analiz **9** (1997), no. 4, 224–248; English transl., St. Petersburg Math. J. **9** (1998), no. 4, 859–879. MR1604318

[Pl6] B. Plotkin, *Algebraic logic, varieties of algebras and algebraic varieties*, Proc. Int. Alg. Conf., St. Petersburg, 1995, St. Petersburg, 1999, 189–271.

[Pl7] B. Plotkin, *Zero divisors in group-based algebras. Algebras without zero divisors* (English, with English, Russian and Moldavian summaries), Bul. Acad. Ştiinţe Repub. Mold. Mat. **2** (1999), 67–84. MR1738244

[Pl8] B. Plotkin, *Algebra, categories and databases*, Handbook of Algebra, Vol. 2, Elsevier/North-Holland, Amsterdam, 2000, pp. 79–148, DOI 10.1016/S1570-7954(00)80028-5. MR1759596

[Pl9] B. Plotkin, *Algebraic geometry in first-order logic* (Russian, with Russian summary), Sovrem. Mat. Prilozh. **22, Algebra i Geom.** (2004), 16–62, DOI 10.1007/s10958-006-0288-2; English transl., J. Math. Sci. (N.Y.) **137** (2006), no. 5, 5049–5097. MR2462070

[Pl10] B. Plotkin, *Infinitary quasi-identities and infinitary quasivarieties* (English, with English and Latvian summaries), Proc. Latv. Acad. Sci. Sect. B Nat. Exact Appl. Sci. **57** (2003), no. 3-4, 111–112. MR2016272

[Pl11] B. Plotkin, *Algebras with the same (algebraic) geometry*, Tr. Mat. Inst. Steklova **242** (2003), Mat. Logika i Algebra, 176–207; translation in Proc. Steklov Inst. Math. 2003, no. 3 (242), 165–196. MR2054494

[PPT] B. Plotkin, E. Plotkin, and A. Tsurkov, *Geometrical equivalence of groups*, Comm. Algebra **27** (1999), no. 8, 4015–4025, DOI 10.1080/00927879908826679. MR1700201

[Pl] T. Plotkin, *Relational databases equivalence problem*, Advances of Databases and Information Systems, Springer, 1996, 391–404.

[PlP] T. Plotkin and B. Plotkin, *Geometrical aspect of databases and knowledge bases*, Algebra Universalis **46** (2001), no. 1-2, 131–161, DOI 10.1007/PL00000333. The Viktor Aleksandrovich Gorbunov memorial issue. MR1835790

[Ra] A. A. Razborov, *On systems of equations in free groups*, Combinatorial and Geometric Group Theory (Edinburgh, 1993), London Math. Soc. Lecture Note Ser., vol. 204, Cambridge Univ. Press, Cambridge, 1995, pp. 269–283. MR1320290

[R] E. Rips, *Cyclic splittings of finitely presented groups and the canonical JSJ-decomposition*, Proceedings of the International Congress of Mathematicians, Vol. 1, 2 (Zürich, 1994), Birkhäuser, Basel, 1995, pp. 595–600. MR1403960

[RS] E. Rips and Z. Sela, *Cyclic splittings of finitely presented groups and the canonical JSJ decomposition*, Ann. of Math. (2) **146** (1997), no. 1, 53–109, DOI 10.2307/2951832. MR1469317

[Ro1] A. L. Rosenberg, *Noncommutative algebraic geometry and representations of quantized algebras*, Mathematics and its Applications, vol. 330, Kluwer Academic Publishers Group, Dordrecht, 1995. MR1347919

[Ro2] A. L. Rosenberg, *Noncommutative schemes*, Compositio Math. **112** (1998), no. 1, 93–125, DOI 10.1023/A:1000479824211. MR1622759

[S] W. R. Scott, *Algebraically closed groups*, Proc. Amer. Math. Soc. **2** (1951), 118–121, DOI 10.2307/2032631. MR0040299

[Se] Z. Sela, *Diophantine geometry over groups I*, IHES **93** (2001), 31–105.

[Sha] I. Shapharevich, *The Foundations of Algebraic Geometry*, Nauka, Moscow, 1972.

[Shi] Yan Shi-jian, *Linear groups over a ring*, Chinese Math.–Acta **7** (1965), 163–179. MR0222185

[ZS] O. Zariski and P. Samuel, *Commutative algebra, Volume I*, The University Series in Higher Mathematics, D. Van Nostrand Company, Inc., Princeton, New Jersey, 1958. MR0090581

Institute of Mathematics, Hebrew University, Jerusalem, Israel
Email address: borisov@math.huji.ac.il

Contemporary Mathematics
Volume **726**, 2019
https://doi.org/10.1090/conm/726/14612

New problems in universal algebraic geometry illustrated by Boolean equations

Artem N. Shevlyakov

ABSTRACT. We discuss new problems in universal algebraic geometry and explain them by boolean equations.

1. Introduction

The process of solving equations is the central part of mathematics. The most general and important problems in this area are the following.

(1) Is a given equation consistent over an algebraic structure (algebra for short) $\mathcal{A}$?
(2) Find all solutions of a given equation over an algebra $\mathcal{A}$.

There are many surveys and papers devoted to equations in various classes of algebras. Let us just mention the survey [**5**] for group equations.

However, recent achievements of universal algebraic geometry (see the papers [**2**,**3**] by E. Daniyarova, A. Miasnikov, V. Remeslennikov, and [**4**] by B. Plotkin) allow us to pose new problems on equations (all required definitions may be found in Section 2 of the present paper).

(3) *Systems of equations vs. algebraic sets.* Let Y be an algebraic set over an algebra $\mathcal{A}$. Obviously, there exists more than one systems of equations (systems for short) with solution set Y. Let us fix a family of systems $\mathbf{S}$, and let $S(Y) \subseteq \mathbf{S}$ be all systems with solution set Y. It turns out that the numbers $|S(Y)|$ have a wide spread of values for almost all natural $\mathbf{S}$ (e.g. in [**8**] this fact was proved for semilattice equations). Thus, the following problem arises: do there exist an algebra $\mathcal{A}$ and a natural family $\mathbf{S}$ such that the variance of the set $\{|S(Y)|\}$ is minimal?

   The meaning of this problem is the following. Suppose we want to generate random algebraic sets over an algebra $\mathcal{A}$ by a random generation of systems from $\mathbf{S}$. If the variance of the set $\{|S(Y)|\}$ is small, the random distribution of algebraic sets becomes close to the uniform distribution.
(4) *Irreducible algebraic sets.* Let Y be an algebraic set over an algebra $\mathcal{A}$. Is there an algorithm that decides whether Y is irreducible or not? If Y is reducible, can we find its irreducible components? Can one find the average number of irreducible components of all algebraic sets in $\mathcal{A}^n$?

2010 *Mathematics Subject Classification.* Primary 03G05, 03C98.
The author was supported by the Russian Science Foundation (project 17-11-01117).

The importance of this problem is the following. According to [**2**, **3**], the structure of irreducible algebraic sets over $\mathcal{A}$ determines the universal theory of $\mathcal{A}$. Moreover, if $\mathcal{A}$ is finite, irreducible algebraic sets over $\mathcal{A}$ correspond to subalgebras of $\mathcal{A}$.

(5) *Isomorphic algebraic sets.* In [**2**] isomorphisms between algebraic sets, namely, isomorphic algebraic sets have the same properties with respect to universal algebraic geometry were defined. For any algebra $\mathcal{A}$ one can pose the following problem: how many non-isomorphic algebraic sets are there in $\mathcal{A}^n$? The solution of this problem allows us to decide about the complexity of the class of all algebraic sets over $\mathcal{A}$.

(6) *Equationally extremal algebras.* Let $\mathcal{A}_n$ be the class of $\mathcal{L}$-algebras of order n (for example, $\mathcal{A}_n$ is the class of all semilattices of order n) and $\mathbf{S}$ a finite set of systems. The problem is the following: find an algebra $\mathcal{A} \in \mathcal{A}_n$ such that the number of consistent systems from $\mathbf{S}$ is maximal (minimal) for $\mathcal{A}$.

Let us cite the papers, where the problems above were solved for some algebras. In [**9**], we describe irreducible algebraic sets and compute the average number of irreducible components of algebraic sets over linearly ordered semilattices. In [**8**], for the class of semilattices of order n, we describe equationally extremal semilattices, which have maximal (minimal) number of consistent equations. Above we mentioned the paper [**8**], where we consider problem (3) for semilattices. The results of all these papers show that problems (3)–(6) are nontrivial even for simple algebras. However, there exists a class of algebras, the class of boolean algebras, where these problems have nice solutions.

Thus, the aim of this paper is the solution of problems (3)–(5) in the class of finite boolean algebras (problem (6) is not relevant for boolean algebras since $|\mathcal{A}_n| \leq 1$ for any $n \in \mathbb{N}$). So the reader may consider this paper as a vast testing ground for examples for these problems.

The paper is organized as follows. In Section 2 we give basic notions of universal algebraic geometry. Section 3 contains the rules of transformations of equations over boolean algebras. Actually, any boolean system $\mathcal{S}(X)$ in n variables X can be transformed to an equivalent *orthogonal* system $\mathcal{S}'(Z)$ in 2^n variables Z. Solving the problem (3), we prove that any algebraic set defined by a system in n variables is isomorphic to the solution set of a unique orthogonal system in 2^n variables.

In Section 4, we describe irreducible algebraic sets over finite boolean algebras and decompose any algebraic set into a finite union of irreducible ones. In Section 5, we compute the average number of irreducible components of algebraic sets over finite boolean algebras. In Section 6, we give the definition of rank of irreducibility $IR(\mathcal{S})$ of a system $\mathcal{S}$ and count the average rank of irreducibility of all orthogonal systems in 2^n variables. Thus, Sections 4–6 solve the problem (4) for finite boolean algebras.

In Section 7, we study problem (5) and directly compute the number of pairs of isomorphic algebraic sets defined by orthogonal systems in 2^n variables.

2. Basic notions

Let $\mathcal{L} = \{\vee^{(2)}, \cdot^{(2)}, {}^{-(1)}, 0, 1\}$ be a language of binary functional symbols $\vee, \cdot$ (join and meet), unary symbol $^-$ (complement) and constant symbols $0, 1$. Clearly, boolean algebras are algebraic structures of the language $\mathcal{L}$ with natural interpretation of functional and constant symbols (see [**1**] for more details).

Recall that for any finite boolean algebra $\mathcal{B}$ there exists a number $r \geq 1$ such that $\mathcal{B}$ is isomorphic to the power set algebra on r elements ($|\mathcal{B}| = 2^r$). The number r is called the *rank* of a boolean algebra $\mathcal{B}$. We assume below that any boolean algebra $\mathcal{B}$ is nontrivial, i.e., $|\mathcal{B}| > 1$.

An element a is an *atom (co-atom)* if $\{ab \mid b \in \mathcal{B}\} = \{0, a\}$ (respectively, $\{a \vee b \mid b \in \mathcal{B}\} = \{a, 1\}$). Note that the rank of a finite boolean algebra is equal to the number of atoms (co-atoms).

Following [**2**], let us give the basic notions of algebraic geometry over boolean algebras.

Let $X = \{x_1, x_2, \ldots, x_n\}$ be a finite set of variables. A term $t(X)$ of the language $\mathcal{L}$ is called an $\mathcal{L}$-term. The set of all $\mathcal{L}$-terms in variables X is denoted by $\mathcal{T}_{\mathcal{L}}(X)$. A *boolean equation* is an atomic formula $\tau(X) = \sigma(X)$ of the language $\mathcal{L}$ (τ, σ are $\mathcal{L}$-terms). Examples of boolean equations are the following expressions: $x_i = x_j$, $x_1 = 1$, $x_1x_2 = x_3 \vee x_4$, $\bar{x}_1 \vee x_2 = \overline{x_3x_4}$.

A *system of equations* (*system* for short) is an arbitrary set of boolean equations. The *set of all solutions (solution set)* of a system $\mathcal{S}$ over a boolean algebra $\mathcal{B}$ is denoted by $\mathrm{V}_{\mathcal{B}}(\mathcal{S})$.

A set $Y \subseteq \mathcal{B}^n$ is *algebraic* over a boolean algebra $\mathcal{B}$ if there exists a system $\mathcal{S}$ such that $Y = \mathrm{V}_{\mathcal{B}}(\mathcal{S})$. A nonempty algebraic set Y is *irreducible* if it is not a finite proper union of other algebraic sets. According to [**2**], any algebraic set $Y \subseteq \mathcal{B}^n$ is decomposable into a finite union of irreducible algebraic sets

$$Y = Y_1 \cup Y_2 \cup \ldots \cup Y_m \ (Y_i \nsubseteq Y_j \text{ for } i \neq j), \tag{1}$$

and decomposition (1) is unique up to a permutation of the sets Y_i. The sets Y_i in (1) are called the *irreducible components* of Y.

Let $Y = \mathrm{V}_{\mathcal{B}}(\mathcal{S})$ be an algebraic set over a boolean algebra $\mathcal{B}$ where $\mathcal{S}$ depends on variables $X = \{x_1, x_2, \ldots, x_n\}$. One can define an equivalence relation $\sim_Y$ on $\mathcal{T}_{\mathcal{L}}(X)$ as follows:

$$t(X) \sim_Y s(X) \Leftrightarrow t(P) = s(P) \text{ for each point } P \in Y.$$

The set of $\sim_Y$-equivalence classes is called the *coordinate algebra* of Y and is denoted by $\Gamma_{\mathcal{B}}(Y)$. By the results of [**2**], $\Gamma_{\mathcal{B}}(Y)$ is a boolean algebra generated by the elements $x_1, x_2, \ldots, x_n$. In other words, all coordinate algebras are finitely generated, and, therefore, *all coordinate algebras of algebraic sets over boolean algebras are finite.*

The following statement describes the properties of coordinate algebras of irreducible algebraic sets.

THEOREM 2.1. *An algebraic set Y is irreducible over a boolean algebra $\mathcal{B}$ iff $\Gamma_{\mathcal{B}}(Y)$ is embedded into $\mathcal{B}$.*

PROOF. In fact, in [**2**, Theorem A] it was proved that $\Gamma_{\mathcal{B}}(Y)$ is discriminated by $\mathcal{B}$ iff the algebraic set Y is irreducible. Since $\Gamma_{\mathcal{B}}(Y)$ is finite, the discrimination is equivalent to the embedding of $\Gamma_{\mathcal{B}}(Y)$ into $\mathcal{B}$. □

There are different algebraic sets with isomorphic coordinate algebras. For example, the algebraic sets

$$Y_1 = \mathrm{V}_{\mathcal{B}}(\{x_1x_2 = x_2\}),\ Y_2 = \mathrm{V}_{\mathcal{B}}(\{x_1x_2 = x_1\})$$

have isomorphic coordinate algebras $\Gamma_{\mathcal{B}}(Y_1), \Gamma_{\mathcal{B}}(Y_2)$ since the second equation is obtained from the first one by the variable substitution (in Example 2.2 we directly compute the coordinate algebras of the sets Y_1, Y_2).

Following [**2**], algebraic sets are *isomorphic* if they have isomorphic coordinate algebras.

EXAMPLE 2.2. Let us compute the coordinate algebra of the algebraic set $Y = \mathrm{V}_{\mathcal{B}}(x_1x_2 = x_2)$, where $\mathcal{B}$ is an arbitrary nontrivial boolean algebra. By definition, $\Gamma_{\mathcal{B}}(Y)$ is generated by the elements x_1, x_2 (we identify here the term x_i with its $\sim_Y$-equivalence class). According to the axioms of boolean algebras, the equality $x_1x_2 = x_2$ gives that the term $\bar{x}_1x_2$ equals 0 in $\Gamma_{\mathcal{B}}(Y)$. Direct computations give that $\Gamma_{\mathcal{B}}(Y)$ consists of 8 elements ($\sim_Y$-equivalence classes)

$$0, 1, x_1, x_2, x_1\bar{x}_2, \bar{x}_1, \bar{x}_2, x_2 \vee \bar{x}_1.$$

Therefore, $\Gamma_{\mathcal{B}}(Y)$ is isomorphic to a boolean algebra of rank 3, and the elements x_2, $x_1\bar{x}_2$, $\bar{x}_1$ (x_1, $\bar{x}_1 \vee x_2$, $\bar{x}_2$) are atoms (respectively, co-atoms) of $\Gamma_{\mathcal{B}}(Y)$.

3. Transformations of boolean equations

Let $X = \{x_1, x_2, \dots, x_n\}$ be a finite set of variables. Let us define new variables $Z = \{z_\alpha | \alpha \in \{0,1\}^n\}$ indexed by all n-tuples $\alpha \in \{0,1\}^n$ ($|Z| = 2^n$). Following [**6**], the variables Z are called *orthogonal*. By $\pi_i(\alpha)$ ($1 \leq i \leq n$), we denote the projection of a tuple $\alpha \in \{0,1\}^n$ onto the i-th coordinate. The substitution of the variables $X = \{x_1, x_2, \dots, x_n\}$ is defined by

$$x_i = \bigvee_{\pi_i(\alpha)=1} z_\alpha. \tag{2}$$

For example, the set $X = \{x_1, x_2\}$ gives $Z = \{z_{(0,0)}, z_{(0,1)}, z_{(1,0)}, z_{(1,1)}\}$ and

$$x_1 = z_{(1,0)} \vee z_{(1,1)},\ x_2 = z_{(0,1)} \vee z_{(1,1)}.$$

According to the axioms of boolean algebras, the variables Z are obtained from X by the following rules:

$$z_\alpha = x_1^{a_1} x_2^{a_2} \dots x_n^{a_2}, \tag{3}$$

where $\alpha = (a_1, a_2, \dots, a_n)$, $a_i \in \{0,1\}$ and

$$x_i^{a_i} = \begin{cases} x_i & \text{if } a_i = 1, \\ \overline{x}_i & \text{if } a_i = 0. \end{cases} \tag{4}$$

For example, the sets $X = \{x_1, x_2\}$, $Z = \{z_{(0,0)}, z_{(0,1)}, z_{(1,0)}, z_{(1,1)}\}$ give $z_{(0,0)} = \overline{x}_1\overline{x}_2$, $z_{(0,1)} = \overline{x}_1x_2$, $z_{(1,0)} = x_1\overline{x}_2$, $z_{(1,1)} = x_1x_2$.

By (2), any system $\mathcal{S}'$ in variables $X = \{x_1, x_2, \dots, x_n\}$ can be written as

$$\mathcal{S} = \{z_\alpha = 0 \mid \alpha \in A\} \cup \bigcup_{\alpha \neq \beta} \{z_\alpha z_\beta = 0\} \cup \Big\{ \bigvee_\alpha z_\alpha = 1 \Big\}, \tag{5}$$

where $A \subseteq \{0,1\}^n$ and $\bigvee_\alpha z_\alpha$ is the join of all variables $z_\alpha \in Z$ (see [**7**] for more details).

Moreover, in [**7**] it was proved that the algebraic sets $\mathrm{V}_{\mathcal{B}}(\mathcal{S}')$, $\mathrm{V}_{\mathcal{B}}(\mathcal{S})$ are isomorphic. A system of the form (5) is called *orthogonal*.

EXAMPLE 3.1. The set $Y = \mathrm{V}_{\mathcal{B}}(x_1x_2 = x_2)$ ($\mathcal{B}$ is an arbitrary boolean algebra) is isomorphic to the solution set of a system

(6)

$$\begin{cases} z_{(0,1)} = 0, \\ z_{(0,0)}z_{(0,1)} = z_{(0,0)}z_{(1,0)} = z_{(0,0)}z_{(1,1)} = z_{(0,1)}z_{(1,0)} = z_{(0,1)}z_{(1,1)} = z_{(1,0)}z_{(1,1)} = 0, \\ z_{(0,0)} \vee z_{(0,1)} \vee z_{(1,0)} \vee z_{(1,1)} = 1 \end{cases}$$

since

$$x_1x_2 = x_2 \Leftrightarrow \bar{x}_1x_2 = 0 \Leftrightarrow z_{(0,1)} = 0.$$

STATEMENT 3.2. The coordinate algebra of the solution set of an orthogonal system $\mathcal{S}$ (5) is isomorphic to the boolean algebra of rank $m-a$, where $m = |Z| = 2^n$ and $a = |A|$.

PROOF. Since all points $P_\alpha = (p_\beta \mid \beta \in \{0,1\}^n)$ $(\alpha \notin A)$,

$$p_\beta = \begin{cases} 1 & \text{if } \beta = \alpha, \\ 0 & \text{otherwise} \end{cases}$$

belong to $Y = \mathrm{V}_{\mathcal{B}}(\mathcal{S})$, the definition of $\sim_Y$-equivalence gives that the elements z_α $(\alpha \notin A)$ are nonzero in $\Gamma_{\mathcal{B}}(Y)$ and $z_\alpha \neq z_{\alpha'}$ for distinct $\alpha, \alpha' \notin A$). The equations $z_\alpha z_\beta = 0 \in \mathcal{S}$ imply that the elements z_α $(\alpha \notin A)$ are exactly the atoms of the boolean algebra $\Gamma_{\mathcal{B}}(Y)$. Since the rank of a boolean algebra is equal to the number of atoms, $\Gamma_{\mathcal{B}}(Y)$ is isomorphic to the boolean algebra of rank $m - a$. □

EXAMPLE 3.3. According to Statement 3.2, the coordinate algebra of the solution set of an orthogonal system $\mathcal{S}$ (6) is isomorphic to the boolean algebra of rank 3 (in Example 2.2 we directly obtained the same result). Using Theorem 2.1, we obtain that the set Y is irreducible over any boolean algebra of rank $r \geq 3$.

If $\mathcal{B}$ is the boolean algebra of rank 2 the solution set of (6) is decomposable into the union of solution sets of the following systems

$$\mathcal{S}_1 = \mathcal{S} \cup \{z_{(0,0)} = 0\},\ \mathcal{S}_2 = \mathcal{S} \cup \{z_{(1,0)} = 0\},\ \mathcal{S}_3 = \mathcal{S} \cup \{z_{(1,1)} = 0\}.$$

For the boolean algebra $\mathcal{B}$ of rank 2 there are no nonzero elements $z_1, z_2, z_3 \in \mathcal{B}$ with $z_iz_j = 0$ $(i \neq j)$. Therefore, for any solution of $\mathcal{S}$, one of the following equalities holds: $z_{(0,0)} = 0$, $z_{(1,0)} = 0$, $z_{(1,1)} = 0$. Thus, $\mathrm{V}_{\mathcal{B}}(\mathcal{S})$ can be decomposed into a union of solution sets of $\mathcal{S}_1, \mathcal{S}_2, \mathcal{S}_3$.

One can prove that for any algebraic set $Y \subseteq \mathcal{B}^n$ there exists a unique orthogonal system $\mathcal{S}$ in $m = 2^n$ variables with solution set isomorphic to Y. Therefore, there arises a one-to-one correspondence between algebraic sets in $\mathcal{B}^n$ and orthogonal systems in $m = 2^n$ variables. It allows us below to identify the class of algebraic sets in $\mathcal{B}^n$ and the class of all orthogonal systems in $m = 2^n$ variables.

4. Irreducible components of algebraic sets

Let Y be the solution set of $\mathcal{S}$ (5) over the boolean algebra $\mathcal{B}$ of rank r. Let $m = |Z| = 2^n$, $a = |A|$.

LEMMA 4.1. *If $m - a \leq r$, then Y is irreducible.*

PROOF. It directly follows from Statement 3.2 and Theorem 2.1. □

LEMMA 4.2. *If $m - a > r$, then Y is a union of solution sets of the following orthogonal systems*

$$\mathcal{S}_B = \{z_\alpha = 0 \mid \alpha \in B\} \cup \bigcup_{\alpha \neq \beta} \{z_\alpha z_\beta = 0\} \cup \Big\{\bigvee_\alpha z_\alpha = 1\Big\} \tag{7}$$

where $B \subseteq \{0,1\}^n$, $B \supseteq A$, $|B| = m - r$. Moreover, the sets $Y_B = \mathrm{V}_{\mathcal{B}}(\mathcal{S}_B)$ are irreducible components of Y.

PROOF. Actually, the statement of this lemma was demonstrated in Example 3.3, where the solution set of $\mathcal{S}$ over the boolean algebra of rank 2 is a union of the solution sets of the systems $\mathcal{S}_1, \mathcal{S}_2, \mathcal{S}_3$. For the systems $\mathcal{S}_1, \mathcal{S}_2, \mathcal{S}_3$ the set B respectively equals $\{(0,0),(0,1)\}$, $\{(1,0),(0,1)\}$, $\{(1,1),(0,1)\}$.

The proof of the lemma follows from the statements below.

(1) *Let us prove* $Y = \bigcup_B Y_B$. Since the systems $\mathcal{S}_B$ contain new equalities $z_\alpha = 0$, obviously $\mathrm{V}_{\mathcal{B}}(\mathcal{S}_B) \subseteq \mathrm{V}_{\mathcal{B}}(\mathcal{S})$ and $\bigcup_B Y_B \subseteq Y$.

Let us prove the opposite inclusion $Y \subseteq \bigcup Y_B$. Let $P = (p_\alpha \mid \alpha \in \{0,1\}^n) \in Y$. Since $p_\alpha p_\beta = 0$ for all $\alpha \neq \beta$, P contains at most r nonzero coordinates (and at least $m-r$ zero coordinates). Therefore, there exists a set $B \subseteq \{0,1\}^n$, $|B| = m-r$, $B \supseteq A$ such that $p_\beta = 0$ for all indexes $\beta \in B$, and therefore $P \in Y_B$.

(2) Statement 3.2 implies that the coordinate algebras of algebraic sets Y_B are isomorphic to $\mathcal{B}$. By Theorem 2.1, *all sets Y_B are irreducible.*

(3) Let us prove that $Y_B \nsubseteq Y_{B'}$ *for distinct sets* B, B'. Let $\beta \in B \setminus B'$. Then the point $P = (p_\alpha \mid \alpha \in \{0,1\}^n)$ with coordinates

$$p_\alpha = \begin{cases} 1 & \text{if } \alpha = \beta \\ 0 & \text{otherwise} \end{cases}$$

belongs to $Y_{B'}$, but $P \notin Y_B$. □

5. Average number of irreducible components

In this section, we obtain a formula for the average number of irreducible components of algebraic sets defined by orthogonal systems (5) over the boolean algebra $\mathcal{B}$ of rank r. Let m be the number of variables in the orthogonal system $\mathcal{S}$ (5) and $a = |A|$. According to Lemmas 4.1, 4.2, the number of irreducible components $\mathrm{Irr}(\mathcal{S})$ of the solution set of $\mathcal{S}$ equals

$$\mathrm{Irr}(\mathcal{S}) = \begin{cases} 1 & \text{if } m - a \leq r, \\ \binom{m-a}{r} & \text{otherwise.} \end{cases}$$

The number of orthogonal systems for fixed m, a equals $\binom{m}{a}$. The number of all orthogonal systems is 2^m; therefore, the average number of irreducible components of algebraic sets defined by orthogonal systems in m variables equals

$$\overline{\mathrm{Irr}} = \frac{1}{2^m}\left(1 \cdot \sum_{a=m-r}^{m} \binom{m}{a} + \sum_{a=0}^{m-r-1} \binom{m}{a}\binom{m-a}{r}\right).$$

We have

$$\sum_{a=0}^{m-r-1}\binom{m}{a}\binom{m-a}{r}=\sum_{a=0}^{m-r-1}\binom{m}{m-a}\binom{m-a}{r}=\binom{m}{r}\sum_{a=0}^{m-r-1}\binom{m-r}{m-r-a}$$
$$=\binom{m}{r}\sum_{a=0}^{m-r-1}\binom{m-r}{a}=\binom{m}{r}(2^{m-r}-1),$$

and, therefore, the average number of irreducible components is

$$\overline{\mathrm{Irr}}=\frac{1}{2^m}\left(\sum_{a=m-r}^{m}\binom{m}{a}+\binom{m}{r}(2^{m-r}-1)\right)$$
$$=\frac{1}{2^m}\left(\sum_{a=m-r+1}^{m}\binom{m}{a}+2^{m-r}\binom{m}{r}\right)=\frac{1}{2^m}\left(\sum_{i=0}^{r-1}\binom{m}{i}+2^{m-r}\binom{m}{r}\right).$$

For a fixed r and $m\to\infty$ we have $\frac{1}{2^m}\sum_{i=0}^{r-1}\binom{m}{i}\to 0$ and

$$\overline{\mathrm{Irr}}\sim 2^{-r}\binom{m}{r}\ \text{for } m\to\infty.$$

6. Ranks of irreducibility

According to Lemmas 4.1, 4.2, the solution set of a system $\mathcal{S}$ (5) may be reducible over the boolean algebra of rank r, but the solution set of $\mathcal{S}$ becomes irreducible over the boolean algebras of higher ranks. We say that a system $\mathcal{S}$ (5) has *rank of irreducibility* $IR(\mathcal{S})$ if the solution set of $\mathcal{S}$ is irreducible over the boolean algebra of rank $IR(\mathcal{S})$, but solution set of $\mathcal{S}$ is reducible over each boolean algebra of rank $r<IR(\mathcal{S})$ (if $\mathcal{S}$ is inconsistent over any boolean algebra we put $IR(\mathcal{S})=0$). Below we compute the average rank of irreducibility of orthogonal systems in m variables.

By Lemmas 4.1, 4.2, we have that $IR(\mathcal{S})$ of a system $\mathcal{S}$ (5) equals $m-a$, $a=|A|$. The number of orthogonal systems in m variables with rank of irreducibility $m-a$ is equal to $\binom{m}{a}$. Therefore, the average rank of irreducibility of orthogonal systems in m variables is

$$2^{-m}\sum_{a=0}^{m}(m-a)\binom{m}{a}=2^{-m}\left(m\sum_{a=0}^{m}\binom{m}{a}-\sum_{a=0}^{m}a\binom{m}{a}\right)$$
$$=2^{-m}\left(m2^m-m2^{m-1}\right)=m/2.$$

7. Pairs of isomorphic algebraic sets

In this section, we compute the number of pairs (Y_1,Y_2) such that the algebraic sets Y_i are isomorphic to each other and Y_i are defined by orthogonal systems in m variables.

Suppose algebraic sets Y_i are defined by the following orthogonal systems

$$\mathcal{S}_i=\{z_\alpha=0\mid \alpha\in A_i\}\cup\bigcup_{\alpha\neq\beta}\{z_\alpha z_\beta=0\}\cup\Big\{\bigvee_{\alpha} z_\alpha=1\Big\},\tag{8}$$

where $A_i\subseteq\{0,1\}^n$

The following lemma is a simple corollary of Statement 3.2.

LEMMA 7.1. *Algebraic sets Y_1, Y_2 defined by orthogonal systems $\mathcal{S}_1, \mathcal{S}_2$ (8) are isomorphic to each other iff $|A_1| = |A_2|$.*

PROOF. In [**2**] (Corollary 5.7), it was proved that that Y_1, Y_2 are isomorphic iff their coordinate algebras are isomorphic. It remains to apply Statement 3.2. □

The number of pairs $(\mathcal{S}_1, \mathcal{S}_2)$ with $|A_1| = |A_2|$ is equal to

$$\sum_{i=0}^{m} \binom{m}{i}\binom{m}{i} = \binom{2m}{m}.$$

Since there are exactly $2^m 2^m = 4^m$ pairs of algebraic sets defined by orthogonal systems in m variables, two random algebraic sets are isomorphic with probability

$$\frac{\binom{2m}{m}}{2^m \cdot 2^m}.$$

Applying the Stirling formula to the expression $\binom{2m}{m}$, we obtain that the required probability asymptotically equals $\frac{1}{\sqrt{\pi m}}$.

References

[1] S. Koppelberg, *Handbook of Boolean algebras. Vol. 1*, North-Holland Publishing Co., Amsterdam, 1989. Edited by J. Donald Monk and Robert Bonnet. MR991565

[2] E. Daniyarova, A. Myasnikov, and V. Remeslennikov, *Unification theorems in algebraic geometry*, Aspects of infinite groups, Algebra Discrete Math., vol. 1, World Sci. Publ., Hackensack, NJ, 2008, pp. 80–111, DOI 10.1142/9789812793416_0007. MR2571513

[3] È. Yu. Daniyarova, A. G. Myasnikov, and V. N. Remeslennikov, *Algebraic geometry over algebraic structures. II. Foundations* (Russian, with English and Russian summaries), Fundam. Prikl. Mat. **17** (2011/12), no. 1, 65–106, DOI 10.1007/s10958-012-0923-z; English transl., J. Math. Sci. (N.Y.) **185** (2012), no. 3, 389–416. MR2898219

[4] B. Plotkin, *Seven lectures on the universal algebraic geometry*, (2002), arXiv: math.0204245[math.GM].

[5] V. Roman'kov, *Equations over groups*, Groups Complex. Cryptol. **4** (2012), no. 2, 191–239, DOI 10.1515/gcc-2012-0015. MR3043434

[6] S. Rudeanu, *Lattice functions and equations*, Springer Series in Discrete Mathematics and Theoretical Computer Science, Springer-Verlag London, Ltd., London, 2001. MR1902657

[7] A. N. Shevlyakov, *Elements of algebraic geometry over Boolean algebras with distinguished elements* (Russian, with English and Russian summaries), Fundam. Prikl. Mat. **18** (2013), no. 4, 197–218; English transl., J. Math. Sci. (N.Y.) **206** (2015), no. 6, 742–757. MR3431841

[8] A. N. Shevlyakov, *Equivalent equations over semilattices* (Russian, with English summary), Sib. Èlektron. Mat. Izv. **13** (2016), 478–490. MR3512691

[9] A. N. Shevlyakov, *On irreducible algebraic sets over linearly ordered semilattices*, Groups Complex. Cryptol. **8** (2016), no. 2, 187–195, DOI 10.1515/gcc-2016-0014. MR3567874

SOBOLEV INSTITUTE OF MATHEMATICS, PEVTSOVA ST. 13, OMSK, 644099 RUSSIA; OMSK STATE TECHNICAL UNIVERSITY, PR. MIRA, 11, OMSK, 644050 RUSSIA

Email address: a_shevl@mail.ru